S0-BQI-269

MODERN ASPECTS OF ELECTROCHEMISTRY

No. 12

LIST OF CONTRIBUTORS

A. COVINGTON
Department of Chemistry
University of Newcastle
Newcastle, England

T. ERDEY-GRÚZ
The Academy of Science
Budapest, Hungary

J.-P. FARGES
Department of Biochemistry
University of Nice
Nice, France

F. GUTMANN
Department of Chemistry
Macquarie University
Sydney, Australia

M.A. HABIB
School of Physical Sciences
The Flinders University of South Australia
Adelaide, Australia

K.A. KINOSHITA
United Technologies Incorporated
Middletown, Connecticut

S. LENGYEL
The Academy of Science
Budapest, Hungary

K.E. NEWMAN
Department of Chemistry
University of Keele
Keele Park
Staffs, England

P. STONEHART
United Technologies Incorporated,
Middletown, Connecticut

MODERN ASPECTS OF ELECTROCHEMISTRY

No. 12

Edited by

J. O'M. BOCKRIS

School of Physical Sciences
The Flinders University
Adelaide, South Australia

and

B. E. CONWAY

Department of Chemistry
University of Ottawa
Ottawa, Canada

PLENUM PRESS • NEW YORK AND LONDON

Library of Congress cataloged the first volume of this title as follows:

Modern aspects of electrochemistry. no. [1]
Washington, Butterworths, 1954-

v. illus. 23 cm.

No. 1-2 issued as Modern aspects series of chemistry.
Editors: no. 1- J. Bockris (with B. E. Conway, no. 3-
Imprint varies: no. 1, New York, Academic Press.—No. 2, London, Butterworths.

1. Electrochemistry—Collected works. I. Bockris, John O'M., ed. II. Conway, B. E. ed. (Series: Modern aspects series of chemistry)

QD552.M6 54-12732 rev

Library of Congress Catalog Card Number 54-12732
ISBN 0-306-37652-0

A Division of Plenum Publishing Corporation
227 West 17th Street, New York, N.Y. 10011

Printed in the United States of America

To

A. N. Frumkin

Preface

The first chapter in the present volume takes up a well-known theme in modern context: the ideas concerning non-Stokesian mechanisms of ion transport. We are happy that one of the great pioneers of modern electrochemistry, T. Erdey-Grúz, in collaboration with S. Lengyel, has consented to write this article for us. Along with it is a solution-oriented article in spectroscopic vein, namely, that by A. Covington and K. E. Newman on the analysis of solution constituents by means of nuclear magnetic resonance studies.

Progress in the electrochemistry of the double layer has perked up, and the advances have been triggered from critical experiments, one showing that fluoride ions are specifically adsorbed, and the other showing that the position of maximum disorder of the water molecules occurs at a charge opposite to that needed for interpretations of capacitance humps in terms of water molecules. M. A. Habib, who has contributed to the theory in this area, reviews the consequences of these changes in information.

The rise in the price of energy toward a situation in which sources other than the fossil fuels become economical implies much for the fuel cell and electrocatalysis. It has long been known that electrocatalysis in real situations was more than a consideration of exchange current densities, and a gap remains in the formulation of the theory of supports for such catalysts, although Boudart has stressed so much the vital nature of them. P. Stonehart and K. A. Kinoshita describe progress in this area.

Lastly, the Bockris and Reddy prediction that the eventual largest area of application of electrochemistry would come in biology and medicine has advanced apace, and some of this intriguing material is described by F. Gutmann and J. P. Farges.

Adelaide, South Australia
Ottawa, Canada

J. O'M. Bockris
B. E. Conway

Contents

Chapter 1

PROTON TRANSFER IN SOLUTION

T. Erdey-Grúz and S. Lengyel

I. Introduction 1
II. Experimental Findings 3
1. Anomalous Mobility of Ions of the Solvent Molecule in a One-Component Solvent 3
2. Effect of Structural Changes on Proton-Transfer Mobility 7
III. Theories of Abnormal Conductance 18
1. Theories Reviewed by Conway 18
2. Recent Suggestions in the Theory of Abnormal Conductance 23
3. A Semiempirical Theory of the Average Size of Clusters in Water 26
4. The Quantum Mechanical Theory of Weidemann and Zundel 28
5. Theories Rejecting Proton Transfer 29
IV. The Theory of the Contribution of Transfer Reactions to Transport Processes 31
V. Concluding Comments 36
References 37

Chapter 2

NMR STUDIES OF THE STRUCTURE OF ELECTROLYTE SOLUTIONS

A. K. Covington and K. E. Newman

I. Introduction 41
II. General Principles 42
1. Shift Measurements 42
2. Linewidth and Relaxation Phenomena 45
III. Studies on Solvents and Solvent Mixtures 48
1. Structure of Water 48
2. NMR Studies of Water 50
3. Interaction of Organic Molecules with Water 56
IV. NMR Studies of Electrolyte Solutions 61
1. Aqueous Solutions 61
2. Nonaqueous Solutions 95
3. Mixed Solvents 108
V. Concluding Remarks 123
References 123

Chapter 3

SOLVENT DIPOLES AT THE ELECTRODE–SOLUTION INTERFACE

M. Ahsan Habib

I. Introduction 131
II. Experimental Information on Solvent Properties 132
1. Solvent Excess Entropy 132
2. Entropy of Formation of the Double Layer 133
III. Models of Water in the Interface 137
1. The Two-State Water Model 137
2. The Three-State Water Model 140
3. Solvent Excess Entropy: Modelistic Approach 150

4. The Four-State Water Model 162
IV. Influence of the Metal on the Solvent Properties 165
1. Introduction 165
2. Dependence of Some Electrochemical Properties on the Nature of the Substrate Metal 165
V. Discussion and Conclusion 173
References 179

Chapter 4

PREPARATION AND CHARACTERIZATION OF HIGHLY DISPERSED ELECTROCATALYTIC MATERIALS

K. Kinoshita and P. Stonehart

I. Introduction 183
II. Electrocatalyst Supports 184
1. Carbon 185
2. Other Supports 189
3. Electrocatalyst–Support Interactions 191
III. Preparation of Electrocatalysts 194
1. Impregnation 194
2. Ion-Exchange and Adsorption 197
3. Colloidal Electrocatalyst Particles 200
4. Adams' Method 204
5. Raney's Method 205
6. Thermal Decomposition 206
7. Freeze-Drying 210
8. Binary Alloy Catalysts 211
IV. Characterization of Electrocatalysts 214
1. Adsorption Methods 214
2. X-Ray Analysis 235
3. Electron Microscopy 244
4. Reaction–Rate Measurements 251
V. Summary 257
References 258

Chapter 5

CHARGE-TRANSFER COMPLEXES IN ELECTROCHEMISTRY

Jean-Pierre Farges and Felix Gutmann

I. Theoretical Introduction ... 267
1. Charge-Transfer Complexes ... 267
2. Donors, Acceptors, and Complexation ... 268
3. Weak Charge-Transfer Complexes ... 270
4. Strong Charge-Transfer Complexes ... 272
II. Solid Charge-Transfer Complexes ... 273
1. Crystal Structures ... 273
2. Electrical and Magnetic Properties ... 275
3. Modes of Charge Transfer in the Solid State ... 276
4. Tetracyanoquinodimethan Salts ... 282
III. Identification of Charge-Transfer Complexes ... 289
1. Spectroscopy ... 289
2. Conductimetric Titrations ... 289
3. Dielectric Properties ... 294
4. Other Electrical and Magnetic Methods ... 298
IV. Biological Applications ... 301
1. Introduction ... 301
2. Charge-Transfer Interactions on Biological Membranes ... 301
3. Charge-Transfer Complexes and Drug Activity ... 303
4. Charge-Transfer Complexes in Biological Energy Transfer ... 305
5. Charge-Transfer Complexes and Carcinogenesis ... 306
6. Enzymatic Charge-Transfer Complex Reactions ... 307
References ... 308

Index ... 315

1

Proton Transfer in Solution

T. Erdey-Grúz and S. Lengyel

Academy of Science, Budapest, Hungary

I. INTRODUCTION

The understanding of the role of proton transfer in the mechanism of transport processes in solution is based on three accepted concepts:

1. In a solution containing ions produced by dissociation of solvent molecules, rapid charge transfer may take place between these ions and solvent molecules when they collide. This charge transfer and the sudden charge displacement (jump) that is involved, e.g., between a hydrogen ion and a water molecule or a CH_3O^- ion and a methanol molecule, may contribute to the transport of current in solution. This possibility was recognized as early as 1905 by Danneel.[1]

2. Successive charge transfers in a particular direction (e.g., under the influence of an external electric field) are necessary for this contribution. A series of successive transfers can take place only if each jump is followed by structural rearrangement in the liquid. As an example of such rearrangements, rotation of solvent molecules was also mentioned by Danneel.[1]

3. No free protons exist permanently in aqueous solutions. The hydrogen ion is the monohydrate of the proton, i.e., the hydronium (or oxonium) ion, H_3O^+. Recognition of this ion was first substantiated by Goldschmidt and Udby[2] in their theory of kinetics of acid-catalyzed esterification reactions. Recognizing the hydrogen ion as H_3O^+ also meant identifying Danneel's charge transfer as proton transfer from H_3O^+ to H_2O.

In liquids in which proton transfer contributes to transport processes, the structure is much influenced by the presence of hydrogen bonds. It was Latimer and Rodebush[3] who first recognized that "a free pair of electrons on one water molecule might be able to exert sufficient force on a hydrogen held by a pair of electrons on another water molecule to bind the two molecules together," or in the more general form, "the hydrogen nucleus held between two octets constitutes a weak bond." This was the first formulation of the hydrogen bond. Latimer and Rodebush also understood the role the hydrogen bond might play in promoting proton-transfer "shifting of hydrogen nuclei from one water molecule to another." Recognition of the hydrogen bond may thus be considered, along with the three concepts enumerated above, as forming the groundwork for theories of proton transfer in transport processes in solution.

The hydrogen bond and proton transfer are decisive factors for a number of phenomena. In many crystalline hydrates, the hydrogen bond plays an important structure-influencing role. On the other hand, in some solids, such as KH_2PO_4, the hydrogen bond is responsible for ferroelectric properties. In biological materials, characteristic structural features are determined by hydrogen bonds; the α-helix structure, myoglobin and lysozyme structures of polypeptide chains, and the double helix of DNA should be mentioned as examples, together with the suggestion[4] that proton tunneling in nucleic acid–base pairs may affect the transmission of genetic information. In liquids, association by hydrogen bonding has long been recognized as related to high dielectric constants. Proton transfer is a part of the mechanism of acid, base, and enzymatic catalysis. In this chapter, we shall focus our attention on the exchange reactions of proton transfer and their contribution to transport processes in solution.

Regarding chemical reactions controlled by proton transfer, the dynamics of the formation and dissociation of hydrogen bonds, and the effect of hydrogen bonding on reaction rates, we refer to reviews and monographs on this topic.[5–8]

Since Conway[9] published an excellent detailed analysis of theories concerning proton-transfer processes in solution in this series in 1964, we restrict ourselves here to referring to well-known findings and theories and focus our attention on the advances described in the literature after 1964.

II. EXPERIMENTAL FINDINGS

1. Anomalous Mobility of Ions of the Solvent Molecule in a One-Component Solvent

The limiting value of the equivalent conductance of atomic or other simple inorganic ions in aqueous solutions at 25°C is about 35–80 Ω^{-1} cm^2 $equiv^{-1}$. The hydrogen and the hydroxide ions are exceptions, with the much higher values of 350 and 192,[10] respectively. If we use a model of spherical charged particles subject to resisting forces proportional to their velocity and moving in a homogeneous fluid of definite viscosity, and apply Stokes's law, we obtain reasonable values for ionic radii in the range 1.2–4.4 Å. They are approximately in accordance with the hydrated radii.[11,12] However, the value expected for the hydrogen ion is unreasonably small, no free protons exist permanently in aqueous solutions, and the chemical species that actually exist, H_3O^+ and OH^-, have dimensions near that of a water molecule, which, if considered to be roughly a sphere, has a radius of about 1.4 Å.

As in the case of hydrogen and hydroxyl ions in aqueous solutions, anomalously high conductances were found for the CH_3O^- ion in methanol by Dempwolff[13] and Tijmstra,[14] for the pyridinium ion ($C_5H_5NH^+$) in pyridine, the formate ion in formic acid, and the acetate ion in acetic acid by Hantzsch and Caldwell[15] in the years 1904, 1905, and 1907, respectively. The generalization that all ions produced by dissociation of the solvent molecule have anomalously high conductance was first suggested by Danneel.[1]

For these ions, in contrast to the hydrodynamic migration of the others, a different mechanism of current transport was proposed as early as 1905.[1] In the case of other ions, ion–solvent interactions (hydration, solvation) had to be considered in order to explain differences between Stokes and crystallographic radii and other facts.

The presence of the H_3O^+ ion in solids as well, namely, in crystal hydrates, could be detected,[16,17] and its molecular geometry was recently determined by Lundgren and Williams[18] by neutron scattering on *p*-toluenesulfonic acid monohydrate single crystal.* This crystal proved to be ionic, with the chemical species H_3O^+ at cation positions. According to this study, the H_3O^+ ion has a pyramidal

*For earlier studies on the structure of H_3O^+, see Ref. 9, pp. 53–56.

structure with three O—H distances of 1.01 Å and H—O—H angles of 110.4°. The oxygen nucleus is at a distance of 0.32 Å above the plane formed by the three protons. The shape and size of the whole ion do not differ too much from those of the water molecule, and taken as a sphere, it can be assumed to be an intermediate between the ions of sodium and potassium, closer to the latter. Thus, we also have crystallographic radius for H_3O^+. A Stokes radius cannot be calculated directly from conductance because of a non-Stokesian mechanism of transport. However, hydrodynamic migration ("bodily transport") by the chemical species H_3O^+ cannot be excluded. Consequently, at least two mechanisms for the transport of current by protons must be supposed.

The hydrodynamic fraction of this conductance can be estimated, starting from assumed values for the radii of ionic spheres. Lorenz[19] used radii of the H atom and the OH^- ion, calculated from atomic volumes by Reinganum's formula. A more adequate way of estimation is to take H_3O^+ as a sphere of a size intermediate between that of Na^+ and K^+ and assume a corresponding hydrodynamic mobility (a value between 50.1 and 73.5 Ω^{-1} cm^2 equiv^{-1}).[10]

A higher value of 85 Ω^{-1} cm^2 equiv^{-1} results if we assume the hydrodynamic mobilities λ of the species H_3O^+ and H_2O to be approximately equal and calculate the latter from the experimental value of the self-diffusion coefficient of the water molecule ($D^* = 2.25 \times 10^{-5}$ cm^2 sec^{-1})[20] by using Nernst's expression in the form

$$D^* = (\lambda/e_0 F)kT$$

Here e_0 denotes electronic charge, F the Faraday constant, and k Boltzmann's constant.

A higher value for the radius of the hydrated H_3O^+ ion is obtained from thermodynamic ionic properties (heat capacities[21,22] and activity coefficients[23]) of solutions of lithium halides and $LiClO_4$ and salts of other alkali metals with those of acid solutions. This comparison shows that the behavior of the Li^+ ion comes closest to that of the H_3O^+ ion.

If we take the hydroxide ion as a sphere of the same radius as that of the isoelectronic fluoride ion, we obtain by Stokes's formula $\lambda = 55$ Ω^{-1} cm^2 equiv^{-1}. A fraction of the conductance of OH^- corresponding approximately to this value is due, then, to hydrodynamic migration.

The differences between the total and the hydrodynamic conductances must be explained by a special conductance mechanism for both the H_3O^+ and the OH^- ion.

The temperature dependence of the total equivalent conductance (relative mobility) of H_3O^+ differs from that of other ions.[9,10,24–27] Since the structure of water changes with temperature, the temperature dependence of the equivalent ionic conductance cannot be fitted by a single apparent activation energy over a longer range of temperature. However, if values of apparent activation energies calculated for the same temperature interval are compared, the values of H_3O^+ and OH^- are smaller than those of any other simple ion. In the interval 0–156°C, the activation energy decreases with increasing temperature for all simple ions. However, this decrease is more rapid for H_3O^+ and OH^-.

As was shown by Hückel,[24] the differences $\lambda_{H^+} - \lambda_{Na^+}$ (or $\lambda_{H^+} - \lambda_{K^+}$) can be taken approximately as the nonhydrodynamic fraction of the conductance of the hydrogen ion. The apparent activation energies calculated from these values are lower than the apparent activation energies of other ions. Their decrease with increasing temperature is very rapid.

In contrast to other ions, the equivalent conductance of the hydrogen ion increases with increasing pressure.[28–33] Correspondingly, the activation volume calculated from the pressure dependence of the abnormal contribution to proton mobility is negative.

Bernal and Fowler assumed[25] that "the effective mobility of the isotope D must be less than that of the ordinary H." Since hydrodynamic and transfer mechanisms are influenced to a different degree by the isotope effect, it is important to know the ratio of the mobilities of H^+ in H_2O to D^+ in D_2O. At 25°C, it is 1.42.[27] This ratio can be derived from measurements of the conductances of hydrochloric acid and potassium chloride in ordinary water and in nearly pure D_2O.[34] (See also the footnote on p. 836 of Ref. 27.) Alternatively, it can be derived from the diffusion coefficients calculated by the corrected form of the Ilkovic equation from the polarographically measured diffusion currents.[35] However, the ratio (1.52) derived from the latter approach[36] differs slightly from the above-mentioned value. This ratio has been studied as a function of the concentration of the supporting electrolyte in solutions of some alkali halides and tetraalkylammonium bromides.[35,36]

Assuming a hydrodynamic mobility mechanism only, we should expect a smaller value for $\lambda_{H^+}/\lambda_{D^+}$, corresponding to the differences between viscosities and dielectric constants of H_2O and D_2O, respectively. Supposing, on the other hand, that in the mechanism involving proton (or deuteron) transfer (usually referred to as the *prototropic mechanism*), it is this transfer, not a structural rearrangement, that is the rate-determining step; we should then expect a higher value for $\lambda_{H^+}/\lambda_{D^+}$ than that observed, due to the mass effect on transfer probability.

The difference between the equivalent conductances of H^+ and OH^- is also important. The conductance of OH^- is less than that of the ion H^+.

Roberts and Northey[36] calculated from their polarographic measurements of the limiting diffusion current a limiting value of $D^0_{H^+} = 9.4 \times 10^{-5}$ cm^2 sec^{-1} for the diffusion coefficient of H^+ at 25°C. According to Nernst's formula, this corresponds to an equivalent conductance of $\lambda_{H^+} = 352\,\Omega^{-1}$ cm^2 equiv^{-1}. This experimental evidence for equal mobilities in diffusion and current transport was preceded by the observations of Woolf[37] that the presence of "supporting" electrolytes decreases tracer diffusion of H^+ to a higher degree than that of any other ion. Correspondingly, Glietenberg *et al.*[38] found a particularly large effect from Li^+ ions. These effects were interpreted by structural changes caused by ion–solvent interactions, i.e., hydration of the ions of the supporting electrolyte. These changes lead to structures less favorable for proton transfer.

A comparison of the equivalent conductances of other ions with those of the hydrogen ion in solvents such as methanol and ethanol[9] also shows abnormal proton mobility. This abnormality is explained by consecutive processes of proton transfer and structural rearrangements. Temporary clusters of solvent molecules, bound by hydrogen bonds, certainly play an important role in this mechanism.

Experimental evidence shows that a contribution to transport processes from a transfer reaction is not limited to proton transfer. Electronic conductivity in solutions was predicted by Frumkin, as noted by Levich in his review.[39] The increase of diffusion by electron transfer was measured by Ruff *et al.* in the systems ferrocene and ferricinium ion [i.e., biscyclopentadienyl complexes of Fe(II) and Fe(III)] in alcohols,[40] ferroin and ferriin [tris1,10-phenantroline complexes of Fe(II) and Fe(III)] ions in water,[41] and ferrous and ferric forms of cytochrome *c* in water.[42] A contribution of the transfer of atoms or molecules to diffusion was also measured in systems containing sulfide and disulfide ions,[43] triiodide and iodide ions,[44] and tribromide and bromide ions[45] in aqueous solution. For systems containing antimony trichloride and aliphatic alcohols, acids, amides of carboxylic acids, and alkyl halides, there is, in addition to proton transfer, a contribution to the electrical conductivity from the transfer of atomic chlorine.[46,47]

2. Effect of Structural Changes on Proton-Transfer Mobility

Hydrogen bonds seem to be a precondition for proton transfer from H_3O^+ to H_2O; at least, quantum mechanical calculations of proton tunneling probability are usually based on hydrogen-bonded models.

The addition of nonelectrolytes to aqueous electrolyte solutions affects the structure of the solvent mainly by dissolving a fraction of the hydrogen bonds and changing the size of clusters held together by these bonds. The addition of electrolytes may have either a structure-breaking or a structure-forming effect on water. However, the added ions affect electrical conductivity by their own current transport. Separating these two influences has been the aim of many authors in studying the effect of the addition of nonelectrolytes on the conductance of the oxonium and the hydroxide ions. In this respect, Wulff and Hartmann[48] were the first to carry out studies on the influence of the dioxane content in hydrochloric acid solutions in water–dioxane mixtures on viscosity, dielectric constant, and equivalent conductance.

Nonelectrolytes that are highly soluble in water usually contain hydroxyl groups, have a characteristic structure, and show current transport by a prototropic mechanism in their pure state.

Several authors studied the effect of varying the concentration of added nonelectrolytes on the electrical conductance of aqueous acid and base solutions. Some investigated the variation of conductance with the nature of the added nonelectrolytes and their concentrations. Thus, the conductance of aqueous hydrochloric acid solutions containing methanol and ethanol has already been investigated, mainly by Goldschmidt and Dahl,[49] Walden,[50] Thomas and Marum,[51] Berman and Verhock,[52] Kortüm and Wilski,[53]

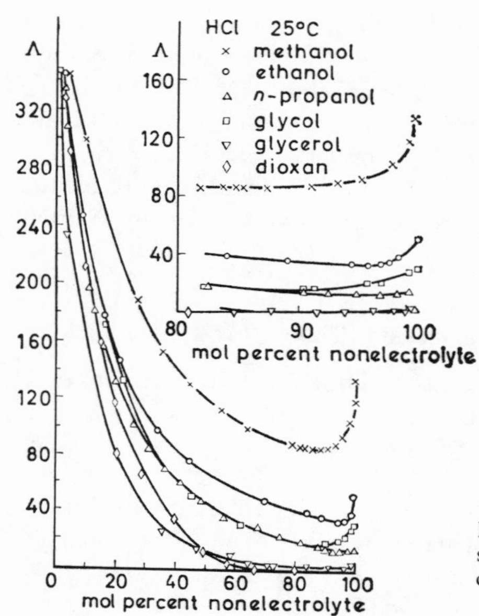

Figure 1. Conductance of 0.01 *m* solutions of HCl as a function of the concentration of various nonelectrolytes at 25°C.

Dnieprov,[54] el Aggan *et al.*,[55] and Tourky and Mikhail.[56] The change in conductance of aqueous hydrochloric acid solutions on the addition of ethylene glycol has been studied by Dnieprov[54] and Kirby and Maass,[57] while the effect of glycerol has been measured by Dnieprov,[54] Conway *et al.*,[27] Woolf,[58] and Accascina *et al.*[59]; further, the effect of butanol has been investigated by Accascina *et al.*[60] It has been found that the equivalent conductance of aqueous hydrochloric acid solutions decreases rapidly on the additions of alcohols; with increasing alcohol concentration, it passes through a minimum and increases again in solutions containing very small amounts of water. According to Tourky *et al.*,[61] and Abdel Hamid *et al.*,[62] the conductance of hydrochloric acid varies similarly in mixtures of propanol, *t*-butanol, and water. The influence of water on proton migration in 1-pentanol has been studied by de Lisi and Goffredi.[63] Figures 1 and 2 show how the equivalent conductances of dilute (0.01 *M*) hydrochloric acid and potassium hydroxide solutions vary with the concentration of the nonelectrolyte (methanol, ethanol, *n*-propanol, ethylene glycol, glycerol, and dioxane) added to aqueous solutions. These studies were made by Erdey-Grúz *et al.*[64,65] The results for HCl coincide qualitatively with those of former authors. However, more detailed information can be derived from recent studies. The conductance minimum occurs at about 90 mol percent alcohol content of the solvent (see detailed figure for the range 80–100 mol percent). On the curve for methanol, a sharp minimum can be seen, while it is smoothed out in the cases of *n*-propanol and glycerol.

The equivalent conductance of KOH solutions falls rapidly with the increasing concentration of the nonelectrolyte (Fig. 2), but a smooth minimum appears only on the curve for methanol, at about 75 mol percent of the nonelectrolyte.

Hydrodynamic mobilities depend on the viscosity of the solution. For the model of spherical charged particles subject to resisting forces proportional to their velocity and moving in a homogeneous fluid of definite viscosity, Walden's rule

$$\Lambda^0 \eta^0 = (0.82 \times 10^{-8})/r = \text{const}$$

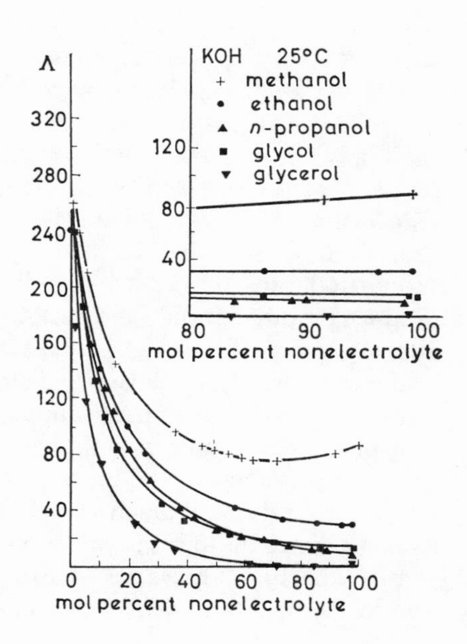

Figure 2. Conductance of 0.01 *m* solutions of KOH as a function of the concentration of various nonelectrolytes at 25°C.

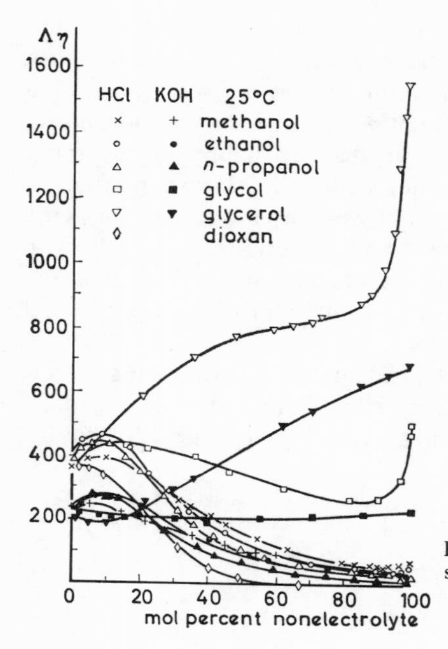

Figure 3. Product $\Lambda\eta$ in HCl and KOH solutions vs. nonelectrolyte concentration at 25°C.

for a given electrolyte in a series of solvents follows from Stokes's law. Here Λ^0 is the limiting conductance, η^0 the viscosity of the pure solvent, and r the hydrodynamic radius of the ion. However, Walden's rule is valid only approximately and only for ions of larger radii. Although the Walden product depends on temperature and on other factors influencing ionic migration, it is less dependent on viscosity than the equivalent conductance, which is, in a very simplified model, proportional to $1/\eta^0$. Thus, in studies of the variations of Walden's product, the direct effect of the variation of the macroscopic viscosity is approximately eliminated. This explains the role that studies of the Walden product have played, and still play, in many investigations. Thus, conclusions concerning liquid structure and mechanism of transport can be drawn from the variations of Walden's product with such parameters as temperature and composition.

Figure 3 shows Walden's product of HCl and KOH solutions in its dependence on the concentration of the nonelectrolyte.[64,65] For comparison, Fig. 4 shows the same functions for KF and KCl,[64,65] i.e., electrolytes without transfer mechanisms. Even in the latter cases, marked variation of the value of Walden's product can be observed due to several effects. Radii of the solvated ions may change with changing composition of the solvent, affecting composition and structure of the first solvation shell. A dependence of the viscosity in the neighborhood of the ion on the distance from the latter may also result. It is not justified to assume the same mechanism of displacement of both the ion and the solvent particles. In an interpretation of these complications, a simplified model of spherical particles moving in a continuum medium of bulk viscosity of the solution (not the solvent!) does not seem adequate.

Figures 3 and 4 both show an increase of Walden's product with the increase of alcohol concentration in the lowest concentration range (glycol and glycerol seem to be exceptions if KOH is the electrolyte). The decrease of mobilities is overcompensated by the increase in bulk viscosity, due to structure-strengthening effects of the alcohols. Their molecules may be assumed to fill the holes in the water

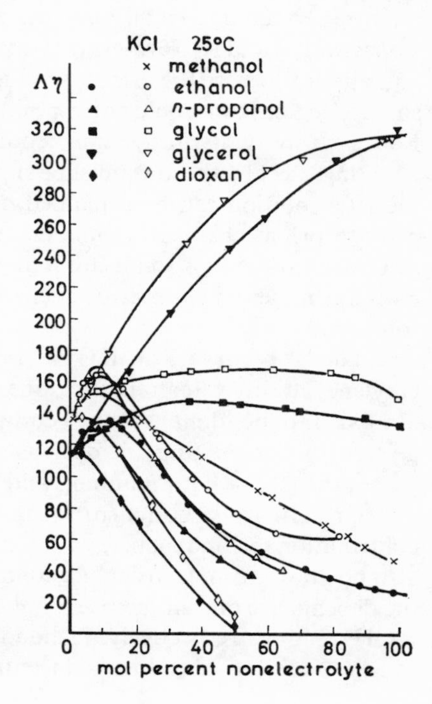

Figure 4. Product $\Lambda\eta$ of KF and KCl in aqueous solutions of nonelectrolytes at 25°C.

structure. This strengthened structure is decomposed by ions to a higher degree than the structure of pure water.

In mixtures of monoalcohols with water, $\Lambda\eta$ shows a maximum between about 4 and 10% alcohol concentration. At higher concentrations, it decreases below the values of solutions in pure water. The maximum is flatter for HCl and KOH than for KF and KCl. The decrease after the maximum is more pronounced than expected from viscosity changes alone. The ions strengthen the structure of the liquid and hinder displacement among its molecules. In addition, in the cases of H^+ and OH^-, the role of prototropic mechanism decreases.

Solutions in dioxane–water mixtures show similar behavior (Figs. 3 and 4). Electrolytes in polyalcohol–water mixtures behave differently (Fig. 3). The curve of the Walden product of HCl solutions increases at low concentrations of ethylene glycol, then passes a flat maximum, and after a minimum of about 85%, shows, finally, a sudden increase. The $\Lambda\eta$ curve for HCl, KF, and KCl in glycerol solutions has a positive slope in the entire range. For HCl, a particularly sudden increase in mixtures containing small amounts of water can be observed. The slopes for KF and KCl are less than that for HCl. In solutions of glycol and glycerol, where the viscosity is high, conductance decreases to a smaller degree than would correspond to an increase in bulk viscosity. For exclusively hydrodynamic conductance, a strong structure-breaking effect by the ions must be supposed. The variant behavior of the HCl solution can be explained by a decrease in the role of the prototropic mechanism at moderate concentrations. However, in solutions containing small amounts of water, the proton transfer mechanism seems to prevail even more than in a pure aqueous solution.

The $\Lambda\eta$ curves of KOH show a different behavior. In ethylene–glycol–water mixtures, the value of Walden's product is almost independent of the composition of the solvent. In glycerol–water mixtures, it decreases at low glycol concentration, passes through a flat minimum, and increases moderately.

Conclusions concerning the variation of prototropic contribution to conductance with the composition of the solvent can also be drawn from transference numbers.[65,66] Figure 5 summarizes the dependence of transference numbers of H^+ and OH^- in HCl and KOH solutions, respectively, on the addition of certain nonelectrolytes to the solvent. At low concentrations of the alcohol, the

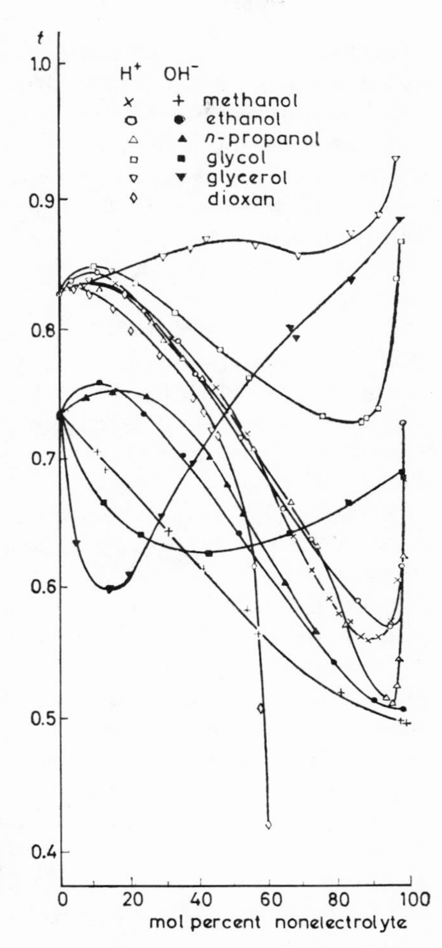

Figure 5. Transference number of hydrogen and hydroxide ions in HCl and KOH solutions as a function of the concentration of nonelectrolytes at 25°C.

transference number of H^+ increases with the alcohol content, passes through a maximum, and, after decreasing and passing through a minimum, increases again in solutions containing small amounts of water. In monoalcoholic solutions, the minimum is sharp, and its position is at about 90–95 mol percent. The latter increase has a steep slope. The first increase may be explained by strengthening of the structure of the liquid. The decrease beyond the maximum, on the other hand, may be due to the structure-breaking effect of the alcohol, which hinders proton transfer between H_3O^+ and H_2O. At the composition corresponding to the minimum transference number, the main contribution to conductance is given by the hydrodynamic migration of the chemical species H_3O^+. In solvents containing small amounts of water, the particular hydrogen-bonded liquid structure of the alcohol prevails, and the proton transfer between hydrogen-bonded ROH_2^+ and ROH contributes to the transport of current.

A particular liquid structure with hydrogen-bonded clusters seems to be a precondition to prototropic conductance. This view is supported by the behavior of HCl solutions in dioxane–water mixtures.[66] Apart from the region of very low concentrations, the transference number of H^+ decreases abruptly with the increasing moles percent of dioxane, and at 60 mol percent reaches a value of 0.42. In this solution, the conductance of H^+ is smaller than that of Cl^-. Dioxane destroys the water structure, and no more proton transfer is possible.

In the range of small water concentrations, the transference number of the H^+ ion increases with a very high slope if the concentration of glycol or glycerol is increased. This is due to a contribution of proton transfer to the conductance that is much higher than in pure water. Presumably, in a polyalcohol, all the hydroxyl groups of the molecule may function in proton transfer. Such double or triple roles of a molecule may have a higher probability than rotation of the molecules in a fluid of high viscosity. Since structural rearrangement consequent to proton transfer is necessary for prototropic conductance, and since molecular rotation seems to be excluded by high viscosity, intramolecular rearrangement in the polyalcohol molecule or its protonated form may be assumed instead.

On addition of methanol to aqueous solutions of KOH, the transference number of the OH^- ion decreases to a value of about 0.5 in pure methanol (Fig. 5). After a slight increase and minor maximum at 10–20 mol percent alcohol, the transference number of OH^- decreases also in water–ethanol and water–*n*-propanol mixtures with increasing concentration of alcohol and approaches the values 0.5 and 0.56, respectively. Consequently, no successive proton transfer can be assumed between OH^- and the monoalcohol molecule. The polyalcohols behave differently. In the range of small polyol concentrations, the transference number of OH^- decreases with increasing concentration of alcohol, particularly in the case of glycerol. After passing a minimum, it rises again. In a solvent containing a small amount of water and a high percentage of glycerol, the transference number of the OH^- ion is higher than in pure water. At low alcohol concentrations, a considerable part of the hydrogen bonds is dissolved by alcohol, and proton transfer is therefore hindered. At higher alcohol concentrations, the molecules of the alcohol may participate in proton-exchange reactions.

Comparison of the conductance $\Lambda\eta$ and transference number curves for H^+ and OH^- leads to the conclusion that different mechanisms for proton transfer must be assumed in acid and base solutions, and that the mechanism also depends on the nature and composition of the solvent. (For further details, see Ref. 67).

Quantitative theories of the anomalous conductance of the H^+ and OH^- ions usually assume infinite (or at least high) dilution and calculate limiting conductance. The presence of acids or bases at higher concentrations influences the structure of the solvent. Consequently, the variation of the equivalent ionic conductance with varying concentration of the electrolyte may give additional information about the mechanism of proton transfer. Transference number studies on acid and basic solutions, together with equivalent conductance data, can be used to obtain such information.

The transference numbers (relative to the solvent) of the H^+ ions in hydrochloric acid solutions in the higher concentration ranges were determined by Szabó,[68] Harned and Dreby,[69] and Lengyel *et al.*[70–72] by measuring the electromotive force of concentration cells, and by Kaimakov and Fiks[73] by the moving boundary method.[74] Konstantinov and Troshin[75] determined the transference number of the H^+ ion relative to the solution in hydrochloric acid solutions over the entire concentration range above 1 mol/liter. All these studies show that the transference number of the H^+ ion (ca. 0.84 at $c = 1$) diminishes markedly at higher acid concentrations, and the mobility ratio $\lambda_{H^+}/\lambda_{Cl^-}$ falls from about 5.39 (at 25°C and 1 m) to about 1.8 (at 12 m).

The transference number of the OH^- ion in alkali hydroxide solutions has also been studied. Lengyel *et al.*[76] determined the transference number of the OH^- ion in aqueous KOH and NaOH solutions at 20°C in the molality ranges 1–17 and 2–17, respectively, by measuring the electromotive force of concentration cells. In the case of KOH, the value (relative to the solvent) of 0.74 found for the lowest and the highest concentrations changed to a maximum of 0.78 (at 3–12 *m*). NaOH behaves differently. The value of 0.94 at 2 *m* increases to 0.98 (at 3–6 *m*), then diminishes to 0.86, increases again, and at $m = 14$ has a maximum of 1.00, after which it decreases to 0.59 (at $m = 17$). The extremely high (apparent) values were explained by formation of complex ions. Troshin and Zvyagina determined the transference number of OH^- relative to the solution in concentrated aqueous solutions of KOH,[77] NaOH[78] and LiOH[79] at 25°C. Their values for KOH increased uniformly from 0.79 (at $c = 1$ equiv/liter) to 0.847 (at $c = 12$). For the transference number relative to the solution of OH^- in NaOH solutions, they obtained a value of 0.905 at $c = 1$, which increases to 0.911 at $c = 2$, then diminishes uniformly to 0.826 at $c = 18$. No second maximum occurs. For LiOH, the high value of 0.947 at $c = 1$ was found. This increases to 0.980 at $c = 4$, then drops to 0.965 at $c = 4.8$. Since this compound is not very soluble in water, values for higher concentrations could not be measured.

From the limiting equivalent conductances,[10] the following limiting values can be calculated for the transference numbers at 25°C: in HCl, 0.823; OH^- in KOH, 0.72; OH^- in NaOH, 0.76. Comparison of these values with those for $m = 1$ or $c = 1$ shows a strong effect of the electrolyte even at this very low concentration.

In almost all theories of abnormal conductance, successive acid–base proton transfer, whether rate-controlling or not, is assumed. The rate of proton transfer is then calculated quantum mechanically or classically. By observation of the broadening of the corresponding nuclear magnetic resonance (NMR) lines, Meiboom *et al.*[80–86] and other authors[87–89] determined this rate constant experimentally for a series of reactions. Table 1 shows a compilation of the experimental values of the second-order rate constant of proton transfer from H_3O^+ to H_2O and H_2O to OH^-. Table 2, on the other hand, gives rate constant values of proton transfers between an ionic and a neutral molecular species for a series of other reactions. If we may rely on these rate constants, then one of the uncertain points in abnormal conductance theories seems to be eliminated. The results of a comparison of theoretical and experimental conductances no longer depends on the uncertainties of calculations of the proton-transfer rate constant.

Table 1
Values of the Second-Order Rate Constants for the Reactions $H_3O^+ + H_2O \rightarrow H_2O + H_3O^+ (k_1)$ and $H_2O + OH^- \rightarrow HO^- + H_2O\ (k_2)$ at 25°C (mol^{-1} liter sec^{-1})

References	$k_1 \times 10^9$	$k_2 \times 10^9$
Meiboom[80]	10.6	3.8
Löwenstein and Szöke[87]	11.0	5.5
Luz and Meiboom[81]		4.8
Luz and Meiboom[82]		3.4
Luz and Meiboom[83]	9.7	2.9
Glick and Tewari[88a]	7.9	4.5
Rabideau and Hecht[89]	8.2	4.6

[a]Calculated from $6.0 \times 10^{11} e^{-2.4/RT}$ and $1.0 \times 10^{11} e^{-2.1/RT}$, as given by the authors, at 25°C.

Table 2
Values of the Second-Order Rate Constants at 25°C for Some Proton Reactions Involving Proton Transfer

Reactants	Products	kmol^{-1} liter sec^{-1}	Temp., C°	Ref.[a]
$H_2O_2 + H_3O^+$	$H_3O_2^+ + H_2O$	1.6×10^7	27	84
$CH_3OH + H_3O^+$	$CH_3OH_2^+ + H_2O$	10^8	22	85
$CH_3O^+H + CH_3OH_2^+$	$CH_3O^+H_2^+ + CH_3OH$	3.6×10^9	24.8	86
$CH_3OH + OH^-$	$CH_3O^- + H_2O$	2.6×10^6	22	85
$CH_3OH + CH_3O^-$	$CH_3O^- + CH_3OH$	7.5×10^8	24	86
$C_2H_5OH + H_3O^+$	$C_2H_5OH_2^+ + H_2O$	2.8×10^6	22	85
$C_2H_5O^+H + C_2H_5OH_2^+$	$C_2H_5O^+H_2^+ + C_2H_5OH$	1.1×10^6	22	85
$C_2H_5OH + OH$	$C_2H_5O^- + H_2O$	2.8×10^6	22	85
$C_2H_5OH + C_2H_5O^-$	$C_2H_5O^- + C_2H_5OH$	1.4×10^6	22	85

[a]Reference 85 deals with solutions containing water; Ref. 86, with solutions in pure methanol.

III. THEORIES OF ABNORMAL CONDUCTANCE

1. Theories Reviewed by Conway[9]

(i) Early Theories, Semiempirical Treatments, and Qualitative Considerations

This chapter begins by reminding the reader that Danneel[1] was the first to recognize the consecutive mechanism of proton transfer combined with structural reorientation (often referred to as *structural diffusion*) in the liquid. Rotation of H_2O dipoles was regarded by Danneel as the actual form of reorientation.

In his attempt at a quantitative treatment, Hückel[24] assumed a combination of proton jumps from H_3O^+ to H_2O with rotation of the resulting new H_3O^+. Neglecting the equivalence of the three protons in H_3O^+ and considering one of them fixed, he calculated the relaxation time of the rotating ionic sphere subjected to a momentum of the force acting on this proton and to the resisting force in the viscous liquid. He also calculated the proton-transfer frequency by classic vibration statistics. The ionic mobility was then related to both relaxation times by assuming a stationary orientation distribution resulting from the equilibrium one by perturbation by the external field.

Instead of this field, the Lorentz field in a spherical cavity in the dielectric was taken into account. The inadequacy of this theory, as well as of Wannier's calculations[90] relating to the same mechanism, was shown by Conway *et al.*[27] and by Conway.[9] The following argument should be added: because of the relatively high inertia of the rotating oxonium ion, a decrease of anomalous mobility would be expected at high frequencies of an alternating current. However, this expectation is not fulfilled.

Assuming a two-step consecutive process involving quantum mechanical tunneling proton transfer, coupled with water molecule rotation, Bernal and Fowler[25] calculated the frequency of proton tunneling between two water molecules in a suitable configuration and the corresponding proton mobility, which is due to an external (Lorentz) field. The rotation frequency estimated for the water molecule as a classic isotropic free rotator had the same order of magnitude as the frequency of proton tunneling. The latter was then proposed as the rate-determining step, because it seemed that orientation of the water molecule suited for proton transfer would be presented to the H_3O^+ ion at least as frequently as this ion could utilize it. Apart from the fact that the predicted decrease of the abnormal conductance with increasing pressure is at variance with experience, the detailed analyses by Conway *et al.*,[27] and again by Conway[9] show that some of Bernal and Fowler's assumptions were not adequate. The tunneling frequency depends greatly on the potential energy profile of the proton between its two alternative sites. Bernal and Fowler's rectangular-well profile and its dimensions result in lower values for the net rate of transfer than the more realistic potential energy curve used by Conway *et al.*[27] and by Conway.[9] Also, the free rotation model of the water molecule is incompatible with the hydrogen-bonded structure of water. For the ratio $\lambda^0_{H^+}/\lambda^0_{D^+}$ (see Section II.2), Bernal and Fowler estimated a value of about 5 (see Ref. 25, p. 548). Their paper[25] was received by the *Journal of Chemical Physics* on April 29, 1933. Lewis and Doody[34] published their experimental value of 1.39 at 25°C in a paper received by the *Journal of the American Chemical Society* on July 24 and published August 5, 1933. (See also Section II.2 and Ref. 35.) The discrepancy between the two values is experimental evidence against the assumption of proton tunneling as the rate-determining step, which would necessarily result in much higher values for $\lambda^0_{H^+}/\lambda^0_{D^+}$.

Huggins, who, independently of Latimer and Rodebush,[3] had proposed the concept of the hydrogen bond,[91] presented reasoning in favor of the hydrogen ion as $(H_2OHOH_2)^+$: "By contemporaneous shifts of two protons such an ion can add OH_2 at one end and lose H_2O at the other, thus shifting the position in the solution of the excess positive charge." Thus, Huggins was the first to suggest the existence of hydrogen-bonded ionic clusters (aggregates, chains) $(H_5O_2)^+$, $(H_7O_3)^+$, $(H_9O_4)^+$, etc., which were established experimentally by Beckey[92] some 20 years later. Huggins assumed rapid simultaneous proton transfers within the hydrogen-bonded chains and a less frequent, consequently rate-limiting, proton transfer between two chains at a position where the hydrogen bond was absent.[93,94] Thus, he proposed a new modification of the "structural diffusion" concept. On the other hand, his calculations and conclusions pertaining to the potential energy profile of the hydrogen bond cannot be accepted (see p. 118 and 119 of Ref. 9).

Using the theory of absolute reaction rates, Stearn and Eyring[26] calculated the free energy of activation for the proton transfer between two water molecules from Hückel's estimate: $(\lambda_{H^+} - \lambda_{Na^+})$ for the abnormal conductance. This activation energy was taken as the highest limit to the energy barrier over which the proton passes in the classic transfer process. Quantum mechanical tunneling was neglected because this tunneling, if it were rate-controlling, would involve a higher value for the $\lambda_{H^+}/\lambda_{D^+}$ ratio than has been found by experiment (see Section II.2). For the abnormal ionic mobility, a two-step mechanism of classic proton transfer combined with water-molecule rotation in the Lorentz field was assumed. The authors seem to have intended to calculate an approximate value (1–3.5 kcal mol^{-1}) of the free energy of activation for the process of proton transfer from H_3O^+ to H_2O, instead of presenting a theory of anomalous conductance. For this reason, the question of which of the two successive steps is rate-controlling was not considered. Reasons against the classic transfer mechanism are given by Conway[9] (his point IV. 4.V). Thus, Conway and Salomon[95] calculated lower (4.6) and upper (6.5) limits to the isotope ratio $\lambda_{H^+}/\lambda_{D^+}$ resulting from the classic proton-transfer step. Without presenting a quantitative calculation, Wulff and Hartmann[48] suggested a structural diffusion mechanism similar to that of Huggins.

In a semiempirical treatment based on Eucken's model[96] of stoichiometric water association, Gierer and Wirtz[97] assumed activated classic proton transfers within the hydrogen-bonded ionic aggregates, one transfer in $(H_5O_2)^+$ and three simultaneous transfers in $(H_9O_4)^+$, contributing to the abnormal conductance. The preexponential factor can be obtained by combination of the Nernst and the Einstein expressions for diffusion. The average square displacement of the transfers was called the *structure factor*, and was linearly composed of Eucken's temperature-dependent mole fractions of $(H_2O)_2$ and $(H_2O)_4$. With these as functions of the temperature, the product of the exponential and preexponential could be well fitted to the experimental temperature dependence of the abnormal conductances $\lambda_{H^+} - \lambda_{Na^+}$ and $\lambda_{OH^-} - \lambda_{Cl^-}$, respectively, over the long range 0–306°C by a single value of the activation energy (2.44 kcal mol^{-1} for H^+ and 3.0 kcal mol^{-1} for OH^-, with the same structure factor for both). Gierer and Wirtz consider a two-step mechanism of structure reorientation (decomposition of hydrogen bonds and formation of new ones) and proton transfer. However, their treatment allows both to be the rate-controlling step. An attempt was made to explain the pressure dependence of the abnormal conductance and the ratio $\lambda_{H^+}/\lambda_{D^+}$, as well. For a criticism of the Gierer and Wirtz theory, see the review by Conway.[9]

To interpret anomalies occurring in the interpretation of the infrared spectra of aqueous solutions, Wicke *et al.*[17] suggest a modification of Gierer and Wirtz's basic assumptions on the structure of ion hydrates. They regard $(H_9O_4)^+$ as a complex (H_3O^+, plus the first hydration shell) stable to about 160°C, surrounded by a second hydration shell of water molecules, hydrogen-bonded to those in the first shell. The number of these bonds decreases with increasing temperature to zero in the lower temperature ranges (below 160°C). Within $(H_9O_4)^+$, the protons are assumed to be mobile. To explain the abnormal H^+ mobility on the basis of this hydration model, Wicke *et al.*[17] and Eigen and De Maeyer[6,98] assumed (in the lower temperature range) very fast protonic charge fluctuation inside the H-bonded $(H_9O_4)^+$ complex, as in Hermans' cage[99] (see below), and considered the formation and decomposition of hydrogen bonds at the periphery of the complex as the rate-limiting process. In the higher temperature range (160°C), decomposition of hydrogen bonds in the first hydration shell (i.e., in the complex $H_9O_4^+$) leads to a further decrease of abnormal mobility. These conclusions were based mainly on analysis of the temperature dependence of the thermal conductivity of water and electrolyte solutions[100] and of the temperature and pressure dependence of the equivalent conductance of the hydrogen ion. Some support for this view is based on the reaction distance (7.5–12.5 Å) of the recombination reaction $H_3O^+ + OH^-$, estimated from the rate constant measured by Eigen and Schoen.[101] "This distance corresponds to the dimensions of the hydration sphere and indicates that the proton is mobile within its H-bonded hydration structure" (Ref. 98, p. 518).

To explain the experimental value of the isotopic ratio $\lambda_{H^+}/\lambda_{D^+}$ (see Section II.2), Darmois and Sutra[102] treated the protons as freely moving particles of an ideal gas with an average kinetic energy proportional to temperature and applied classic metallic conduction equations. For the inadequacy of this theory, see Conway *et al.*,[27] and Conway.[9] Earlier, Hermans[99] had suggested the application of the cage theory of liquids to free protons oscillating in cells of the liquid and making classically activated jumps between the cells.

(ii) Theory of Conway, Bockris, and Linton

Rejecting H_3O^+ rotation because of the equivalence of the three protons and the flatness of the ion, and neglecting the possibility of simultaneous proton transfers within hydrogen-bonded ionic aggregates, Conway *et al.*,[27] gave a thoroughgoing analysis of classic and quantum mechanical (tunneling), proton transfer from H_3O^+ to H_2O, pure thermal rotation, and rotation of water molecules, accelerated by the field of the H_3O^+ ion. Classic transfer was found to be about three times, and tunnel transfer two orders of magnitude, faster than corresponds to the observed abnormal proton mobility. Thermal reorientation of water molecules is insufficiently rapid. Rotation of the water molecule hydrogen-bonded to an H_3O^+ ion and to three other H_2O molecules (i.e., rotation of one water molecule in the complex $H_9O_4^+$) was found to be the rate-determining reaction, which is preceded and succeeded by fast proton transfers. The calculated value of the abnormal mobility of H^+ ($\lambda_{calc} = 270\ \Omega^{-1}\ cm^2\ mol^{-1}$) is in good agreement with the experimental mobility ($\lambda^0_{H^+} - \lambda^0_{Na^+}$ or $\lambda^0_{H^+} - \lambda^0_{K^+}$). For the isotopic ratio $\lambda_{H^+}/\lambda_{D^+}$, the square root mass ratio $\sqrt{m_D/m_H} = \sqrt{2}$ was obtained, also in good agreement with the experimental value (see Section II.2). In the case of the OH^- ion, the rotation of the water molecule is accelerated by the average OH^- ion–dipole force, because the hydrogen-bonded $(H_7O_4)^-$ hydrate structure contains two protons fewer than the $(H_9O_4)^+$ structure, and consequently, proton–proton repulsion forces acting during rotation of a water molecule in $(H_9O_4)^+$ do not occur in $(H_7O_4)^-$. Hence, the ratio of the abnormal conductances $\lambda_{OH^-}/\lambda_{H^+} = \sqrt{1/4} = 0.5$ was calculated, also in good agreement with the experiments.

Temperature and pressure dependence were qualitatively explained. In *a priori* calculations such as those of Conway, Bockris, and Linton, the model used is of decisive importance. It comprises the geometry of the hydrogen-bonded systems H_3O^+–H_2O, H_2O–OH^-, $(H_9O_4)^+$, and $(H_7O_4)^-$, bond lengths, bond energies, potential energy profiles of the proton transfer and of the rotation of the water molecule, and other parameters. For details of the models and calculations, the reader is referred to the original paper[27] and Conway's review.[9] It should perhaps be added here that according to the calculations of Conway, Bockris, and Linton, "the time taken in the transfer process is about 10^{-14} sec, while the time an H_3O^+ ion has to wait to receive a favorably oriented water molecule is about 2.4×10^{-13}," i.e., 24 times more. This is evidence not only for the existence of the species H_3O^+ but also for its hydrodynamic contribution to the conductance, because the time necessary to accelerate the ionic sphere (H_3O^+) in the viscous medium to final velocity was estimated by Hückel (Ref. 24, p. 550) to be about 10^{-14} sec, i.e., a small fraction of its lifetime.

The paper of Conway, Bockris, and Linton was soon followed by Eigen and De Maeyer's review of self-dissociation and protonic charge transport in water and ice.[98] In this review, the authors criticize some approximations made by Conway *et al.*[27] In calculating tunneling frequency of the protons instead of Maxwell–Boltzmann energy distribution as used in Eqs. (12) and (17) of Ref. 27, a summation of the discrete quantum states is recommended by Eigen and De Maeyer (Ref. 98, p. 516). However, this does not affect the order of magnitude of proton tunneling frequency relative to that of structure reorientation frequency, as pointed out by Conway and Bockris[103] in their reply (cf. p. 118 of Ref. 9). Thus, the use of quantum statistics equally proves structure reorientation to be the rate-controlling step.

Despite the reasons given[9,27,103] for the potential barrier used and the proton jump distance chosen by Conway *et al.*,[27] there is no direct experimental or theoretical evidence for the correctness of this choice, as was emphasized by Eigen and De Maeyer.[98,104]

Pertaining to Conway, Bockris, and Linton's calculations on water-molecule rotation accelerated by the electrical field of the H_3O^+ ion, Eigen and De Maeyer[98,104] pointed out the lack of direct experimental evidence "for the conclusion that H-bonded H_2O molecules in the liquid phase have the same thermal orientational relaxation time as in a complete ice lattice," as was assumed by the former authors. Eigen and De Maeyer asked for a more rigorous treatment of this rotation, including the liberation of H_2O from water structure, on the basis of more detailed knowledge of the water structure.

2. Recent Suggestions in the Theory of Abnormal Conductance

From the effect of nonelectrolytes dissolved in water on prototropic conduction, it can be concluded that the presence of isolated water molecules is not sufficient for the transport of electricity by this mechanism. It requires the presence of some complexes (aggregates, associated molecular or ionic species, polymers, latticelike regions) formed from a certain number of water molecules by hydrogen bonds through which easy transfer of protons may take place. The decrease of the size of such complexes hinders prototropic conduction, and monomer water molecules are unsuitable for this mechanism. This is supported by the fact that the ratio $(\lambda^0_{HCl} - \lambda^0_{KCl})/\lambda^0_{KCl}$, which provides some approximate empirical information on the extent of prototropic conduction, is 2.26, 1.84, and 1.07 at 0°, 25°, and 100°C, respectively. Thus, the share of prototropic mechanism in the transport of current is strongly diminished by the increse of temperature, which is probably due to the loosened connection between water molecules at elevated temperatures; i.e., the size of the regions bound together by hydrogen bonds decreases, and the liquid structure of water is destroyed.

With increasing pressure, however, the share of prototropic conduction increases. According to Hamann,[29,30] the value of the above-mentioned ratio at 25°C increases from 1.85 under atmospheric pressure to 2.15 under 3000 atm. Thus, pressure has a structure-forming effect on water in the sense that it assists the proton exchanges by providing some of the repulsion energy needed to bring the oxygen atoms close enough for a proton switch to occur. A different interpretation is given by Horne and Courant[105] and Lown and Thirsk,[106] however—namely, that "high pressure reduces the extent of H-bonded association in water, thereby facilitating the rotation of water molecules which is regarded as the rate-determining step in the proton transfer mechanism." According to Horne *et al.*,[107] who measured the specific conductances of aqueous KCl, KOH, and HCl solutions as functions of hydrostatic pressure and temperature over the ranges 5–45°C and 1–6900 bar, and calculated the variation of the corresponding apparent activation energies with pressure, "at lower pressures the rotation of water molecules appears to be the rate-controlling step, but above about 1380 bar the proton jump becomes rate-determining," particularly because of the breaking of H-bonds (cf. Ref. 105).

The water-structure-destroying effect of dioxane was mentioned in Section III.1(i). Other phenomena also indicate that H_2O molecules separated from the normal liquid structure of water do not participate in prototropic conduction. For example, Gusev and Palei[108] and Gusev[109] investigated the effect of nonelectrolytes on the conductance of aqueous solutions of electrolytes. They have inferred that water molecules, being in the hydrate spheres of ions (i.e., to a certain extent separated from the original liquid structure of water), do not participate in the prototropic conduction mechanism.

With respect to the role of monomeric water molecules in a prototropic conduction mechanism, Horne and Courant[105] have drawn dissimilar conclusions from their conductance measurements in aqueous hydrochloric acid solutions in the temperature range −1–10°C. They found the apparent activation energy of protonic conduction to be approximately constant, unlike the activation energies of other ions, which exhibit a maximum at the temperature of the density maximum. According to these authors, vacancies determining the hydrodynamic migration can be formed in both the Frank–Wen type[110] molecular clusters and the monomeric water around them, while the rotation of water molecules required for prototropic conduction can occur only in the monomeric state. In general, this latter mechanism is assumed to predominate, but, below 2°C, the share of hydrodynamic conductance increases.

The ratio of prototropic to hydrodynamic conductance depends in a complicated way on temperature, pressure, and concentration of the base or acid. Lown and Thirsk[106,111] measured the electric conductance of aqueous solutions of KOH, NaOH, and LiOH within the ranges 25–200°C, 1–3000 atm and 0.1–6.68 *m* and of orthophosphoric acid at 25°C within the ranges 1–3000 atm and 1–15.7 *m*. The conductance due to proton transfer is reduced progressively with increasing concentration in the 1–6.8 mol liter^{-1} range in LiOH, NaOH, and KOH solutions, and hydrodynamic migration increases considerably. This has been explained by the fact that an increasing portion of the water molecules become bound to the ions in the hydrate spheres of the ions with increasing ionic concentration, and these molecules thus become unsuitable for the conduction via the proton transfer of hydroxide ions. The prototropic conduction mechanism breaks down at lower concentrations in KOH solutions than in LiOH solutions, which can be ascribed to ionic association in the latter case. In KOH solutions, the Walden product becomes more and more independent of temperature and pressure with increasing concentration. This indicates the suppression of the prototropic mechanism. At low concentrations, the conductance of phosphoric acid solutions increases with increasing pressure. In medium concentrations, it decreases, and in the range 11.7–15.7 mol liter^{-1}, it becomes independent of pressure. This behavior can be attributed to the fact that water molecules bound in the hydrate spheres do not participate in the prototropic mechanism. Lown and Thirsk point out that both prototropic and hydrodynamic mechanisms are accompanied by rupture of hydrogen bonds. This may explain why the Debye–Hückel–Onsager theory holds in good approximation for dilute solutions of both KCl and HCl.

Roberts and Northey[36] pointed out that ions such as, for example, K^+, increase fluidity and the self-diffusion of water, but reduce proton mobility. They compiled values of the orientation time of the water molecule in pure water and in diamagnetic salt solutions, as estimated by McCall and Douglass[112] and Hertz and Zeidler[113] from NMR studies. KBr and KI were found to reduce the relaxation time of orientation. The facts pointed out by Roberts and Northey are in accordance with the statement of Conway *et al.*[27] that the *thermal* (as opposed to the field-induced) rotation of the water molecules cannot be the rate-determining step in abnormal conductance of the hydrogen ion.

Assuming proton transfer to be the rate-determining step in the abnormal diffusion of the H^+ ion, Ruff and Friedrich[114] calculated the diffusion coefficient from the value of Luz and Meiboom[83] for the second-order rate constant. The value obtained for the abnormal fraction of the diffusion constant turned out to be equal to the directly measured value of the total diffusion coefficient. From this, Ruff and Friedrich drew the conclusion that H_3O^+ does not move hydrodynamically at all. However, the basic equation of their theory[115] contains a factor due to erroneous averaging, as was shown by Lengyel.[116]

3. A Semiempirical Theory of the Average Size of Clusters in Water

Naberukhin and Shuiskii,[117] in their semiempirical theory, assumed water molecule aggregates (clusters) in aqueous solutions. Corresponding to the H^+ and OH^- concentration, some clusters contain an excess (or defect) proton. The average linear dimension of the clusters can be taken as n times the diameter d_0 of the water molecule. For the motion of the excess proton within the cluster two alternative possibilities were assumed:

a. Oscillation through the hydrogen bonds in the entire volume of the cluster with the frequency experimentally determined for proton transfer between an H_3O^+ ion and a water molecule (or H_2O and OH^-).

b. Oscillation in a potential minimum at the boundary of the cluster.

The frequency ν_0 of this oscillation was assumed to be of the order of magnitude of the frequencies in diffusion-controlled processes. The excess (or defect) proton may be transferred to the neighboring cluster either by diffusion of an H_3O^+ ion from one cluster to the other or by rotation of a water molecule on the boundary. For both alternatively assumed steps, as well as for (b) and for the hydrodynamic diffusion of an ordinary ion of the same size, the same frequency, $\nu_0 \approx 10^{13}\ \text{sec}^{-1}$, was chosen. Proton transfer from cluster to cluster involves the mean square displacement $n^2 d_0^2$; hydrodynamic diffusion of an ordinary ion, on the other hand, d_0^2. Comparison of the relationship $D = \delta^2 \nu/6$ between diffusion coefficient D, mean square displacement δ^2, and frequency ν for proton transfer between clusters and for hydrodynamic diffusion, respectively, leads to

$$n = (D\nu_0/D_h k_1[H_2O])^{1/2} = (\lambda_{H^+}\nu_0/\lambda_{K^+} k_1[H_2O])^{1/2} \qquad \text{(a)}$$

in case (a) and

$$n = (D/D_h)^{1/2} = (\lambda_{H^+}/\lambda_{K^+})^{1/2} \qquad \text{(b)}$$

in case (b). Here, D is the total diffusion coefficient of the hydrogen ion and D_h is the diffusion coefficient of an ion of the same size moving only hydrodynamically. Case (a) was regarded by Naberukhin and Shuiskii as referring to a Grotthus mechanism.

Formula (a) yielded higher values ($n = 11.5$ at 0°C and $n = 4.6$ at 100°C for H^+), and its application to water–nonelectrolyte mixed solvents leads to contradictions. With expression (b), $n = 2.4$ was calculated at 0°C. Its use for the mixed solvents water–ethanol, water–dioxane, and water–acetone, allowed the evaluation of how the addition of a nonelectrolyte decomposes water clusters.

These calculations start from a layer-structure model of the mixed solvent, according to which the solvent is composed of microscopic—though relative to molecular dimensions, macroscopic—regions of its components. An average mobility is calculated from the relative dimensions of these regions. Finally, for the parameter n characterizing the size of the water clusters, the expression

$$n = \frac{(\lambda_{H^+}/\lambda_{K^+})^{1/2}}{1 - p(x_{nonel}/x_{water})[(\lambda_{H^+}/\lambda_{K^+})^{1/2} - 1]}$$

was derived. Here, p is a parameter ($0 < p \leqslant 1$), and x denotes the respective mole fractions. With increasing mole fraction of the nonelectrolyte, the water clusters gradually decompose. In water–acetone mixtures at $x_{nonel} = 0.6$, decomposition is complete.

Some basic assumptions of the theory of Naberukhin and Shuiskii must be criticized. In the case of pure water as a solvent, the model and the assumptions are too general to allow conclusions on the real mechanism of proton transfer and electrical conduction. In solvent mixtures, the molecular distribution assumed for the components (assuming alternating microcrystalline regions of the two components) may be justified in exceptional cases only.

Proton transfer cannot be excluded in such nonelectrolytes as ethanol, as was shown by, for example, Lin *et al.*[118–120]

A weak point of the theory is that although one assumption leads to good results with respect to one phenomenon, it is found necessary to change it in the case of another phenomenon. Thus, Walden product values in good agreement with experiment were obtained by assuming $p = 1$, but this assumption led to unreasonable n values. Reasonable n values required the assumption $p = 0.1$.

4. The Quantum Mechanical Theory of Weidemann and Zundel[121]

In their quantum mechanical theory, Weidemann and Zundel used a model based on IR spectroscopic studies.[122–125] The kernel of the model is an ionic grouping $(H_5O_2)^+$ in which the H_3O^+ ion is bound to an H_2O molecule by a hydrogen bond. Proton transfer between them occurs, and a symmetrical, double minimum potential energy curve of the proton, similar to that of Conway *et al.*,[27] is assumed. The $(H_5O_2)^+$ grouping, on the other hand, is linked to four water molecules by hydrogen bonds. These hydrogen-bonds are, normally, bent, and thus hinder proton tunneling from the $(H_5O_2)^+$ to neighboring H_2O molecules. However, for short periods, they will be linearized by thermal motion. During these periods, two of the three protons of the H_3O^+ [in the $(H_5O_2)^+$] become equivalent, and in addition to the original transfer, transfer to one of the H_2O molecules bound to $(H_5O_2)^+$ may occur. If, finally, the bond within the original $(H_5O_2)^+$ bends, it will be found that $(H_5O_2)^+$ has moved. This process is the structural diffusion defined by Weidemann and Zundel. This structural diffusion is influenced by an external field that polarizes $(H_5O_2)^+$ in the sense that the tunneling (excess) proton will be found with higher probability in one of its alternative sites. The site lying in the field direction will be preferred. This increases the probability of structural diffusion in the field direction and results in net charge transfer.

For the symmetrical double minimum potential well in a hydrogen bond, the shift of the split zero level was quantum mechanically calculated. With the value $\nu = 10^{13}\,\text{sec}^{-1}$ for the tunneling frequency,[6] and the transition dipole moment calculated using a linear dimension of 2 Å and a charge of $\frac{1}{2}e_0$ Zundel[125] and Weidemann and Zundel,[126] an extremely small shift was obtained as a function of the external electrical field. Then, from this shift, the shift of the weights of the two alternative sites ("proton boundary structures") of the proton and the induced dipole momentum and polarizability of the hydrogen bond were evaluated. Finally, by relating the abnormal mobility to the induced dipole moment and polarizability of the hydrogen bond, averaging over all orientations, and using the value $s = 2.5\,\text{Å}$ for the mean displacement by a structural diffusion step, Weidemann and Zundel obtained the value $v_s = 1.3 \times 10^{12}\,\text{sec}^{-1}$ for the number of structural diffusion steps per second if no external field is acting. This value agrees reasonably with that of Eigen,[6] who, comparing mechanisms of proton transport in ice and water, estimated a value of about $0.3 \times 10^{12}\,\text{sec}^{-1}$. The conclusion is made that the frequency of the structural diffusion step is about one order of magnitude smaller than the proton tunneling frequency and that the small shift of the weights of the alternative sites of the proton in the hydrogen bond, due to the external field, is the rate-determining factor in abnormal conductance.

5. Theories Rejecting Proton Transfer

Perrault[127] assumed that hydrogen ions exert exclusively hydrodynamic motion. However, he considered that only a small (activated) fraction C^*/C of the ions take part in this motion. He regarded the activation energy E_{H^+} in $C^*/C = e^{-E_{H^+}/kT}$ as the desolvation energy of the proton, and evaluated this, semiempirically, from the apparent activation energies of the electrical conductance and viscosity of pure water and the ionic product and the limiting conductances of the ions H^+ and OH^-. He obtained a value of $E_{H^+} = 0.260$ eV. Consequently, the concentration C^* is $e^{-10.1} = 4 \times 10^{-5}$, which is less than the stoichiometric concentration of the hydrogen ion. This very low value, of course, must be combined with a very high value of mobility in order to obtain the experimental value of the limiting specific conductance.

Thus, the mobility must be increased by about $1/(4 \times 10^{-5})$. A Stokes's law radius of 1.55×10^{-13} cm appears to follow from the model.

According to this theory, H_3O^+ does not move under the influence of the external field, and only the free protons undergo hydrodynamic motion.

Perrault's theory is at variance with all the important experimental and theoretical evidence. The value estimated for the standard free energy of the protonation reaction $H_2O + H^+ \rightarrow H_3O^+$ is some 30 times Perrault's value (see, for example, point III.1. in Conway's review[9] and Ref. 128). Also, it is doubtful whether the use of Stokes's law is justified for particles of the dimensions of atomic nuclei. What forces prevent the H_3O^+ species from moving under the influence of the external field?

Zatsepina[129] assumed that the H_3O^+ ion is not linked to water molecules; i.e., she assumed that no hydrogen bonds between the H_3O^+ ion and water molecules occur in water or in aqueous acid (or base) solutions. In Zatsepina's model, the "hydrophobic" H_3O^+ species exert translational motion in the holes of the hydrogen-bonded structure of water, between collisions, with average velocity v corresponding to the kinetic energy of an ideal gas molecule. A theoretical development is made that is similar to the classic theory of ionic conduction in gases. The mean free path is found to be $l = 0.8 \times 10^{-8}$ cm. The cube root of the free volume of liquid water at 0°C is of the same order. The diffusion coefficient of the H_3O^+ ion in water proves to be 1.5×10^{-4} cm sec^{-1} by the gas-kinetic formula $D = (1/3)lv$. This value was compared with the experimental value, i.e., $D_{H_3O^+} = 6 \times 10^{-5}$ cm^2 sec^{-1}. Also, the rate constant k_d (the reciprocal of the lifetime τ_{H_2O} of the H_2O molecule) of the dissociation reaction

$2H_2O \rightleftarrows H_3O^+ + OH^-$ was estimated. A recombination radius R was taken as the distance at which the potential energy of the attracting electrical force is equal to the kinetic energy (of ideal gas molecules) $3kT/2 = e^2/\varepsilon R$. The value $R = 10^{-7}$ cm was found. The average distance L between the point where an H_3O^+ ion is created by dissociation and the point where it will disappear because of recombination was obtained by dividing the average volume containing an ion by πR^2. The value 1.6 cm was obtained. This divided by the average gas kinetic velocity gives a lifetime $\tau_{H_3O^+}$, of 2.8×10^{-5} sec. From the relationship $C_{H_3O^+}/C_{H_2O} = \tau_{H_3O^+}/\tau_{H_2O} = k_d \tau_{H_3O^+}$, $k_d = 2 \times 10^{-5}$ sec^{-1} was obtained, in good agreement with Mayer's value of 2.5×10^{-5} sec^{-1}.

The corresponding values for ice were calculated ($\tau = 2 \times 10^{-11}$ sec, $l = 1.1 \times 10^{-6}$ cm, $D \approx 2 \times 10^{-2}$ cm^2 sec^{-1}, $R \approx 10^{-6}$ cm, $L = 17$ cm, $k_d = 2 \times 10^{-9}$ sec^{-1}).

Despite the agreement between some experimental findings and the conclusions from Zatsepina's theory, there is undeniable evidence of strong interaction between the H_3O^+ ion and water. Thus, it is difficult to imagine free translational motion with collisions in a condensed phase. It is also difficult to regard a notion in which the size of the particle and the mean free path length are of the same order of magnitude.

IV. THE THEORY OF THE CONTRIBUTION OF TRANSFER REACTIONS TO TRANSPORT PROCESSES

Proton transfer is an exchange reaction in which the proton is transferred, for example, from an H_3O^+ ion to a neighboring (colliding or encountering) water molecule.

By transfer of the proton between two encountering particles, the particles change places. Consequently, a displacement of the ion and molecule (in the opposite or approximately opposite directions) results. In contrast to hydrodynamic motion ("bodily transport") of the ion, displacement by proton transfer is not preconditioned by change of the translational coordinates of the surrounding particles (e.g., the solvent molecules). In the case of fast exchange reactions, the displacement by transfer may have a higher velocity than hydrodynamic movement.

In general, by the exchange reaction

$$AX + B \rightleftarrows A + XB \tag{1}$$

the particle X (electron, proton, atom, radical, molecule, ion, or energy quantum) will frequently be transferred between two colliding (encountering) particles.

If the rate of the forward reaction

$$k_f C_{AX} C_B$$

differs from that of the reverse reaction

$$k_r C_A C_{XB}$$

then the exchange reaction *may* contribute to the fluxes of the transport processes in the solution. However, the additional condition of continuous repetition of excess transfer in one direction must be fulfilled. Thus, to make successive transfer possible, a second fast elementary process is needed. Rotation of XB, for example, may be such a process. A consecutive mechanism of dissociation–association of ion pairs (under the influence of the external field) followed by their rotation was proposed as a general mechanism for electrolytic conduction in solutions by Grotthuss. Similar double mechanisms are usually referred to as Grotthuss *chain mechanisms*, even if they differ somewhat from the original model.

Theories of proton-transfer contributions to transport processes in liquids were reviewed above. Electron transfer between two atoms was examined in a quantum mechanical theory of electronic conduction by McCrea.[130] The same method of calculation was used by Bernal and Fowler[25] for proton transfer in aqueous solutions. Marcus,[131–135] Laidler,[136] and Levich[39] made well-known contributions to electron-transfer processes between ions in solution. In addition to the effect of the electrical field, Dahms[137] calculated the contribution to electronic conduction due to local concentration gradients.

Ruff and Friedrich[115] generalized the theory of the contribution of exchange reactions (if rate-controlling) to transport processes for transferred particles (atoms, radicals, ions, electrons, energy quanta) and for isothermal diffusion, electric conductance, and thermal diffusion. However, an erroneous factor arises in the equation connecting flux and reaction rate.[116]

The contribution J_X of the exchange reaction to the flux of particle X can be calculated from the difference between the rates of the forward (f) and reverse (r) reactions. Consider an exchange reaction between the triiodide and iodide ions, or the tribromide and the bromide ions, respectively, as studied by Ruff *et al.*[44,45] The triiodide ion has a linear structure,[138] and the same can be assumed for the tribromide ion. Let us restrict ourselves to linear collisions only, and assume that transfer proceeds exclusively at the distance of closest approach of the ions. In such a case, the forward reaction in (1) (with B identical to A) displaces the mass center of AX by the vector $\boldsymbol{\delta}_{AX}$ and that of A by $\boldsymbol{\delta}_A$. $\boldsymbol{\delta}_{AX}$ and $\boldsymbol{\delta}_A$ are antiparallel, and directions coincide with that of the collision (linear collision) (Fig. 6). The reverse reaction rearranges the original configuration.

In such cases, because of the symmetry of AX, successive transfer is possible without a second elementary process other than collision of two ions. Thus, no rotation of AX is needed, and the rate of the exchange reaction is the only elementary process on which the contribution of transfer to transport processes (e.g., diffusion) is dependent.

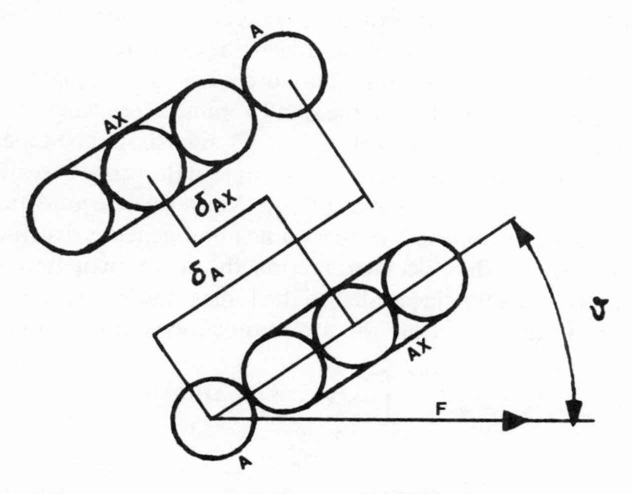

Figure 6. Geometric model for the exchange reaction $AX + A \rightleftarrows A + XA$.

If external forces F (e.g., for charged particles in an electric field) act in the solution and the concentrations vary with the coordinates, then each factor in the product $kC_{AX}C_A$ depends on the position, and for the contribution of the exchange reaction to the flux of AX, one obtains

$$J_{X,\theta} = \boldsymbol{\delta}_{AX}(k_f C_{AX} C_B - k_r C_A C_{XB})$$
$$= kC_{AX}C_A\boldsymbol{\delta}_{AX}[1 - e^{-\alpha\cos\theta}(1 + \beta\cos\theta)] \tag{2}$$

with

$$\alpha = (\pm\delta_{AX}F_{AX} \pm \delta_A F_A \pm \delta_{AX}|\mathrm{grad}\,\mu_{AX}| \pm \delta_A|\mathrm{grad}\,\mu_A|)/RT \tag{3}$$

and

$$\beta = \pm\frac{\delta_{AX}}{C_{AX}}|\mathrm{grad}\,C_{AX}| \pm \frac{\delta_A}{C_A}|\mathrm{grad}\,C_A| \tag{4}$$

Here β is a measure of concentration changes and $\alpha RT\cos\theta$ is the difference of the potential energy due to the external field and to the thermodynamic forces for the configurations before and after transfer of X. Correspondingly, $e^{-\alpha\cos\theta}$ is the factor by which the values of the rate constants differ in the two configurations, if a Boltzmann type of distribution is taken into account. The chemical potentials are denoted by μ. θ is the angle between the directions of collision and the forces that are assumed to be parallel or antiparallel with respect to each other, k is the rate of the exchange reaction if no forces act, and C_{AX} and C_A are some average concentrations in the small volume in the reaction takes place. Finally, $\delta = |\boldsymbol{\delta}|$ and $F = |\mathbf{F}|$. The "+" sign should be taken if $\boldsymbol{\delta}$ and the corresponding force subtend an angle in the range 0 to $\pi/2$; the "−" sign, for an angle in the range $\pi/2$ to π. Any term of second order in values of δ has been omitted. If we assume homogeneous distribution of linear collision with respect to direction, the direction of the average contribution to the flux is parallel to the forces, and by averaging the component in this direction, the absolute value is found to be

$$J_X = \frac{1}{2\pi}\int_0^{\pi/2}\int_0^{2\pi} J_{X,\theta}\cos\theta\sin\theta\,d\theta d\psi$$

$$= kC_{AX}C_A\delta_{AX}\frac{\alpha - \beta}{3} \tag{5}$$

Integration leads to a sum of exponential functions. These have been expanded, and terms higher than those of the first order in α and β have been neglected. We obtain the same result, i.e., Eq. (5), if we do not involve a Boltzmann distribution and assume that the rate constants in the two positions differ by a factor f. Here, $f = f(\alpha \cos \theta)$ is any analytical function with $f(0) = 1$. Expansion in series and neglect of terms higher than the first order before integration lead to Eq. (5).

The gradients of the activity coefficients may be neglected. Therefore, we obtain

$$J_X = \frac{kC_{AX}C_A\delta_{AX}^2}{3}\left(\pm F_{AX} \pm \frac{\delta_A}{\delta_{AX}}F_A \pm \frac{2}{C_{AX}}|\text{grad } C_{AX}| \pm \frac{2}{C_A}\frac{\delta_A}{\delta_{AX}}|\text{grad } C_A| \right) \tag{6}$$

For electric conductance and isothermal diffusion, we have

$$J_X^{\text{cond}} = \frac{kC_{AX}C_A\delta_{AX}^2 eE}{3}\left(Z_{AX} + \frac{\delta_A}{\delta_{AX}}Z_A \right) \tag{7}$$

and

$$J_X^D = \frac{2k\delta_{AX}^2}{3}\left(\pm C_A|\text{grad } C_{AX}| \pm \frac{\delta_A}{\delta_{AX}}C_{AX}|\text{grad } C_A| \right) \tag{8}$$

respectively. Here e denotes the positive elementary charge, Z_{AX} and Z_A the (positive or negative or zero) valency of the particles, and E the electric field intensity.

The conditions for the validity of Eqs. (6)–(8) can be summarized as follows.

a. The exchange reaction controls the rate of the transport process.
b. Transfer occurs only at a definite distance (close contact) of the colliding particles.
c. The collision is linear, and the displacements of the particles are parallel.
d. External and thermodynamic forces have the same directions or their opposites.

e. The orientation distribution of the direction of displacements and collision with respect to the forces is homogeneous.

If conditions (b)–(e) do not hold, then more complicated averaging becomes necessary, and in the relationships (6)–(8), other numerical constants will be obtained.

The theory outlined above could be applied to the anomalous conductance of the H^+ and OH^- ions insofar as (a) is fulfilled; i.e., proton transfer controls the overall ionic mobility. However, probably, this is not the case.

V. CONCLUDING COMMENTS

The existence of relatively long-lived H_3O^+ and OH^- species in water, a strong interaction between these ions and the solvent, and the role of proton transfers through hydrogen bonds not only between the water molecules but also between H_3O^+ and H_2O as well as the liquid structure, should be accepted and taken as basic assumptions in further theories.

New experimental results, e.g., the values obtained experimentally (by NMR line broadening) for the rate constants of proton transfer reactions (cf. Tables 1 and 2), should be related to the results of classically and quantum mechanically calculated proton-transfer frequencies and variations of these frequencies by external fields. Further, the results of recent theoretical investigations on hydrogen-bonded systems and proton-transfer processes in these systems should be taken into account. The model treatments by Brickmann and Zimmermann,[139,140] Singh and Wood,[141] Janoschek *et al.*,[142] and Anderson and Lippincott[143]; the investigations of proton transfer in a symmetrical potential well by Sussmann,[144] Gosar,[145] and Fischer *et al.*[146]—all consider proton–lattice couplings. Rosenstein's classical time correlation theory[147] and Rösch's quantum mechanical strong coupling scheme[148] may contribute to a solution of the kinetic problem. Finally, the study of Fang *et al.*,[149] calculating protonic Hartree states for the $H_{21}O_{10}+$ complex and proton–proton correlations, should be mentioned here, as well as the SCF, MO, and LCGO studies of Kraemer and Diercksen[150] on hydrogen bonding in the system $(H_2OHOH_2)^+$. Such studies lead to a linear symmetrical hydrogen bond with an equilibrium oxygen–oxygen distance of 2.39 Å. Some *ab initio* studies on the structure and energetics of the systems $H_3O^+(H_2O)_n$, and $OH^-(H_2O)_n$, with $n = 0$–4, by Newton and Ehrenson,[151] may also provide a basis for further work.

REFERENCES

[1] H. Danneel, *Z. Electrochem.* **11** (1905) 125.
[2] H. Goldschmidt and O. Udby, *Z. Phys. Chem.* **60** (1907) 728.
[3] W. M. Latimer and W. H. Rodebush, *J. Am. Chem. Soc.* **42** (1920) 1419.
[4] P. O. Löwdin, *Adv. Quantum Chem.* **2** (1966) 213.
[5] S. N. Vinogradov and R. H. Linnell, *Hydrogen Bonding*, Van Nostrand, New York, 1971, Ch. 10.
[6] M. Eigen, *Z. Angew. Chem.* **75** (1963) 489; *Angew. Chem.* (*Int. Ed. Engl.*) **3** (1964) 1.
[7] E. S. Amis, *Solvent Effects on Reaction Rates and Mechanism*, Academic Press, New York, 1966.
[8] G. C. Pimentel and A. L. McClellan, *Ann. Rev. Phys. Chem.* **22** (1971) 347.
[9] B. E. Conway, *Proton Solvation and Proton Transfer Processes in Solutions*, in *Modern Aspects of Electrochemistry* (1964), Vol. 3, p. 43.
[10] *Handbook of Chemistry and Physics*, 54th Ed., Chemical Rubber Co., Cleveland, Ohio, 1973, p. D-132.
[11] V. M. Goldschmidt, *VII Skritf Norsk. Vidensk. Akad. Oslo* **1** (1926).
[12] L. Pauling, *J. Am. Chem. Soc.* **49** (1927) 765.
[13] C. Dempwolff, *Phys. Z.* **5** (1904) 637.
[14] S. Tijmstra, *Z. Phys. Chem.* **49** (1904) 345.
[15] A. Hantzsch and K. S. Caldwell, *Z. Phys. Chem.* **58** (1907) 575.
[16] M. Volmer, *Ann. Chem.* **440** (1024) 200.
[17] E. Wicke, M. Eigen, and T. Ackermann, *Z. Phys. Chem. N.F.* **1** (1954) 340.
[18] J. O. Lundgren and J. W. Williams, *J. Chem. Phys.* **58** (1973) 788.
[19] R. Lorenz, *Z. Phys. Chem.* **73** (1910) 252.
[20] J. Tamás, S. Lengyel, and J. Giber, *Acta Chim. Acad. Sci. Hung.* **38** (1963) 225.
[21] T. Ackermann, *Discuss Faraday Soc.* **27** (1957) 180.
[22] E. Wicke, M. Eigen, and T. Ackermann, *Z. Phys. Chem. N.F.* **1** (1954) 340.
[23] E. Högfeldt, *Acta Chem. Scand.* **14** (1960) 1597.
[24] E. Hückel, *Z. Elektrochem.* **34** (1928) 546.
[25] J. D. Bernal and R. H. Fowler, *J. Chem. Phys.* **1** (1933) 515.
[26] A. E. Stearn and H. Eyring, *J. Chem. Phys.* **5** (1937) 113.
[27] B. E. Conway, J. O'M. Bockris, and H. Linton, *J. Chem. Phys.* **24** (1956) 834.
[28] A. J. Ellis, *J. Chem. Soc. London* **1959** (1959) 3689.
[29] S. D. Hamann, *Physico-chemical Effects of Pressure*, Butterworth, London, 1957.
[30] S. D. Hamann and W. Strauss, *Trans. Faraday Soc.* **51** (1955) 1684.
[31] S. Glasstone, K. Laidler, and H. Eyring, *The Theory of Rate Processes*, McGraw-Hill, New York, 1941, p. 570.
[32] Zisman, *Phys. Rev.* **39** (1932) 151.
[33] Taumann and Tofante, *Z. Anorg. Chem.* **182** (1929) 353.

[34]G. N. Lewis and T. C. Doody, *J. Am. Chem. Soc.* **55** (1933) 3504.
[35]N. K. Roberts and H. L. Northey, *J. Chem. Soc. Faraday Trans. 1* **68** (1972) 1528.
[36]N. K. Roberts and A. L. Northey, *J. Chem. Soc. Faraday Trans. 1* **70** (1974) 253.
[37]L. A. Woolf, *J. Phys. Chem.* **60** (1960) 481.
[38]D. Glietenberg, A. Kutschker, and M. Stackelberg, *Ber. Bunsenges. Phys. Chem.* **72** (1968) 562.
[39]V. G. Levich, *Present state of the theory of oxidation–reduction in solution (bulk and electrode reaction)*, in *Advances in Electrochemistry and Electrochemical Engineering*, Ed. by P. Delahay, 1961, Interscience, New York, Vol. 4, pp. 249–371.
[40]I. Ruff, V. J. Friedrich, K. Demeter, and K. Csillag, *J. Phys. Chem.* **75** (1971) 3303.
[41]I. Ruff and M. Zimonyi, *Electrochim. Acta* **18** (1973) 515.
[42]I. Ruff, M. Zimonyi, and P. Kovács, private communication.
[43]I. Ruff, A. Hegedüs, and V. J. Friedrich, *J. Phys. Chem. Magy. Kém. Foly.* **79** (1973) 133.
[44]I. Ruff, V. J. Friedrich, and K. Csillag, *J. Phys. Chem.* **76** (1972) 162.
[45]I. Ruff and V. J. Friedrich, *J. Phys. Chem.* **76** (1972) 2957.
[46]Yu. Ya. Fialkov and Yu. A. Karapetyan, *Ukr. Khim. Zh.* **37** (1971) 398.
[47]V. P. Basov, Yu. A. Karapetyan, and A. D. Krysenko, *Elektrokhimiya* **10** (1971) 420.
[48]P. Wulff and H. Hartmann, *Z. Elektrochem.* **47** (1941) 858.
[49]H. Goldschmidt and P. Dahl, *Z. Phys. Chem.* **108** (1924) 121; **114** (1925) 1.
[50]P. Walden, *Z. Phys. Chem.* **108** (1924) 341.
[51]L. Thomas and E. Marum, *Z. Phys. Chem.* **143** (1929) 191.
[52]I. I. Berman and F. H. Verhock, *J. Am. Chem. Soc.* **67** (1945) 1330.
[53]G. Kortüm and H. Wilski, *Z. Phys. Chem. N.F.* **2** (1954) 256.
[54]G. F. Dnieprov, *Chem. Zap. LGU* **40** (1939) 19.
[55]A. M. el Aggan, D. C. Bradley, and W. Wardlaw, *J. Chem. Soc.* **1958** (1958) 2092.
[56]A. R. Tourky and S. Z. Mikhail, *Egypt. J. Chem.* **1** (1958) 1, 13, 187.
[57]P. Kirby and O. Maass, *Can. J. Chem.* **36** (1958) 456.
[58]L. A. Woolf, *J. Phys. Chem.* **64** (1960) 500.
[59]F. Accascina, A. D'Aprano, and M. Goffredi, *Ric. Sci.*, **34** (1964) II, 443.
[60]F. Accascina, R. de Lisi, and M. Goffredi, *Electrochim. Acta* **15** (1970) 1209.
[61]H. R. Tourky, S. Z. Mikhail, and A. A. Abdel Hamid, *Z. Phys.* **252** (1973) 289.
[62]A. A. Abdel Hamid, M. F. Ragaii, and I. Z. Slim, *Z. Phys. Chem.*, **254** (1973) 1.
[63]R. de Lisi and M. Goffredi, *J. Chem. Soc. Faraday Trans. 1* **72** (1974) 787.
[64]T. Erdey-Grúz, E. Kugler, K. Balthazár-Vass, and I. Nagy-Czakó, *Acta Chim. Acad. Sci. Hung.* **79** (1973) 169.
[65]T. Erdey-Grúz, E. Kugler, and L. Majthényi, *Electrochim. Acta* **13** (1968) 947.
[66]T. Erdey-Grúz and I. Nagy-Czakó, *Acta Chim. Acad. Sci. Hung.* **67** (1971) 283.
[67]T. Erdey-Grúz, *Transport Phenomena in Aqueous Solutions*, A. Hilger, London, 1974.
[68]Z. Szabó, *Magy. Kém. Foly.* **40** (1935) 52.
[69]H. S. Harned and E. C. Dreby, *J. Am. Chem. Soc.* **61** (1939) 3113.
[70]S. Lengyel, J. Giber, and J. Tamás, *Acta Chim. Acad. Sci. Hung.* **32** (1062) 429.
[71]J. Giber, S. Lengyel, J. Tamás, and P. Tahi, *Magy. Kém. Foly.* **66** (1960) 170.
[72]S. Lengyel, in *Electrolytes*, Pergamon, (London, 1962), p. 208.
[73]E. A. Kaimakov and V. B. Fiks, *Zh. Fiz. Khim.* **35** (1961) 1777.
[74]B. P. Konstantinov and V. P. Troshin, *Zh. Prikl. Khim.* **35** (1962) 2420.
[75]B. P. Konstantinov and V. P. Troshin, *Izv. Akad. Nauk SSSR* (1966) 2104.
[76]S. Lengyel, J. Giber, Gy. Beke, and A. Vértes, *Acta Chim. Acad. Sci. Hung.* **39** (1963) 357.
[77]V. P. Troshin and E. V. Zvyagina, *Elektrokhimiya* **8** (1972) 505.
[78]V. P. Troshin and E. V. Zvyagina, *Elektrokhimiya* **8** (1972) 1712.

[79]V. P. Troshin and E. V. Zvyagina, *Elektrokhimya* **8** (1972) 586.
[80]S. Meiboom, *J. Chem. Phys.* **34** (1961) 375.
[81]Z. Luz and S. Meiboom, *J. Chem. Phys.* **39** (1963) 366.
[82]Z. Luz and S. Meiboom, *J. Am. Chem. Soc.* **86** (1964) 4766.
[83]Z. Luz and S. Meiboom, *J. Am. Chem. Soc.* **86** (1964) 4768.
[84]M. Anbar, A. Loewenstein, and S. Meiboom, *J. Am. Chem. Soc.* **80** (1958) 2630.
[85]Z. Luz, D. Gill, and S. Meiboom, *J. Chem. Phys.* **30** (1959) 1540.
[86]E. Grunwald, C. F. Jumper, and S. Meiboom, *J. Am. Chem. Soc.* **84** (1962) 4664.
[87]A. Loewenstein and A. Szöke, *J. Am. Chem. Soc.* **84** (1962) 1151.
[88]R. E. Glick and K. C. Tewari, *J. Chem. Phys.* **44** (1966) 546.
[89]S. W. Rabideau and H. G. Hecht, *J. Chem. Phys.* **47** (1967) 544.
[90]G. Wannier, *Ann. Phys.* **24** (1935) 545, 569.
[91]M. L. Huggins, Undergraduate thesis (1919).
[92]*Proceedings of the Fourth International Congress on Electron Microscopy*, Springer-Verlag, Berlin, 1958; *Z. Naturforsch.* **15a** (1960) 822.
[93]M. L. Huggins, *J. Phys. Chem.* **40** (1936) 723.
[94]M. L. Huggins, *J. Am. Chem. Soc.* **53** (1931) 3190.
[95]B. E. Conway and M. Salomon, in *Chemical Physics of Ionic Solutions*, Ed. by B. E. Conway and Barradas, Wiley, New York 1966, p. 541.
[96]A. Eucken, *Gött. Nachr., Math-Phys. Kl.* **1946** (1946) 39; *Z. Elektrochem.* **52** (1948) 255.
[97]A. Gierer and K. Wirtz, *Ann. Phys. 6 F* **6** (1949) 257.
[98]M. Eigen and L. De Maeyer, *Proc. Roy. Soc. A* **247** (1958) 505.
[99]J. J. Hermans, *Rec. Trav. Chim. Pays-Bas* **58** (1939) 917.
[100]M. Eigen, *Z. Elektrochem.* **56** (1952) 176.
[101]M. Eigen and J. Schoen, *Z. Elektrochem.* **59** (1955) 483.
[102]Darmois and Sutra, *C. R. Acad. Sci. Paris* **222** (1946) 1286.
[103]B. E. Conway and J. O'M. Bockris, *J. Chem. Phys.* **31** (1959) 1133.
[104]M. Eigen and L. De Maeyer, *J. Chem. Phys.* **31** (1959) 1134.
[105]R. A. Horne and R. A. Courant, *J. Phys. Chem.* **69** (1965) 2224.
[106]D. A. Lown and H. R. Thirsk, *Trans. Faraday Soc.* **67** (1971) 132.
[107]R. A. Horne, B. R. Myers, and G. R. Frysinger, *J. Chem. Phys.* **39** (1963) 2666.
[108]N. I. Gusev and P. N. Palei, *Zh. Fiz. Khim.* **45** (1971) 1164.
[109]N. I. Gusev, *Zh. Fiz. Khim.* **45** (1971) 1164, 2238, 2243.
[110]H. S. Frank and W -Y. Wen, *Discuss. Faraday Soc.* **24** (1957) 133.
[111]D. A. Lown and H. R. Thirsk, *Trans. Faraday Soc.* **67** (1971) 149.
[112]D. W. McCall and D. C. Douglass, *J. Phys. Chem.* **69** (1965) 2001.
[113]H. G. Hertz and M. D. Zeidler, *Ber. Bunsenges. Phys. Chem.* **67** (1963) 774; **68** (1964) 621.
[114]I. Ruff and V. Friedrich, *J. Phys. Chem.* **76** (1972) 2954.
[115]I. Ruff and V. Friedrich, *J. Phys. Chem.* **75** (1971) 3297.
[116]S. Lengyel, *Magy. Kém. Foly.* **80** (1974) 187.
[117]Yu. I. Naberukhin and S. I. Shuiskii, *Zh. Strukt. Khim.* **11** (1970) 197.
[118]J. Lin and J. J. DeHaven, *J. Electrochem. Soc.* **116** (1969) 805.
[119]W. G. Breck and J. Lin, *Trans. Faraday Soc.* **61** (1965) 2223.
[120]J. Lin, *J. Electrochem. Soc.* **116** (1969) 1708.
[121]E. G. Weidemann and G. Zundel, *Z. Naturforsch.* **25a** (1970) 627.
[122]G. Zundel and H. Metzger, *Z. Phys. Chem. Frankfurt* **58** (1968) 225.
[123]G. Zundel and H. Metzger, *Z. Phys. Chem.* **59** (1968) 225.
[124]G. Zundel and H. Metzger, *Z. Naturforsch.* **22a** (1967) 1412.
[125]G. Zundel, *Hydration and Intermolecular Interaction*, Academic Press, New York, 1969.

[126] E. G. Weidemann and G. Zundel, *Z. Physik* **198** (1967) 288.
[127] G. Perrault, *C. R. Acad. Sci. Paris* **252** (1961) 3779.
[128] J. Long and B. Munson, *J. Chem. Phys.* **53** (1970) 1356.
[129] G. N. Zatsepina, *Zh. Strukt. Khim.* **10** (1969) 211.
[130] W. H. McCrea, *Proc. Carub. Phil. Soc.* **24** (1928) 438.
[131] R. A. Marcus, *J. Chem. Phys.* **24** (1956) 966.
[132] R. A. Marcus, *J. Chem. Phys.* **26** (1957) 867, 872.
[133] R. A. Marcus, *Can. J. Chem.* **37** (1959) 155.
[134] R. A. Marcus, *Discuss. Faraday Soc.* **29** (1960) 21.
[135] R. A. Marcus, paper presented at the Moscow CITEE Meeting, 1963.
[136] K. J. Laidler, *Can. J. Chem.* **37** (1959) 138.
[137] A. Dahms, *J. Phys. Chem.* **72** (1968) 362.
[138] *Tables of Interatomic Distances*, Chemical Society, London, 1960.
[139] J. Brickmann and H. Zimmermann, *Ber. Bunsenges. Phys. Chem.* **70** (1966) 157, 521.
[140] J. Brickmann and H. Zimmermann, *J. Chem. Phys.* **50** (1969) 1608.
[141] T. R. Singh and J. L. Wood, *J. Chem. Phys.* **48** (1967) 4567.
[142] R. Janoschek, E. G. Weidemann and G. Zundel, *J. Chem. Soc. Faraday Trans. 2* **69** (1973) 505.
[143] J. R. Anderson and R. E. Lippincott, *J. Chem. Phys.* **55** (1971) 4077.
[144] J. A. Sussmann, *Mol. Cryst. Liq. Cryst.* **18** (1972) 39.
[145] P. Gosar, *Novo Cimento* **30** (1963) 931.
[146] S. Fischer, G. L. Hofacker, and J. R. Sabin, *Phys. Kondens. Mater.* **8** (1968) 268.
[147] R. A. Rosenstein, *Ber. Bunsenges. Phys. Chem.* **77** (1973) 493.
[148] N. Rösch, *Chem. Phys.* **1** (1973) 220.
[149] J. K. Fang, K. Godzik, and G. L. Hofacker, *Ber. Bunsenges. Phys. Chem.* **77** (1973) 980.
[150] W. P. Kraemer and G. H. F. Diercksen, *Chem. Phys. Lett.* **5** (1970) 463.
[151] M. D. Newton and S. Ehrenson, *J. Am. Chem. Soc.* **93** (1971) 4971.

2

NMR Studies of the Structure of Electrolyte Solutions

A. K. Covington

Department of Physical Chemistry, University of Newcastle upon Tyne, England

and K. E. Newman*

Department of Chemistry, University of Keele, Keele Park, Staffs, England

I. INTRODUCTION

The importance of NMR to the study of solution effects was first noted as early as 1951 by Arnold and Packard.[1] In the past 25 years there has been a spectacular growth of literature concerned with this broad topic. In particular, electrolyte solutions have been widely studied, and knowledge of the structure of these solutions has been considerably advanced by the use of this technique. It is unfortunate that much of this work is fragmentary. Although there have been a number of reviews from different aspects, little attempt has been made to bring all the various approaches together in a critical way in order to "assess the state of the art," and to suggest where further work might be fruitful. The purpose of this chapter is to attempt such an assessment.

Excellent discussions of the theory of NMR are available in the standard works on the subject.[2–4] Advances in particular areas of the subject are frequently surveyed in the various review series now available.[5–8] It has been necessary to limit the scope of this review to electrolyte solutions, but because of the structural nature of such solutions, studies of electrolytic solvents must also be included.

* Present address: Department of Molecular Sciences, University of Warwick, Coventry, England.

Hydrogen-bonding interactions in solvent mixtures have been very widely studied (see, for example, the excellent review of Laszlo[9]), but unless the system studied is relevant to electrolyte solutions, such work will not be discussed. The technique of nuclear magnetic relaxation has in recent years received much attention, particularly from Hertz's group.[10] Unfortunately, the interpretation of the results is theoretically very complicated. However, the field was reviewed by von Goldammer[11] in a previous volume of this series. Thus, the relaxation studies will be mentioned only when appropriate. Chemical shift and linewidth studies of paramagnetic salt solutions yield valuable information on both the nature of bonding and kinetic exchange parameters between ion and solvent, but they will be discussed only cursorily, and the extension to ligand exchange and complex formation will be mentioned only in passing.

This chapter is essentially complementary to that of von Goldammer,[11] and is written primarily for solution physical chemists and electrochemists, who require an appreciation of the advances made in the understanding of the structure of electrolyte solutions through the use of NMR techniques, much of which information has not yet reached the standard texts.

II. GENERAL PRINCIPLES

1. Shift Measurements

In principle, it is possible to perform NMR experiments on any magnetic (nonzero nuclear spin quantum number) nucleus. In practice, there are many limitations. The most widely studied nucleus is ^1H, which resonates approximately 42.6 MHz at a field of 1T (10^4 G). It has the highest magnetogyric ratio γ, where $\gamma = 2\pi\nu/H_0$ (H_0 is the applied field and ν is the resonance frequency) of any nucleus with the exception of ^3H and a free electron.[12] Chemical shifts are conventionally expressed in terms of the dimensionless quantity

$$\frac{\delta}{\text{ppm}} = 10^6 \frac{(H - H_r)}{H_r}$$

where H and H_r are the resonant fields of the signal of interest and the reference. Upfield shifts are therefore positive.

NMR sensitivity depends on the population difference between the energy levels through the Boltzmann distribution; thus, for maximum sensitivity, one needs to work at the highest frequency possible. This frequency is nearly always limited through the magnetogyric ratio by the size of the magnet available. It has proved impossible to construct either permanent magnets or electromagnets of sufficient stability and homogeneity with a field much larger than 2.4 T due to the enormous forces operating on the pole pieces. Recently, it has been shown possible to use superconducting magnets,[13] and a commercial instrument with a field strength of approximately 7.0 T is now available.[14] In general, however, with field strengths usually the limiting factor, nuclei with the highest magnetogyric ratio are the most sensitive. Thus (at constant field) the reactive sensitivity of ^{14}N (3.08 MHz at 1.0 T), compared with ^{1}H, is only 1.01×10^{-3}. This is the principal reason for the preponderance of ^{1}H results, with less sensitive nuclei seeming less important. The recent advent of Fourier-transform NMR[15] as a technique for improving signal-to-noise ratios means that low-sensitivity nuclei, even when at only natural abundance, become feasible. In particular, the spectacular growth of ^{13}CNMR (relative sensitivity 1.59×10^{-2}, abundance 1.1%) over the past few years demonstrates this admirably. There are, however, instrumental difficulties concerned with NMR measurements on different nuclei. In order to work at the highest field possible and also to allow field-locking procedures, it is necessary to work at a fixed field, and it is then necessary to have a separate oscillator, transmitter, probe, and receiver for each of the widely separated nucleus frequencies. The use of frequency synthesizers may help to alleviate this problem, but broad-band transmitters and receivers still present problems. The only other alternative is to work at various spot frequencies and wind down the field strength of the magnet until resonance is reached. This means that sensitivity is reduced and that it is not possible to use a field lock, since a variable frequency field lock is not yet possible due to electronic problems.

In comparison with other nuclei, proton chemical shifts are small (0–12 ppm). Referencing, in addition to imposing stringent conditions on magnet stability and homogeneity, becomes in itself an important problem. Two approaches are available, although neither is without disadvantages. The first is to use an external reference in a capillary inserted into the NMR tube or, in order to eliminate field distortion due to the capillary, to use precision glass coaxial tubes, the reference being placed in the annulus and the sample in the inner tube. Unfortunately, the macroscopic field experienced by the inner tube is affected by the bulk volume susceptibility of the solution in the annulus. It is therefore necessary to correct the measured shifts[16] for susceptibility effects by means of the formula

$$\delta_{\text{true}} = \delta_{\text{obs}} + 2\pi(\chi_{v(\text{ref})} - \chi_v)/3 \tag{1}$$

where χ_v and $\chi_{v(\text{ref})}$ are the volume susceptibilities of sample and reference, respectively. The measurement of volume susceptibilities of diamagnetic solutions to the required degree of accuracy is no easy matter. Although several NMR methods have been suggested,[17–19] none is very accurate. Thus, it is doubtful whether susceptibilities are readily measurable to $\pm 0.005 \times 10^{-6}$, whereas shift measurement to ± 0.1 Hz, or 0.001 ppm, at 100 MHz is very easy. For mixtures, Weidemann's additivity law is often used[20] when the susceptibilities of the components are known, although the accuracy of this law to NMR limits is uncertain. When conventional magnet systems are used, the field is applied perpendicularly to the long axis of the sample tube, whereas for superconducting magnets, the field is applied along this axis. The susceptibility correction then becomes[21]

$$\delta_{\text{true}} = \delta_{\text{obs}} + 4\pi(\chi_{v(\text{ref})} - \chi_v)/3 \tag{2}$$

By measuring shifts using both magnet systems, the effects of susceptibility corrections may be eliminated on combination of Eqs. (1) and (2). It is possible that imperfections in NMR tubes and in tube alignment may limit the accuracy of these two formulas.

The alternative referencing procedure is to use an internal reference. This standard may be added to the solution of interest; most organic chemists use tetramethyl silane (TMS), while for aqueous solutions, the use of both the sodium salt of trimethylsilylpropionic acid[22] and the Me_4N^+ ion[23] has been advocated. It is, however, dubious when studying solvent effects to assume that the presence of standard will not perturb the solvent, and that the solvent itself will not perturb the standard. By working at varying standard concentrations and extrapolating back to zero concentration, the former, but not the latter, effect may be removed. Some recent work comparing internal and external standards in organic solvents and using the conventional magnet–superconducting magnet technique to estimate susceptibility corrections has shown that very large solvent effects are exhibited by

TMS.[21] The internal standard may in some cases be a proton resonance of one of the species in solution; e.g., in MeOH + H_2O mixtures, the methyl protons are used as a reference, whereas for EtOH, the methylene group is most suitable. Although small solvent effects for these groups have been observed,[24] they are generally much smaller than the shifts of the protons of interest, viz., the hydroxyl protons in MeOH + H_2O mixtures.

The considerations outlined above limit the accuracy of proton-shift measurements. Indeed, in some cases, susceptibility changes can completely alter the direction of measured shifts. For nearly all nuclei except ^1H, the range of chemical shifts is very much larger and susceptibility corrections are usually negligible. An extreme example is ^{205}Tl where the shift between saturated aqueous $TlNO_3$ and TlOH is of the order of 1700 ppm.[25]

2. Linewidth and Relaxation Phenomena

The width of NMR lines is generally dictated by the lifetime of both ground and excited states through the uncertainty principle of Heisenberg. The relaxation rate from excited to ground state is called the spin-lattice relaxation rate ($1/T_1$), since it involves loss of spin energy to its surroundings. However, there is another possible relaxation process by which adjacent spins can adiabatically exchange energy. Thus, phase coherence between spins is lost after resonance has been passed (T_2). This mechanism is important in exchange phenomena, where the phase coherence may be lost rapidly. It is T_2 that controls the linewidth of the absorption. In inhomogeneous fields, different parts of the sample are at different magnet field strengths, and the observed signal is artificially broadened. In solids, adjacent magnetic nuclei exert an additional magnetic field on the nucleus of interest. With approximately equally distributed Boltzmann energy levels, this additional field may be of either sign. Thus, nuclei experience an apparently inhomogeneous field. In liquids and gases, reorientation of adjacent dipoles is so fast that this relaxation mechanism is unimportant. The relaxation process of loss of phase coherence dictates linewidths. However, for liquids and gases in general, T_1 equals T_2 unless exchange processes occur. It is only for viscous liquids and solids that T_2 is less than T_1.

NMR experiments involve very small energy differences between ground and excited states, so it is important to consider how the excited state may return to the ground state (relaxation). If the rate at which relaxation occurs ever becomes less than the rate of excitation, the excess population of the lower state will fall and the signal intensity will decrease; this effect is known as *saturation.* The probability of spontaneous return to the ground state is negligible. It is necessary for the excited state to couple to its surroundings in some way in order to lose energy.[26] For nuclei with spin $\frac{1}{2}$, e.g., ^{1}H, ^{13}C, ^{19}F, the principal mechanism is for the excited state to couple with a fluctuating magnetic field, the extent of coupling being proportional to the square of the component of the magnetic field fluctuating at the resonant frequency. Using time-dependent perturbation theory and the assumptions inherent in the theory of Brownian motion, it has been shown[27] that

$$\frac{1}{T_1} = \frac{2}{3}\gamma^2\overline{H^2}\frac{\tau_c}{1 + 4\pi^2\nu^2\tau_c} \qquad (3)$$

where $1/T_1$ is the rate of relaxation from excited to ground state, $\overline{H^2}$ is the mean-squared value of the magnetic fluctuations, and τ_c, the correlation time, is a measure of the time taken for a molecule to move a distance comparable to its size or rotate through 1 rad.[28] In diamagnetic solutions, the principal cause of fluctuating magnetic fields experienced by a magnetic nucleus will be the magnetic moments of adjacent "magnetic" ($I \neq 0$) nuclei. For liquids like water, there are two different types of adjacent nuclei, the first being the proton on the same molecule as the relaxing species. In this case, $\overline{H^2}$ depends on the molecular proton separation, and the correlation time depends on the molecular rotation. The other adjacent nuclei are those on different molecules. In this case, the fluctuating magnetic field depends on the distance of closest approach, while the correlation time is related to the diffusion coefficient. For non-proton-exchanging solvents (not water), it is possible to separate out the two effects using isotope dilution techniques. It is interesting that for nonabundant nuclei such as ^{13}C (1.1%), there are instances in which fluctuating magnetic field relaxation is very inefficient.[29] Thus, for a ^{13}C tertiary carbon atom, the probability of finding an adjacent magnetic nucleus, i.e., another ^{13}C carbon atom, is very low, and the nearest magnetic nucleus will be at least two bonds away. Relaxation times in excess of 60 sec have been observed. This poses severe problems, particularly with the use of Fourier-transform techniques, when successive pulses in a pulse train should be sufficiently far apart to allow all nuclei to relax completely.

In paramagnetic solutions, the unpaired electron with its very large magnetic moment is generally the most effective fluctuating magnetic field. Thus, solutions containing paramagnetic solutes generally give very short relaxation times, and hence very broad lines (often so wide as to be unobservable). For a full discussion, the reader is referred to Refs. 2–4.

For nuclei with spin $>\frac{1}{2}$, the majority of nuclei, an important relaxation mechanism, and usually the predominant one, is that of quadrupole relaxation. In this, an electric field gradient across a quadrupolar ($I > \frac{1}{2}$) nucleus perturbs the nuclear energy levels. By using time-dependent perturbation theory and Brownian motion theory, it can be shown[30] that

$$\frac{1}{T_1} = \left(\frac{3}{40}\right)\frac{(2I+3)}{I^2(2I-1)}\left(\frac{e^2Qq}{h}\right)^2 \tau_c \tag{4}$$

where Q is the quadrupole moment, q is the electric field gradient, and τ_c is the correlation time characteristic of the fluctuating field gradient. This relaxation mechanism can be very rapid indeed, depending on the field gradient. Thus, for the polar covalently bonded molecule CH_3Cl, the ^{35}Cl relaxation time is 0.025 msec, which corresponds to a linewidth of 13 kHz,[31] whereas in very symmetrical environments, e.g., Cl^- in very dilute solution, the linewidth is only 15 Hz. Increasing concentration is likely to increase the field gradients due to direct ion–ion encounters. Thus concentration-dependence measurements can yield information as valuable as that from conventional shift measurements.

By inspection of either Eq. (3) or (4), it may be seen that there are two causes for change in relaxation rate: either the interaction energy (dipolar or quadrupolar) may change or the correlation time may change. Thus, in viscous solutions, τ_c is long and lines are broad. Indeed, fairly accurate correlations are usually found between either viscosity or diffusion coefficients and correlation times.

The relative merits and demerits of the various nuclei in NMR studies may be summarized as follows: ^1H measurements are relatively straightforward due to high sensitivity and narrow lines, but the range of shifts is only 10 ppm, which makes accurate studies, particularly of solvent shifts, very difficult. Other spin $I = \frac{1}{2}$ nuclei give narrow lines, but (with the exception of fluorine) sensitivity is generally very low, and in several cases, small natural abundance (e.g., ^{13}C, ^{15}N) may make the techniques difficult. Fourier-transforms NMR is likely to reduce these problems considerably. For spin $I > \frac{1}{2}$ nuclei, accurate shift measurements are often limited by the broadness of the lines, but the linewidths themselves may yield useful information. Sensitivity problems are often important as well.

III. STUDIES ON SOLVENTS AND SOLVENT MIXTURES

Water has been the most studied of all solvents, largely because of its importance, but also because the difference between liquid- and gas-phase shifts is very much larger than for other solvents.

1. Structure of Water

Water may be regarded as an atypical liquid from the point of view of its physical and thermodynamic properties. Thus, compared with comparable nonmetal hydrides, it has much higher melting and boiling points and entropy of vaporization, its relative permittivity (dielectric constant) is very high, and it has a larger molar volume than expected from known nearest-neighbor separations (the molar volume also exhibits a minimum value at 3.98°C). These properties have been widely studied; for a full review of these and other spectroscopic, quantum mechanical, thermodynamic, and transport properties of ice, water, and steam, the reader is referred to either of the recent excellent monographs on the subject.[32,33] These properties may be rationalized in terms of molecules with strong intermolecular forces, the resulting liquid having a very open structure. Increase of temperature (or pressure) will tend to break up this open structure. The observed properties may be interpreted in terms of a competition between these two extreme states (dense and less dense). Early work centered on the idea that water was a mixture of "polyols" [H_2O, $(H_2O)_2$, $(H_2O)_3$, etc.], but such ideas were soon discounted following the X-ray diffraction study of Morgan and Warren.[34] Their results confirmed the idea proposed by Bernal and Fowler[35] that liquid water retained short-range tetrahedral order similar to that of ice. It has therefore been common practice to consider water as a mixture of icelike and nonicelike structures. An interesting special case is the "flickering cluster" model proposed by Frank and Wen.[36] They concluded that the formation of hydrogen bonds with some covalent character was a cooperative process whereby large clusters of hydrogen-bonded molecules would form following a single hydrogen-bond formation. Subsequent rupture of a single bond would cause the entire cluster to "dissolve." This model has been widely adopted, although it depends on the covalent nature of hydrogen bonding. However, the extent of covalent bonding is by no means certain; indeed, Coulson[37] considers that in terms of resonance structures, a covalent bond between adjacent water molecules has a weighting of only 4%. The lattice energy of ice was estimated as early as 1933 by a simple electrostatic model,[35] and recent molecular dynamics calculations have predicted the thermodynamic properties of liquid water using a simple electrostatic pair potential.[38]

Another special mixture model is the "interstitial" model proposed by Samoilov,[39] whereby free water molecules reside in cavities that are known to be present in ice. Most approaches, unfortunately, need a simple structural approach in order to make the statistical mechanics tractable. Even then the approximations, of doubtful validity, often contain several disposable parameters that are constrained to reproduce the experimental thermodynamic, diffraction, or spectroscopic data.

Pople,[40] however, has managed to explain adequately the thermodynamic properties of water in terms of a fully hydrogen-bonded liquid structure that allows for the effects of hydrogen-bond bending. Initially, this model does seem conceptually unrealistic, involving, as it does, no hydrogen-bond breaking at all. It does have something in common with the recent molecular dynamics approach of Rahman and Stillinger.[38] In this model, the pairwise-additive potential of Ben-Naim and Stillinger[41] was used. This potential consists of a repulsive R^{-12} and an attractive R^{-6} term with weighting coefficients corresponding to the isoelectronic neon atom with an array of four point charges arranged to permit tetrahedral hydrogen-bonding interaction. The dynamics calculations agreed well with respect to energy of vaporization, diffusion coefficients, radial distribution functions, and rotational correlation functions. Also, the apparent activation energy of many of these processes was approximately 2.6 kcal mol^{-1}, which has often been interpreted as the excitation between the two species of the mixture model. The Pople model is one of the few available that does not treat the hydrogen-bonding interaction as either on or off, which is also the essence of the electrostatic nature of the molecular dynamics calculations. It is unfortunate that it is not yet possible to study quantum effects from a molecular dynamics viewpoint and consequently not yet feasible to assess the importance of covalent effects of three-body interactions.

2. NMR Studies of Water

In variable-temperature NMR shift studies, one of the principal problems is in obtaining a standard that is known to be temperature-independent or the temperature dependence of which is known. Schneider *et al.*,[42] in one of the earliest studies, used cyclopentane as an external standard to calculate susceptibilities with density data. These workers noted that the chemical shift of water vapor was temperature- and pressure-independent, whereas for liquid water, a linear temperature dependence of 0.0095 ppm K^{-1} and a gas-to-liquid shift of 4.58 ppm was observed at 0°C. Malinowski *et al.*[43,44] used ethane gas as a reference, obtaining a linear temperature dependence of 0.00956 ppm K^{-1} and a gas-to-liquid shift of 4.53 ppm at 0°C. Hindman,[45] in a very thorough study, noted slight nonlinearity in the shift-temperature dependence ranging from 0.00986 ppm K^{-1} at 0°C to 0.00875 ppm K^{-1} at 100°C (Fig. 1).

The range over which temperature measurements can be made has recently been extended by containing the water in narrow capillaries (10–20 or 80–120 μm) or by keeping the water in a hexane emulsion. It proved possible[46] to make measurements by these means down to −38°C. Where the data overlap, they are concordant with Hindman's results. Below 0°C, the curvature mentioned by Hindman is enhanced (Fig. 1). Credence has been given to this method by ^{2}H relaxation measurements of D_2O + hexane emulsions[47] down to −37°C, where again the data are concordant with conventional measurements where they overlap.

Although there are now several studies on the pressure dependence of relaxation parameters of water[48] and theoretical interpretation of these results,[49] it is unfortunate that there are no very high pressure or temperature measurements of proton shifts of pure water, where it is thought that hydrogen-bonded structure is becoming

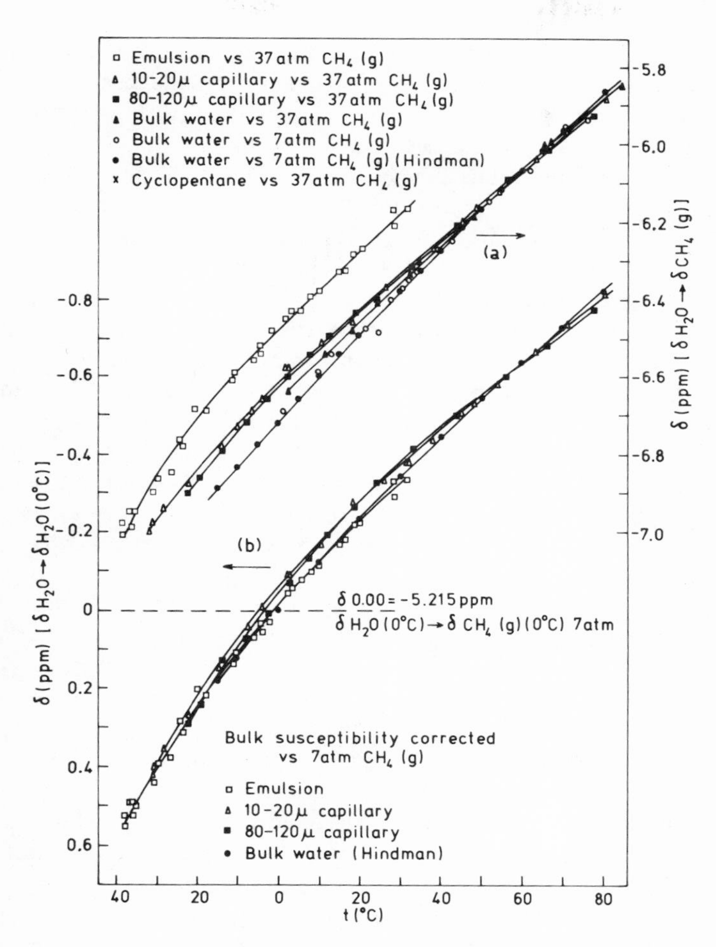

Figure 1. (a) Uncorrected ^1H shifts for bulk water, water in fine capillaries, and in emulsion form over the temperature range -38 to 80°C; (b) Corrected shift values referred to liquid water at 0°C (reproduced from Angell *et al.*[46] with permission of the American Chemical Society).

extensively broken down. Thus, Franck and Roth[50] observed that only at water densities of about 0.04 g ml^{-1} and less (400°C and <100 bar) is it possible to observe IR spectra similar to those observed in dilute water vapor (corresponding to non-hydrogen-bonded water). Such extensive hydrogen-bond rupture would be readily observable from NMR measurements. However, the magnet field homogeneity requirements are very much more stringent for shift measurements than for relaxation, and the difficulties of constructing a high-pressure-shift apparatus are formidable.

^{17}O shifts have not been extensively studied, although Florin and Alei[51] have reported a study on a 30% enriched water sample. They found a linear temperature dependence of 0.0474 ± 0.005 ppm K^{-1} and a gas-to-liquid shift of 37 ± 1 ppm at 0°C. Instead of a temperature-independent standard, the authors used a proton magnetometer, a small NMR probe that fits in the spectrometer sample holder, which enables changes in the magnetic field with temperature to be assessed. Both Fister and Hertz[52] and Luz and Yagil[53] have reported highly nonlinear temperature dependences. In neither study, however, is it apparent whether temperature-independent references have been made. If they have not, as seems likely, it is feasible that the NMR probe assembly can show large susceptibility changes with temperature, as has been reported by Richards and co-workers.[54]

High-resolution shift measurements are not currently feasible with solids due to the large dipolar broadening of lines caused by the presence of neighboring magnetic nuclei of fixed orientation. (In liquids, rapid intermolecular motion averages this effect to zero.) Recent developments[55] suggest that if the sample is rotated around the "magic angle" of 54° 44′ to the applied magnetic field at a frequency faster than the linewidth expressed in hertz, then narrowing of the signal will occur. Such rotation may be performed mechanically, or the r.f. magnetic field may be rotated around the sample using an r.f. pulse sequence.

Using a conventional high-resolution spectrometer, Akitt and Lilley[56] observed a broad signal at −12°C at approximately the same frequency as that observed for water vapor (a truly unexpected result). This problem was resolved by Clifford,[57] who used a broad-line spectrometer and observed a very broad line with a shift that could not be determined and a narrow line (presumably the one found by Akitt and Lilley) at about the shift expected for water vapor. It is likely that the broad component is due to fully hydrogen-bonded water in the ice lattice (strongly dipolar broadened), and the narrow component is due to either freely rotating water molecules in interstices in the ice lattice or free –OH groups relaxing by a rotational process but unable to exchange rapidly with lattice protons.

The interpretation of the shift measurements is not unambiguous, but clearly, in the final analysis, any proposed water structure model must be compatible with the shift measurements. It is generally accepted that the principal reason for a hydrogen-bonded proton resonating downfield from a free proton is an electrostatic effect[42,58,59] whereby the electron cloud around the proton is polarized by the presence of an adjacent oxygen atom that possesses a residual negative charge. Hasted[60] has shown that because the hydrogen bond is so weak, excited vibrational states with different interatomic distances are occupied at room temperatures. Thus, the polarization effect is temperature-dependent, and it has been estimated that the hydrogen-bond shift from vapor should be between 6.6 and 3.9 ppm, with a linear liquid temperature dependence between 2 and 8×10^{-3} ppm K^{-1}.

However, in spite of the work by Hasted, most interpretations of proton shifts involve the idea of H-bond breaking with increase of temperature, a concept in accord with the upfield shift with temperature. In terms of a mixture model, we may thus write

$$\delta_{\text{obs}} = X_F\delta_F + X_B\delta_B \tag{5}$$

where X_F and X_B are the fractions of free and bound protons of shifts δ_F and δ_B, respectively. Unfortunately, we do not know the temperature dependences of any of the quantities in Eq. (5). The most extensive discussion of this problem is due to Hindman,[45] who estimated δ_F from studies of very dilute solutions of water in organic solvents and theoretical estimates using the approach of Buckingham *et al.*[61] to be between 0.21 and 0.43 ppm downfield of the gas-phase shift due to dispersion forces. Using a simple statistical mechanical mixture model, Hindman estimated X_F to be 0.155 at 0°C, and thus obtained a value of δ_B of 5.50 ± 0.07 ppm downfield from the gas-phase value. Hindman,[45] however, also showed that the NMR shift data could be explained quantitatively in terms of Pople's hydrogen-bond bending theory.[40] As already mentioned, such ideas are not unrelated to the molecular dynamics approach of Rahman and Stillinger,[38] and they also relate to the approach of Hasted[60] on the temperature dependence of the hydrogen bond.

Muller[62] has recently employed a much simpler analysis in order to appraise the success of various water structure models. Taking a value of 0.0 for δ_F relative to the gas phase and a value for δ_B of the order of 5 ppm, he compared his results for X_F with the structure models of Nemethy and Scheraga,[63] Marchi and Eyring,[64] and Davis and Litovitz.[65] The Nemethy–Scheraga model consists of a mixture of five species forming 4, 3, 2, 1, or 0 hydrogen bonds. The Marchi–Eyring model is essentially identical to that of Samoilov,[39] consisting of a fully hydrogen-bonded network with freely rotating water molecules in some cavities. The Davis–Litovitz model is a two-species mixture model consisting of an open icelike arrangement of hexagonal rings and a close-packed arrangement of hexagonal rings in which each water is hydrogen-bonded to only two others. Ruterjans and Scheraga[66] have employed a similar analysis comparing estimates of X_F with those from various theoretical models. Both groups conclude that of the models studied, only the Davis–Litovitz model is in accord with the NMR shift measurements. Comparisons are shown in Table 1. Although there are currently no pressure-dependence measurements of chemical shifts, O'Reilly[67] has made an estimate of the magnitude of this effect. He pointed out that in terms of an electrostatic polarization model, it is necessary to take into account the pressure dependence of the hydrogen-bond length as well as the number of bonds present. He estimates that at 100°C, $(\delta\delta/\delta P)_T = -0.25$ ppm kbar^{-1}.

It is to be hoped that any future proposed model for water structure will be shown to be compatible with as many experimental data as are currently available, be they from NMR, IR, Raman, thermodynamic, or scattering experiments, before it is accepted.

The success of treating hydrogen bonding simply in terms of electrostatic interactions that are long-range and nondirectional in nature means that there may be a very wide range of interaction energies depending on both distance and orientation, whereas a

Table 1
Comparison of Estimates of Fractions of Hydrogen Bonds Broken at Different Temperatures for Various Models

Temp. °C	Muller[62]	Hindman[45]	Davis and Litovitz[65]	Nemethy and Scheraga[63]	Marchi and Eyring[64]
0	0.094	0.155	0.181	0.472	0.018
25	0.142	0.21	0.232	0.552	0.051
100	0.284	0.35	0.345	0.675	0.351

mixture model treats the interaction as either on (hydrogen bonded) or off (non-hydrogen-bonded). A definitive test of the electrostatic nature of hydrogen bonding and of the electrostatic cause of proton shifts in water would require modifying the Rahman–Stillinger molecular dynamics calculations[38] to calculate the electrostatic field parameters at a proton nucleus used in Buckingham's shift theory[59] averaged over a long time. Hence an estimate of the chemical shift of that proton could be made.

The ^{17}O shifts, however, are not sufficiently well characterized to allow complete analysis. However, from Eq. (5) we find

$$\delta - \delta_F = X_B \delta_B \tag{6}$$

for both ^{17}O and ^{1}H shifts. Thus

$$\frac{^{17}(\delta - \delta_F)}{^{1}(\delta - \delta_F)} = \frac{^{17}\delta_B}{^{1}\delta_B} \tag{7}$$

where the superscripts 17 and 1 refer to ^{17}O and ^{1}H shifts, respectively. Assuming that δ_F may be identified with the gas-phase shift, a plot of the ^{17}O shifts[51] vs. ^{1}H shifts (both relative to their gas-phase values) should give a straight line through the origin. It is clear from Fig. 2 that

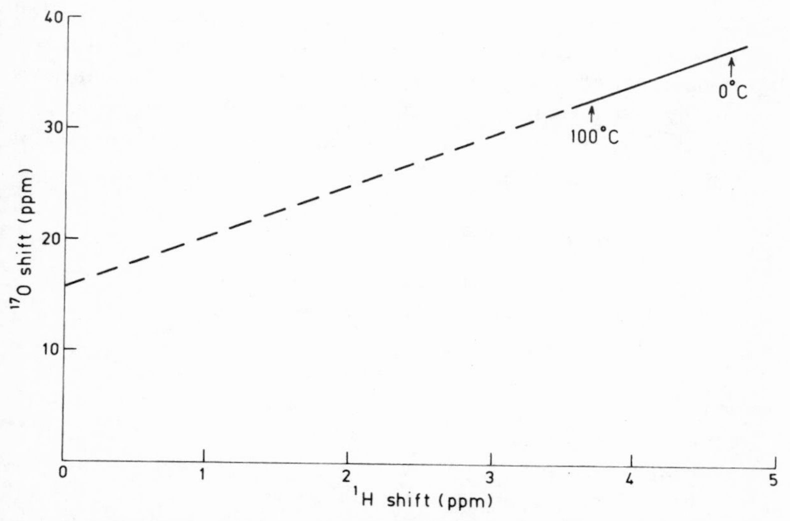

Figure 2. Plot of ^{17}O against ^{1}H shifts for liquid water, with reference to the respective gas-phase values, at different temperatures.

such behavior is not observed. This means that ^{1}H-derived hydrogen-bond fractions will be different from ^{17}O-derived fractions. Clearly, much more work is needed to resolve this problem. Even though ^{17}O shifts for a variety of organic and inorganic compounds range[3] through 1600 ppm, the gas–liquid shift for water is only 38 ppm. For ^{1}H shifts, the range is approximately 12 ppm, with a gas-phase shift for water of 4.6 ppm. It therefore seems apparent that ^{17}O and ^{1}H shifts arise through different mechanisms. This conclusion will be corroborated later during the discussion of the effect of salts on ^{17}O shifts, where it is apparent that an electrostatic (polarization) shift mechanism does not predominate. It will be shown that the most likely cause of ^{17}O shifts is the repulsive overlap between the orbitals of the oxygen electrons restricting the rotation of these electrons, which gives rise to a downfield shift.

3. Interaction of Organic Molecules with Water

The interactions between organic molecular species and water are clearly of very great interest, not only because of the fundamental biological importance of water, but also because of interest in solvation and solvent effects generally. This area of interest is extremely large, and it is beyond the scope of this review to give a full account of all the work done in this field. Instead, the various approaches will be mentioned, and certain important consequences will be discussed. Work has centered on two areas: (i) studies of dilute solutions of water in organic solvents, and (ii) studies that extend right through the solvent composition range.

(i) Dilute Water Solutions

This area has recently been reviewed extensively by Christian *et al.*,[68] and only a few brief observations will be made. It is generally accepted[69] that in pure water, no "free" molecules exist at normal temperatures and pressures, but in very dilute solutions in inert solvents such as CCl_4, water molecules exist almost exclusively as the monomer. Thus, Hindman[45] found the shift of water in saturated CCl_4 (0.0061 mol liter^{-1} at 30°C[70]) to be only 0.77 ppm downfield of its gas-phase value. This finding is borne out by other spectroscopic and thermodynamic techniques.[68] In solvents that can accept a hydrogen bond from a water molecule, the downfield water shift in very dilute solution from its gas-phase value will clearly be much larger (e.g., ethylene oxide, where $\delta = 2.6$ ppm).[71] The very low solubility of water in the various organic solvents that are incapable of hydrogen-bonding does suggest that water molecules that are not hydrogen bonded at all are of very unfavorable energy, an idea in close accord with the view expressed by Frank.[69] The present authors believe that herein lies the reason for the apparently strange structural behavior of aqueous solutions of nonelectrolytes (see below).

In slightly more polar solvents such as 1,2 dichloroethane, in which water solubility is higher, there is reasonable evidence to suggest some degree of water polymerization, most of the data being analyzable in terms of either monomer–dimer or monomer–trimer equilibria.[68] It is likely that the monomer has some hydrogen-bonding interaction with the organic solvent, but neither NMR[72] nor IR[73] measurements shed any light on this.

Considerable work has also been done on the association of water molecules with polar molecules in inert solvents. Li and co-workers[74] have made an extensive study of various polar molecules and have interpreted their NMR shift data in terms of the equilibria

$$\begin{aligned} H_2O + A &\rightleftharpoons H{-}O{-}H\cdots A \\ H{-}O{-}H\cdots A + A &\rightleftharpoons A\cdots H{-}O{-}H\cdots A \end{aligned} \qquad (8)$$

Much of the earlier work has also been reviewed by Christian *et al.*[68]

The solvation of dilute solutions of water in strongly polar solvents is only poorly understood. Mavel[75] and Holmes *et al.*,[76] have estimated apparent dimerization constants of water from the initial slopes of proton water shift vs. concentration plots. Such treatments do, however, belie the undoubted complexity of such solutions. The limiting value of the water shift can itself yield valuable information concerning the state of the water molecule in such solutions. Kingston and Symons[77] have noted that the upfield shift of water from its bulk value is very much larger for triethylamine than for the lower alcohols, even though the amine is a very much stronger base than the alcohols. These authors noted that the IR spectra suggested free –OH groups in the amine solutions, but not in the alcohol solutions. It seemed likely that the ability of a water molecule to form a second hydrogen bond is affected by its ability to donate a lone pair to an acceptor. Thus, a single hydrogen bond tends to "deactivate" the molecule to further hydrogen bonding unless one of the lone pairs is itself involved in hydrogen bonding, in which case the second hydrogen is then "reactivated." The triethylamine is unable to accept a lone pair; thus, free –OH groups are observed. Tertiary butanol behaves like the amine solutions, but the cause of free –OH groups is likely to be steric in origin.

(ii) Polar Organic Solute Solutions

The effect of organic solutes on the proton chemical shift of water is very much harder to interpret. With the exception of the lower alcohols, which are believed to be anomalous due to proton exchange and will be considered separately, most organic solutes and ions shift the water resonance upfield, which implies hydrogen-bond cleavage.[78] However, the NMR relaxation data show that the proton relaxation rate of water decreases with increasing solute concentration, which suggests some form of structural enhancement.[79] This enhancement is well established. It has been noted before from both thermodynamic and kinetic experiments.[80] It appears that the proton shift results are anomalous. There is as yet no generally accepted explanation for this structural enhancement (often known as *hydrophobic interaction*). The most widely quoted is that due to Frank and Wen.[36] This water structure model has been discussed above, but these authors believed that the presence of an inert solute somehow buffered a cluster against initial hydrogen-bond breakage by removal of excess energy. In addition to the shortcomings discussed earlier in this section, the cluster stabilization is conceptually hard to accept.

In view of the three-dimensional nature of water, it is clear that a plane section through water cuts through a certain number of hydrogen bonds; similarly, the formation of a hole (larger than an interstitial site in a water lattice) will break hydrogen bonds as well. This, then, is a possible cause for the hydrogen-bond breakage observed by NMR. It is, unfortunately, very difficult to quantify such ideas, particularly for small molecules. As the hole size increases, the number of broken hydrogen bonds will not necessarily increase regularly, since the water lattice is likely to distort in order to minimize hydrogen-bond breakage. Krishnan and Friedman[80] have noted that the enthalpies of solvation of alcohols in water are not linear with alcohol size, in accord with such ideas. For noninteracting particles much larger than water molecules, it might be expected that hydrogen-bond breakage would be proportional to solute size. The situation will clearly become very much more complicated when the solute molecule contains a polar group. In addition to the hydrogen-bond breakage due to hole formation for the nonpolar end of the molecule, there may be hydrogen-bond accepting by, or donation from, the polar group. The properties observed will depend on the relative importance of the two effects. For molecules with more than one functional group, whether the groups are compatible with water structure will determine whether hydrogen-bonding interactions will occur. An interesting example is glucose,[81] in which all six –OH groups are so orientated as to be compatible with an icelike structure. Consequently, glucose shows no hydrophobic properties.

This interpretation, however, does not explain the structural enhancement involved in such solutions; indeed, the hydrogen-bond cleavage caused by hole formation, although it explains the NMR, would be expected to cause structural disruption. To return to consideration of the interface between solute and solvent, it is likely that each interface water molecule will have two hydrogen bonds broken. It has already been shown that free nonbonded water molecules do not exist in water, presumably because of their unfavorable energy. Thus, if any interface water molecule breaks only one further hydrogen bond, the final bond will remain intact until such time as the second bond is restored. Clearly, this restriction will lead to a marked reduction in the mobility of all solvent molecules adjacent to the solute. Such a description bears some similarity to surface tension, where unsatisfied intermolecular forces cause a macroscopic structuring effect on the surface of liquids.

It is unfortunate that it is extremely difficult to suggest experiments that could add weight to the hypothesis described above. It may be necessary to await molecular dynamics calculations on the interaction of nonpolar solutes and water to see whether such ideas are valid.

As already mentioned, the alcohol plus water results appear anomalous: they give initial low-field shifts that reach a minimum at approximately 0.1 mol fraction alcohol and then shift upfield.[77,78,82] This behavior is particularly pronounced at low temperatures. Various authors have put forward involved arguments to explain such results, but it is possible to explain them simply in terms of the fast exchange between alcohol and water –OH protons. A typical example, *t*-butanol plus water, taken from the work of Kingston and Symons,[77] is shown in Fig. 3.

At 273°K, above an alcohol mol fraction of approximately 0.3, separate signals for both –OH groups are observed, whereas below 0.3 mol fraction, only a single peak is observed. As may be seen, the water peak is closely linear in mol fraction and extrapolation passes close to the origin through which it is constrained to pass. Similar behavior was observed for all the other systems studied by Kingston and Symons.[77] It is reasonable to suppose that the extrapolation to

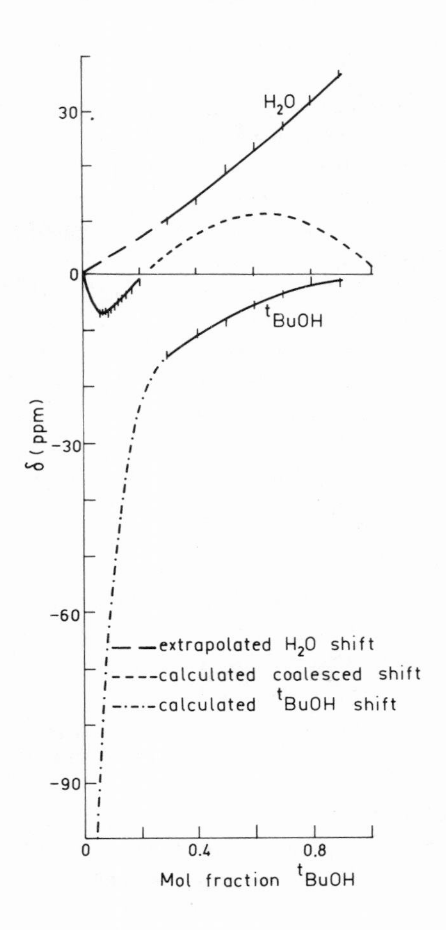

Figure 3. ^1H shift of hydroxyl proton in mixtures of water and *t*-butanol (after Kingston and Symons[77]).

zero alcohol content is straightforward. If such an extrapolation is used, then the alcohol shift can be estimated from the observed coalesced resonance,[83] which is also shown in Fig. 3 at both temperatures. It is initially surprising to find that the shift of the alcohol in pure water is apparently nearly 6 ppm downfield of its value for the pure alcohol at 273°K. If this is so, then tertiary butanol is very extensively hydrogen-bonded in water, but very much less so in the pure state. This conclusion finds some support in the analysis of enthalpy of mixing data given by Franks and Ives[84] in their review of alcohol–water mixtures. It would be interesting to have further data on alcohol–water mixtures in view of these somewhat unexpected results.

IV. NMR STUDIES OF ELECTROLYTE SOLUTIONS

1. Aqueous Solutions

(i) ^{1}H Shifts

(*a*) *Salts.* The effect of salts on the chemical shift of protons in water has been widely studied for many years. Such measurements have been made by two different experimental techniques. However, until recently, the link between them has not been exploited. The first approach involves the study of the concentration-dependent shift of the water signal, the observed shift being the weighted average perturbed by cation, water perturbed by anion, and bulk water. The second approach involves the observation of separate signals from water bound to multicharged ions and from free water at low temperatures.

The first approach has been widely used, but it is in this work that the problems associated with measuring very small shifts become of paramount importance. It is likely that the different results obtained by various authors arise from different internal referencing procedures or susceptibility corrections. It is to be hoped that the superconducting magnet procedure discussed in Section II.1 will help solve this problem.

Shoolery and Alder[85] were the first to make a detailed study of the effect of electrolytes on proton shifts. They concluded that small, multicharged ions produced a large downfield shift due to an electrostatic solvent polarization effect, whereas the large, singly charged ions caused upfield shifts due to hydrogen-bond breaking (Fig. 4). Subsequent studies[86–93] have not materially affected the conclusions reached by these authors. Various authors have sought correlations with other properties sensitive to hydrogen bonding and electrostatic polarization. Thus, Fabricand and Goldberg[87] obtained a linear correlation between molar shift and Pauling crystal radius for both anions and cations, which has not been substantiated in later measurements (see Fig. 6). Glick *et al.*,[90] and Hartman[91] used IR studies in order to assess hydrogen-bond cleavage. The former authors attempted to subtract the effect of this from the chemical shifts, whereas the latter author noted a direct correlation between chemical

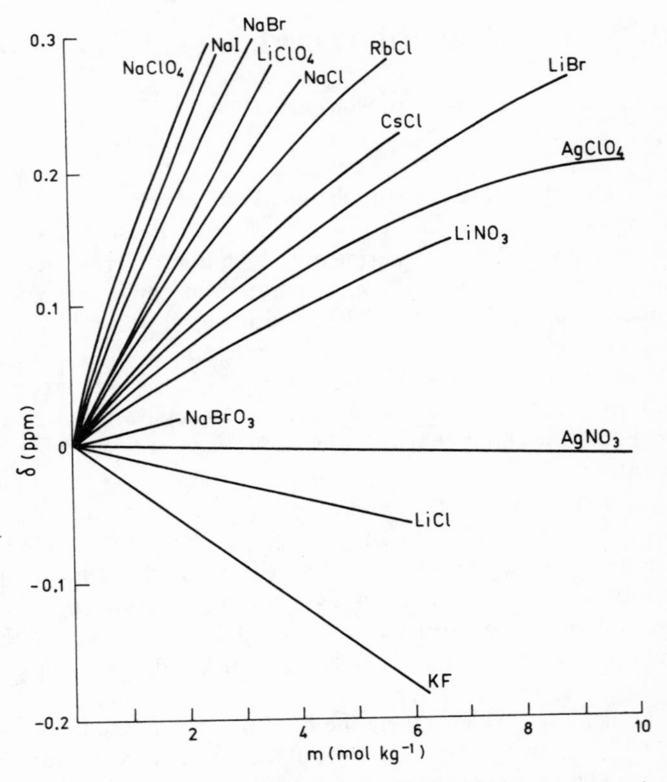

Figure 4. ^{1}H shifts in aqueous electrolyte solutions as a function of concentration (after Hindman[88]).

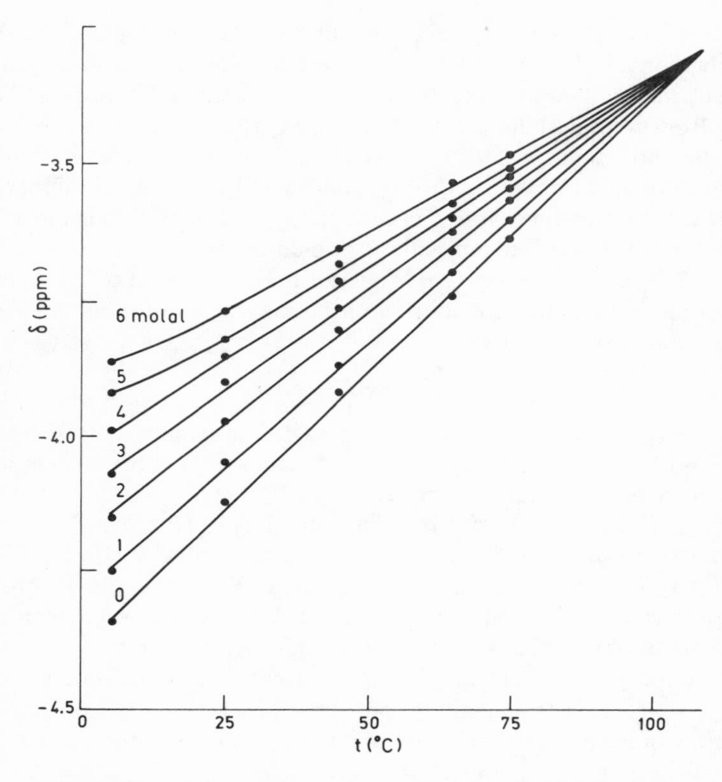

Figure 5. ^1H shift (relative to ethane gas) of aqueous sodium chloride solutions as a function of temperature at various molalities.[43]

shift and HDO IR shift. In concentrated solutions, the salt-effect shifts have been correlated with acid strengths of the anions.[9]

However, the most detailed and extensive treatment of proton-shift measurements is that of Hindman.[88] He discussed the shift data in terms of the Frank and Wen solvation model,[36] viz., a region of strongly dipole-orientated water molecules close to the ion (region A) and beyond it a region B of structural mismatch in which the orientation of the water molecules in the solvation shell and those in bulk water is incompatible, and thus this connecting region (region B) must be hydrogen-bonded to some extent. Region C corresponds to bulk water beyond region B. Hindman identified four contributions to the chemical shift.

1. δ_{bb}, a bond-breaking term due to the organization of the solvation shell water molecules from a hydrogen-bonded water structure configuration to the dipole orientated configuration. The hydrogen-bond shift was evaluated from the gas-to-liquid shift of water and estimates of the hydrogen bonds broken in liquid water. δ_{bb} was estimated from the hydrogen-bond shift, the solvation number of the ion, and considerations of how many hydrogen bonds must be broken when a water molecule is orientated by the ion.

2. The polarization term δ_p was estimated using Buckingham's treatment[59] for the contribution of the electric field to the shielding constant. It is given by

$$\Delta\sigma_p = -AE_{\parallel} - BE^2 \tag{9}$$

where $E_{\parallel}$ is the component of the field E that is parallel to the line joining the hydrogen atom to the charge center. By use of the solvation number of the ion, δ_p can be estimated.

3. The structural contribution, according to Hindman,[88] can be estimated only by difference in a way similar to that whereby the estimates of structural entropy were made by Frank and Evans.[94] However, he worked out maximum limits for the structural term using estimates of the maximum number of possible hydrogen bonds that could form to the orientated water molecules in the solvation shell.

4. From electronegativity considerations, Hindman[88] concluded that with the possible exception of the silver ion, any nonelectrostatic contribution is negligible.

Using these ideas, he estimated solvation numbers, first, by ignoring the structural contribution, which in some cases leads to negative numbers, and also by estimating maximum contributions. The values obtained compare favorably with those estimated from activity coefficients, ion exchange, mobility, entropy, compressibility, density, diffusion, and dielectric data. In particular, for the alkali cations, all methods show a decrease in solvation number with increase in atomic number. For the halide ions, all methods give a very small solvation number, less than 1.

From the viewpoint of distribution of solvent around a solute, the approach discussed in Section IV.3, such results are extremely difficult to reconcile. It is generally agreed that the radial distribution of solvent around a solute is quantitatively very similar to the r.d.f. of pure solvents, viz., an oscillating function that gradually tends to unity.

The area under the first maximum corresponds to the average

number of nearest neighbors, although there may be practical problems delineating the exact extent of this "solvation shell." For the chloride ion in water, this number has been measured[95] as 8.0 ± 0.1. Using this approach, the concept of zero solvation number and decreasing solvation number with increasing solute size is extremely difficult if not impossible to reconcile.[96]

A potential method for separating out polarization and structural contribution lies in variable-temperature studies.

Malinowski *et al.*[43] measured the chemical shift of solutions from 0 to 6 *m* NaCl and from 5 to 75°C. They found that all solutions studied showed a linear shift–temperature dependence, and all extrapolated to the same point (Fig. 5). They used a simple two-site model

$$\delta = X_N \delta_N + X_s \delta_s \tag{10}$$

where X_N and X_s are the fractions of free and bound water of shift δ_N and δ_s, respectively. Noting that

$$x_s = \frac{hm}{55.51} \tag{11}$$

where h is the number of molecules of water bound to each solute, and m is the molality of salt, the authors obtained

$$\delta = \frac{hm}{55.51}(\delta_s - \delta_N) + \delta_N \tag{12}$$

They concluded from the linearity of the shift–temperature plots and the extrapolation to a single point that δ_s is independent of temperature, and at the point of intersection, $\delta_s = \delta_N$. Thus, h was found for NaCl to be 4.0. This work was further extended by Malinowski and co-workers[97,98] for $Al(NO_3)_3$, $NaClO_4$, HCl, and $HClO_4$, for which values of 13.4 ± 0.9, 3.0 ± 0.2, 3.4 ± 0.4, and 2.6 ± 0.4 were obtained. The very high value for $Al(NO_3)_3$ was discussed in terms of either second-sphere effects around the aluminum giving a total coordination of 18 but with some water molecules replaced by nitrate ions, or a coordination number of 6 for aluminum and one of 2.5 for the nitrate ion. The former effect is less likely, since the ion association and hence the effective coordination number would be concentration-dependent, which would affect the linearity of the plots. It is worrying that the coordination number should be so low. Indeed, it has been suggested that anions have a very small coordination number, but such an explanation is conceptually very hard to accept

Table 2
Solvation Numbers of Salts Derived from Variable-Temperature ^{1}H NMR Studies[100]

Salt	Salt concentration (mol kg^{-1})	$h^a(\pm 0.2)$
LiCl	3.08	3.4
LiCl	4.62	3.2
NaCl	3.08	4.6
NaBr	3.01	4.4
$NaClO_4$	2.98	3.0
Na *p*-toluene sulphonate	2.73	2.8
KCl	2.99	4.6
RbCl	3.20	4.0
CsCl	3.02	3.9
$MgCl_2$	2.92	8.2
$CaCl_2$	2.37	9.4
$(CH_3)_4NCl$	2.66	0.6
$(CH_3)_4NBr$	2.92	1.0

^{a}Number of molecules of water bound to each solute; see Eq. (11).

when solvation is considered in terms of the distribution of solvent around solute.* Another possibility is that δ_s is not a constant, but varies linearly with temperature. Satisfactory agreement with ^{1}H-line area measurements can be obtained on this assumption.[99]

Creekmore and Reilley[100] repeated the work of Malinowski and co-workers, and noted that δ vs. temperature plots for different concentration solutions do not always intersect at the same point. This they conclude arises from ion association reducing the solvation number. Their solvation numbers are given in Table 2. From these values they concluded that for the doubly charged ions, second-sphere effects are important. For alkali metal ions, the solvation number is 4.0, whereas for anions it is small, e.g., for Cl^- it is approximately 0.6.

Malinowski *et al.*[101] have also studied the concentration dependence of solvation number in a number of salts. They found that the solvation numbers decreased with salt concentration, as would be

* The most likely explanation is by Akitt[118] (see later) that at room temperature bound-free exchange averaging is not complete. In acidified solutions two signals are observed in solutions cooled to $-10°C$. If this effect is taken into account, then the data are compatible with $h = 6.0$.

expected from considerations of direct ion–ion encounters "squeezing" out solvent molecules. Their general assignment of individual hydration numbers is in broad agreement with those of Creekmore and Reilley.[100]

Swift and Sayre[102] attempted to use a ^{1}H NMR technique to measure hydration numbers. They used as a working definition the number of water molecules residing around an ion for a time longer than the diffusion time. They argued that the ratio of bound to free water could be estimated from a kinetic basis utilizing the fact that a reagent would react with the two water species at different rates:

$$H_2O_{\text{bound}} + Y \xrightarrow{K_b} \text{product} \tag{13}$$

$$H_2O_{\text{free}} + Y \xrightarrow{k_f} \text{product} \tag{14}$$

The pseudo-first-order reaction rate is

$$\frac{-d[Y]}{[Y]\,dt} = k_b[H_2O_{\text{bound}}] + k_f[H_2O_{\text{free}}] \tag{15}$$

For a probe reagent Y, the use of the paramagnetic ion Mn^{2+} was suggested, the reaction rate being water exchange as measured by the line-broadening technique to be discussed later. Unfortunately, the observed reaction rates showed pronounced ionic strength and anion dependence, and no meaningful hydration numbers were obtained. However, using a comparative technique with Al^{3+} or Be^{2+} as standard, and comparing linewidths of both aluminum salts and salts of cations of interest with the same anion and of the same concentration, they thus established an empirical relationship between linewidth and hydration numbers.

In later work, Swift *et al.*[103] obtained a value of 6.0 ± 0.1 for the hydration number of Ga^{3+}. This work, however, was condemned by Meiboom,[104] who concluded that the method was theoretically unsound and the effects observed were likely to be due to viscosity changes affecting linewidths.

As already mentioned, the second NMR approach is involved with the observation of separate bound and free solvent signals for multicharged cations. Thus, Schuster and Fratiello,[105] using very concentrated $AlCl_3$ solutions (1 mol $AlCl_3$/18 mol water), were able to observe separate signals for bound and free water at -47°C. They calculated a hydration number of 6.0 ± 0.1 and, from linewidths, made a very rough estimate of the activation energy of exchange (24 $\pm$ 3 kcal mol^{-1}). It is interesting to note that exchange studies can also be made at much higher temperatures when coalescence has occurred. Thus, Wong and Grunwald[106] have made a study of the pH dependence of proton NMR linewidths of Al^{3+} solutions in water and estimated parameters for proton exchange between the hydration sphere of ion and bulk water.

Although they are not strictly connected with this section, since they are really concerned with mixed-solvent NMR studies, Schuster and Fratiello,[107] using acetone–water mixtures to give a very low freezing point, were able to observe separate bound and free water molecules for $Mg(ClO_4)_2$, and estimated a hydration number of 5.8 $\pm$ 0.2 for Mg^{2+}. The rationale behind this approach was that the acetone is inert and simply acts to lower the freezing point of the water. In water-poor solutions, there will be competition between acetone and water for the solvation sites. Studies of such solutions will be discussed in Section IV.3. Fratiello *et al.*[108] have extended this work to various salts of Al^{3+}, Be^{2+}, Ga^{3+}, In^{3+}, and Mg^{2+}, with and without acetone addition, and have obtained solvation numbers of 6, 4, 6, 6, and 6 ($\pm$0.2), respectively, regardless of acetone concentration or anion. Matwiyoff and Taube,[109] using the acetone addition method, also obtained a solvation number of 6.0 for Mg^{2+} in both water and methanol.

Fratiello *et al.*[110] have since extended their methods to the UO_2^{2+} cation. In perchlorate solutions, a value of 4.0 $\pm$ 0.1 is always found regardless of added perchlorate ion or acetone content. For nitrate solutions, as the acetone concentration is increased (and the dielectric constant falls), the solvation number decreases, which suggests ion association, each nitrate ion acting as a bidentate ligand and replacing two water molecules. Addition of HCl or HBr suggests that complexing by Cl and Br can also occur. For Sn(IV) salts, the solvation numbers are very low[111] ($\sim$2), and suggest some form of ion association, but even ^{119}Sn NMR does not shed much light[112] on the nature of the complexing. Similar measurements have also been made for lanthanide salts.

The link between the two proton NMR techniques was first explored in detail by Davies *et al.*,[113] who measured molal shifts of various electrolytes at different temperatures. In order to resolve one of the classic problems of such shift measurements, that of separating out cation and anion effects, they suggested two approaches. The first is due to the fact that at very low temperatures, separate signals could be observed for bound and free water at high concentrations. They assumed that the bound-water shift was temperature-independent. For methanolic solutions (see below), it has been shown[113,114] that both free and bound solvent show the same concentration dependence. They assumed such behavior occurred also in water. They thus readily obtained the molal shifts of the highly charged cations at the temperature of the bound–free shift measurement. In order to estimate molal shifts at different temperatures, the temperature dependence of the pure-water chemical shift is required. It was obtained by extrapolating Hindman's data to low temperatures. The second approach was to plot the molal shifts of the chlorides against charge–radius. Two smooth curves were obtained (Fig. 6), one for the singly charged ions, the other for the divalent ions. The curves were parallel (where the data overlapped), and it was argued that the correct choice of molal shift should make the two curves concordant. A value of 0.076 ppm $(\text{mol kg}^{-1})^{-1}$ fitted the data very well. It is interesting to note that both methods give closely similar values for individual ion

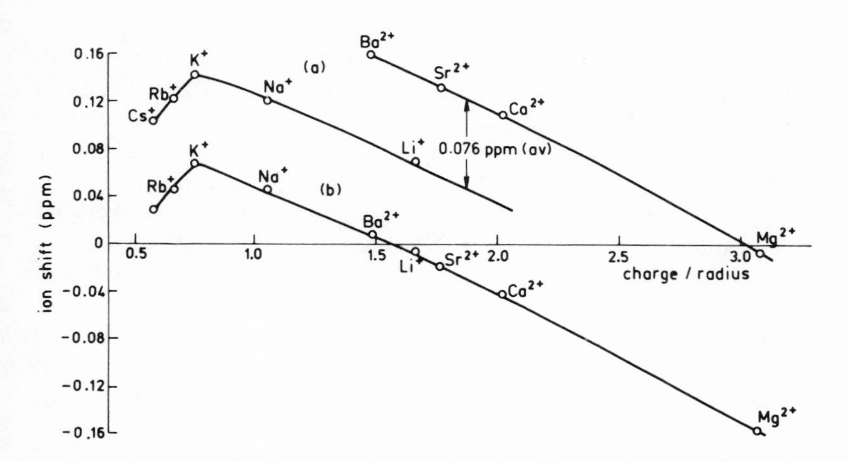

Figure 6. (a) ^{1}H cation shifts in water at 273°K based on $\delta_{Cl^-} = 0$; (b) ^{1}H cation shifts based on $\delta_{Cl^-} = 0.076$ ppm plotted against charge–radius ratio (reproduced from Davies *et al.*,[113] with permission of The Chemical Society, London).

shifts. The second approach clearly has less theoretical justification than the first, which, although it has some theoretical shortcomings, does give further insight into the problem. Apart from the problem of concentration dependence of the bound water and the need to know the temperature dependence of "normal" (?) water well below its freezing point, the principal problem is the assumed temperature independence of the "bound"-water shift. This solvated ion consists of a highly charged central ion surrounded by a fixed number of water molecules firmly held (lifetime ~ 0.1 sec). The alignment of the water dipoles will mean that at the surface of the solvated ion there will be several hydrogen atoms, all capable of hydrogen bonding. Not all will be able to bond simultaneously due to structural mismatch effects. The extent of hydrogen bonding that will in fact occur is certain to be temperature-dependent. Clearly, such a simple model as that described above is unable to take these effects into account. Akitt[115,116] has also explored the connection between shift–salt concentration measurements and observation of separate free and bound signals. Initially,[115] he studied the effects of temperature and concentration on the shifts of both bound and free water in $AlCl_3$ solutions. He found that although the free water showed large temperature dependence, the bound water did not. This value for bound shift is 2.36 ppm to low field of the value estimated by Malinowski and Krapp,[97] and if the revised value of Akitt is used for δ_s [see Eq. (12)], a solvation number of 6 is obtained at 5°C. At 85°C, this value has apparently increased to 7.5. This is unlikely: the most reasonable explanation is that anion effects (and possible second-sphere cation effects) will cause δ_N to deviate from the value for pure water. It thus appears from Akitt's work that the long extrapolations sometimes necessary in the work of Malinowski and co-workers are not valid. In later work, Akitt[116] measured the chemical shifts of bound water of Sn(IV), Al(III), Ga(III), In(III), Be(II), and Mg(II). For all except Mg(II), where an aqueous acetone solution was used, concentrated aqueous solutions were measured. He used Buckingham's theory[59] [Eq. (9)] to estimate the electrostatic contributions to the shift of his solutions using reasonable values for the constants and geometric terms derived from X-ray diffraction data of solid crystal hydrates. It is interesting that he obtained very good linear correlation between calculated and measured shifts of slope of unity. However, the measured shifts (referred relative to gas-phase water) are approximately 2 ppm to low field of the calculated values, which suggests some form of hydrogen bonding between bound and free water. This hydrogen bonding should show temperature dependence. By extrapolation of this line and calculation of $\Delta\delta_p$ for cations, where separate bound signals could not be obtained, he estimated cation shifts for fast-exchanging cations. It is interesting to note that according to Akitt's extrapolation, a "cation" of zero charge would have a shift somewhat downfield from that of gaseous water, but upfield of that of liquid water, which suggests net hydrogen-bond breakage due to structural mismatch between solvation shell and bulk water. It is slightly disturbing that according to Akitt, this structural effect is temperature-independent, which is at variance with most previous ideas on hydrogen bonding. Anion molal shifts were derived by subtracting the estimated cation shifts from the measured shifts for solutions in which exchange between bound and free is rapid and only one signal is observed. The estimated anion shifts are all upfield and temperature-dependent, and Akitt therefore assumes that rapid rotation of water molecules close to the ion averages the electrostatic term to zero. If it is assumed that each anion breaks a given number of hydrogen bonds, it is then possible to fit the variable temperature-dependent molal shifts for a wide variety of salts between 0 and 100°C with remarkable accuracy. There are, however, two important shortcomings to this anion treatment. It is obvious that the orientation of protons toward the anion will mean that these protons will be unable to hydrogen bond to the surrounding solvent, and this hydrogen breakage will be temperature-independent (unless free rotation of the solvating water molecule occurs). However, the structural mismatch between bulk and solvation shell water may be expected to be temperature-dependent. It must therefore be supposed that the hydrogen-bond-breaking effect of anions is temperature-dependent. Recent calculations[117] suggest that free rotation of a water molecule beside an anion will not reduce the first term of Buckingham's equation [Eq. (9)] to zero.

The criticisms that have been leveled against the anion treatment are also valid when applied to singly charged cations with labile solvation shells. It can be shown[117] for cations that the first term of the Buckingham equation changes sign if free rotation occurs increasing in magnitude by a factor of 5.

This indeed may be the origin of upfield shifts caused by large, singly charged cations. As already mentioned, it is difficult to see how hydrogen-bond breakage can be temperature-independent, particularly, since it is well known that "structure-breaking" effects are highly temperature-dependent. It is indeed remarkable, however, that despite its shortcomings, Akitt's model fits the available data with uncanny accuracy.

Akitt has since extended these ideas to include a reassessment of the molal shift data for Al,[118] electric field effects of small anions such as F^- and OH^-,[119] and the Ag^+ ion.[120] For the latter ion, a simple electrical field approach fails, and Akitt invoked a highly unsymmetrical solvation complex in which induced dipoles are important. It is interesting that Hindman[88] mentioned that there may be covalent forces operating in the Ag^+ ion solvation.

(*b*) *Acids.* One of the earliest fruitful quantitative applications of shift measurements was to the study of strong acids, which in concentrated solution are incompletely dissociated. The original observations were made by Gutowsky and Saika,[121] who noted that the shift of the single-proton resonance peak in aqueous hydrochloric acid solutions relative to pure water was a linear function (Fig. 7) of the fraction p of protons on the hydronium ion H_3O^+ present in a solution of acid mol fraction x, where Gutowsky's p function is given by

$$p = \frac{3x}{3x + 2(1 - 2x)} = \frac{3x}{2 - x} \qquad \text{for } x \leq 0.5 \tag{16}$$

relating to the reaction

$$HA + H_2O \rightarrow H_3O^+ + A^- \tag{17}$$

and then $\delta = p\delta_{H_3O^+}$.

For HNO_3 and $HClO_4$, the initially straight lines become curved (Fig. 7) at high concentrations, which was interpreted as due to incomplete dissociation.

Thus

$$\delta = \alpha p \delta_{H_3O^+} + \tfrac{1}{3}(1 - \alpha)_p \delta_{HA} \tag{18}$$

cf. equation (10).

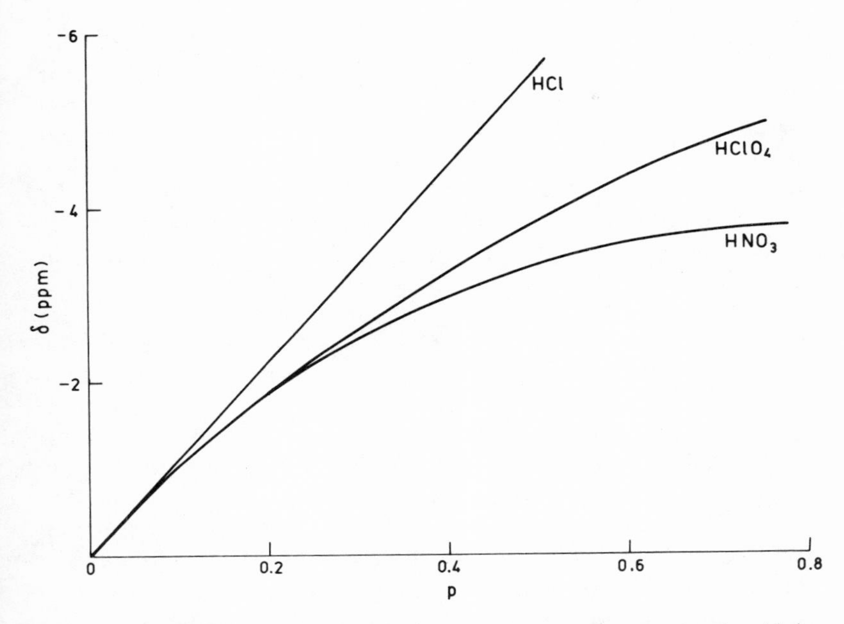

Figure 7. Concentration dependence of ^1H shift of some strong comoprotic acids in aqueous solutions. p is the fraction of protons present in H_3O^+ assuming complete conversion of the acid proton to H_3O: HCl (ref. 88, 121); HNO_3 (ref. 130); $HClO_4$ (ref. 129).

If the intrinsic shifts $\delta_{H_3O^+}$, δ_{HA} can be regarded as independent of environment, then the former can be obtained from the straight-line region and the latter from measurements on the pure acid, if available, or by extrapolation. The potential of the method was quickly realized by Redlich and co-workers,[122–126] who applied it to nitric, perchloric, and some halocarboxylic acids, obtaining results in good agreement with those from other methods such as Raman spectroscopy. The only dibasic acid studied was sulphuric acid, in which there is an additional unknown intrinsic shift parameter, which can be found only by recourse to the results from Raman spectroscopy.

The simple model should yield $\delta_{H_3O^+}$ values that are the same for all acids, but this is not found to be so. This finding does not affect the validity of values for the degree of dissociation, provided the value of $\delta_{H_3O^+}$ used is that for the acid of interest. Several explanations for the different $\delta_{H_3O^+}$ values observed have been advanced, including anion hydration.[127,128] The treatment will also fail if other acid species such as dimers exist. In a later series of papers, Redlich and co-workers[129–131] have modified the Gutowsky and Saika[121] treatment to take into account different degrees of hydration of the proton depending on the mol fraction of the acid, thus involving additional intrinsic shift parameters. These parameters are evaluated by assuming that the chemical shift of each hydrated proton species was directly proportional to the number of exchangeable protons it contains. The treatment was extended to include deuterated acids and mixed protonated–deuterated systems. Other acids have been studied[132–134] or reinvestigated. One report describes the application of the method to nonaqueous solvent systems.[135] Applications to base equilibria are more recent. Study of some pyridine derivatives forms the basis of a simple and instructive student experiment.[136]

A few confirmatory studies of acids have been made using resonances other than proton, such as ^{35}Cl in $HClO_4$[127,137] and ^{15}N in HNO_3.[138] A recent study records the dissociation constant of $B(OH)_4^-$ from ^{11}B chemical shift measurements.[139] Computer fitting of individual ^{13}C titration curves for amino acids, titrated directly into the NMR tube with HCl or NaOH using a microcombination glass electrode, enables precise determination of pK values even when overlap occurs.[140]

(*c*) *Large Organic Ions.* NMR studies have been made on aqueous solutions of organic ions, particularly of the tetraalkylammonium ions. Relaxation studies, employed in order to explore the structural enhancement of the water in these solutions, have attracted attention ever since the pioneering work of Frank and Wen.[36] Much of this work has been reviewed by Zeidler[141] and very recently by von Goldammer.[11] Chemical-shift measurements are almost entirely restricted to studies of the proton shift of water. The alkyl protons have been found to exhibit almost negligible concentration dependence; indeed, they are generally used as an internal reference. Water-proton-shift measurements should be very interesting for large ions, since electrostatic effects should be very small and structural effects should predominate. The earliest studies were those of Hertz and Spalthoff.[142] These studies showed that contrary to expectation, as the size of the tetraalkylammonium ion increased, the shifts became increasingly upfield, which implied increased structure-breaking, whereas thermodynamic, spectroscopic, and kinetic (viscosity and

NMR relaxation) suggested structural enhancement. Subsequent studies[143] have resolved certain anomalies in the early data; in particular, Davies *et al.*[144] noted that the concentration dependence is highly nonlinear, and in the region of measurement can best be represented by two intersecting straight lines (Fig. 8). Such behavior can be rationalized in terms of micelle formation. Figure 9 shows the charge–radius and temperature dependence of the molal shifts of alkali

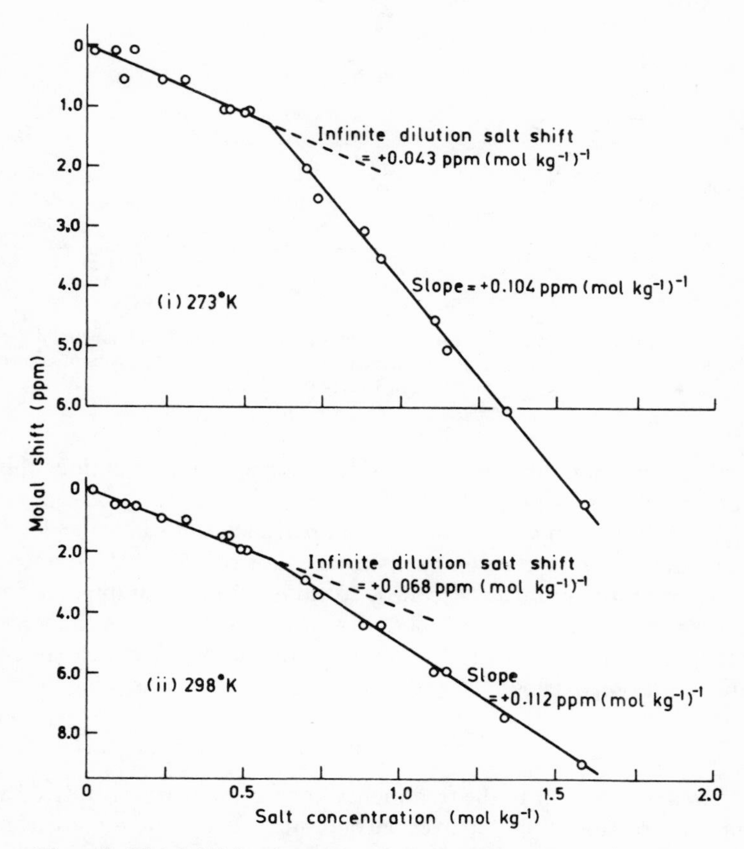

Figure 8. Molal shift for the proton resonance of water induced by tetra-*n*-butylammonium bromide at 273°K (i) and at 298°K (ii), showing a marked discontinuity in the 0.6 *m* region (reproduced from Davies *et al.*,[144] with permission of The Chemical Society, London).

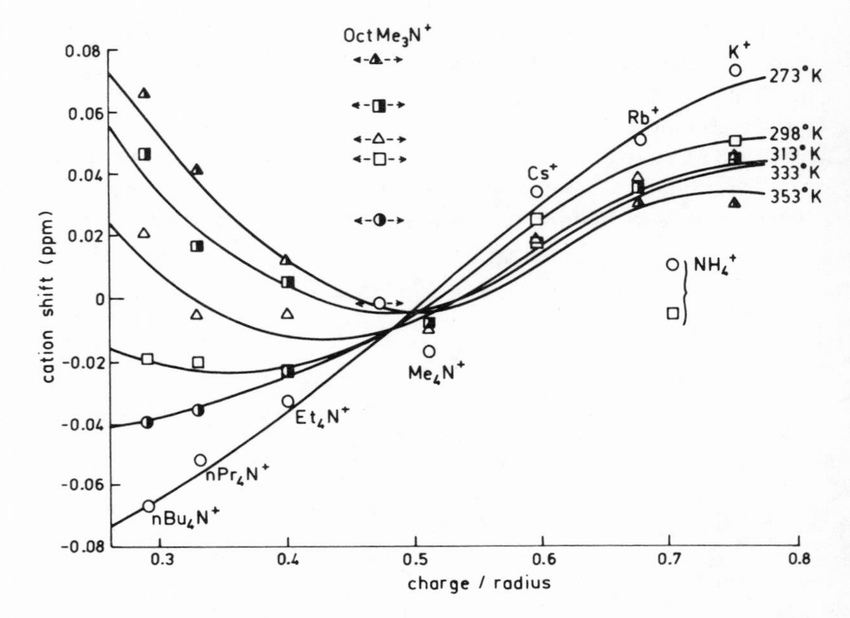

Figure 9. Molal ^1H shifts for tetraalkylammonium and alkali metal ions as a function of charge–radius at various temperatures (derived from Davies *et al.*,[144] with permission of The Chemical Society, London).

and tetraalkylammonium ions, using the separation of cation shifts suggested by Davies *et al.*[113]

It is interesting to explore the cause of such shifts in some detail. This analysis bears some similarity to the study by Akitt,[145] but since it is only semiquantitative, it can be formulated more simply. At any temperature, the molal shift $\delta(m)$ of a solute is defined as the water chemical shift of a 1 mol kg^{-1} solution of the solute δ_M less the shift of pure water (δ_W), i.e.,

$$\delta(m) = \delta_M - \delta_W \tag{19}$$

In a way similar to the treatment above, δ_M may be formulated as the mol-fraction-weighted average of perturbed (δ_P) and unperturbed (δ_W) water. For a 1 m solution, Eq. (20) is obtained:

$$\delta_M = n\delta_P/55.51 + (55.51 - n)\delta_W/55.1 \tag{20}$$

Combining Eqs. (19) and (20) gives

$$\delta(m) = n(\delta_P - \delta_W)/55.51 \tag{21}$$

and (assuming n is constant), we have

$$\frac{d\delta(m)}{dT} = n\left(\frac{d\delta_P}{dT} - \frac{d\delta_W}{dT}\right)\bigg/55.51 \tag{22}$$

The last differential is approximately 0.01 ppm K^{-1} (see Section III.1), and if the interaction between an ion and water is entirely electrostatic, $d\delta_P/dT$ is approximately 0. Therefore,

$$\frac{d\delta(m)}{dT} = -0.00018\, n/\text{ppm K}^{-1} \tag{23}$$

From the tabulation of Davies *et al.*,[144] by choosing appropriate values for n (generally 4–6), it is clear that the electrostatic term predominates for the multiply charged cations. (This conclusion must be tempered by the facts that the cation–anion split was made on the basis of the shift of separate bound and free water from multicharged ions and that cation second-sphere structural effects would be included in the anion, as discussed above.) It is clear, however, that for the larger divalent ions and the alkali metal ions, structural effects become progressively more important. It is immediately obvious that for the tetraalkylammonium ions, structural effects are (as expected) predominant. Furthermore, for the symmetrical ions studied, plots of molal shift vs. temperature all pass through the same point. Also, at all temperatures studied, plots of molal shift vs. charge–radius also pass through the same point, in both cases close to the value $\delta(m) = 0$. To explore this behavior, it will be assumed, as is generally found in practice (see above), that both δ_P and δ_W are linearly dependent on temperature, i.e.,

$$\delta_{P,W} = A_{P,W} + B_{P,W}T \tag{24}$$

From Eqs. (21) and (24), we obtain

$$\delta(m) = \frac{n}{55.51}[A_P - A_W + (B_P - B_W)T] \tag{25}$$

where n, A_P, and B_P should be functions of the ion. The only simple way in which the facts can be rationalized is if A_p and B_p are independent of the ion. Thus, at low temperatures, the water close to tetraalkylammonium (and presumably this holds for other "hydrophobic" solutes) is more hydrogen-bonded than the bulk water, but the perturbed water is more readily thermally disrupted than the bulk. The magnitude of the NMR effect simply depends on the ratio perturbed–bulk water, which depends on the size of the solute through n. A possible mechanism for this low-temperature structural enhancement by hydrophobic groups has already been discussed above.

Recently, Lindenbaum and Levine[146] studied the influence of tetrabutylammonium butyrate on the proton shift of water. The shift variation with concentration, highly curved, shows a shallow minimum at 9.5°C. The shifts are downfield below 33°C and highfield above 56°C. The authors qualitatively analyzed the results in terms of cation–anion, anion–anion, and cation–cation cosphere overlap in accord with the model of Ramanathan *et al.*[147] For tetrabutylammonium bromide, however, anion–anion cosphere terms could be neglected.

(*d*) *Effect of Paramagnetic Species.* Paramagnetic species have a profound effect on NMR spectra. The study of paramagnetic ions in water has thus received much attention. Although this area will not be treated in great depth, a brief review of theory and results will be given.

When there is a finite probability of finding unpaired electron density at a proton, the electron spin and nuclear spin interact to produce a large chemical shift (called a *contact shift*). It is similar in origin to the Knight shift found in metals due to the interaction of metal nucleus with metallic bonding electrons, and is given by[148]

$$\frac{\Delta \nu_c}{\nu} = -\frac{A}{h}\frac{\gamma_e g S(S+1)}{\gamma_N 3kT} \tag{26}$$

where $\Delta \nu_c$ is the shift, ν is the operating frequency; A/h is the nuclear spin–electron spin coupling constant expressed in rad^{-1}; γ_e and γ_N are the electron and nuclear magnetogyric ratios, respectively; g is the Lande splitting factor; and S is the electron spin of the ion of interest. In the limit of fast exchange, the measured shift is then given by the weighted average of protons coupled to the electron and free protons or in terms of a simple two-site model:

$$\delta = X_B \delta_B + X_F \delta_F \tag{27}$$

where X_B and X_F are the fractions of bound and free solvent of shifts δ_B

and δ_F, respectively. Thus, measuring the shift relative to the bulk solvent resonance gives

$$\delta = X_B \frac{\Delta \nu_c}{\nu} \tag{28}$$

If, however, exchange between bound and free water is sufficiently slow (of the same order of time as the separation in shift of the two species), then

$$\frac{\Delta \nu}{\text{Hz}} = \frac{X_B \Delta \nu_B}{1 + 4\pi^2 \Delta \nu_B^2 \tau_B^2} \tag{29}$$

where $\Delta\nu$ is the observed shift and τ_B is the average lifetime of a solvent proton in the bound state. It has been found[149] that for water, the proton chemical shift measurements obey fast-exchange conditions for Mn^{2+}, Co^{2+}, and Cu^{2+}, whereas for Cr^{3+}, Fe^{3+}, and (possibly) Ni^{2+} exchange contributions are important between 0 and 100°C.

An example of recent work is that of Tzalmona and Loewenthal[150] on the effects of Ni^{2+} on the ^1H and ^2H shifts and linewidths in H_2O and D_2O, where in addition to the exchange parameters, hyperfine coupling constants and unpaired electron densities were obtained. Similar parameters have been obtained for V(III) by ^{17}O and ^1H measurements by Chmelnick and Fiat.[151]

It must be emphasized that for some solvents, particularly dipolar aprotic ones, the exchange rate between bound and free solvent is much slower, and fast-exchange conditions hold less frequently. Also, study of other nuclei, such as ^{17}O for water, where the magnetogyric ratio is much smaller than ^1H, will mean that slow-exchange contributions will occur at a lower temperature. Indeed, most exchange studies in water have used ^{17}O.

The presence of unpaired electrons in the solution also has a profound effect on the relaxation rates of the magnetic nuclei. For the observation of an NMR signal, either or both the following conditions must be met.[152]

$$\frac{1}{T_e} \gg \frac{A}{h}, \qquad \frac{1}{\tau_B} \gg \frac{A}{h} \tag{30}$$

where T_e is the spin relaxation time of the unpaired electron. The paramagnetic species in solution affect nuclear relaxation rates via two separate mechanisms. The spinning electron has a high magnetogyric ratio and is thus very effective at producing a fluctuating magnetic field at adjacent magnetic nuclei, as mentioned earlier. The other mechanism arises from the scalar coupling that can arise between nuclear and electron spin mentioned earlier with regard to the contact shift [Eq. (26)]. This interaction gives rise to a fluctuating magnetic field at the magnetic nucleus with a correlation time equal to τ_s, the electron spin–lattice relaxation time. The relaxation rate for bound nuclei is then given by

$$\frac{1}{T_{1,2}} = \frac{1}{T_{1,2\,\text{dipolar}}} + \frac{1}{T_{1,2\,\text{scalar}}} \tag{31}$$

In the limit of fast exchange, the observed rate will be simply the weighted average of both bound and free solvent rates. When the fast-exchange limit does not hold, the observed relaxation rate will be a complex function of the exchange time between sites as well. In particular, T_2 will be a very exchange-sensitive parameter and will involve the shift between bound and free solvent as well. Thus, we have[152]

$$\frac{1}{T_2} - \frac{1}{T_{2_F}} = \frac{X_B(1/T_{2B}^2 + 1/T_{2B}\tau_B + 4\pi^2\Delta\nu_B^2)}{\tau_B[(1/T_{2B} + 1/\tau_B)^2 + 4\pi^2\Delta\nu_B^2]} \tag{32}$$

and

$$\frac{1}{T_1} - \frac{1}{T_{1_F}} = \frac{X_B}{T_{1_B} + \tau_B} \tag{33}$$

It is a generally accepted practice to assume that all the relaxation and exchange parameters show Arrhenius temperature behavior. Schematically, the variation of T_1 and T_2 with temperature is as shown in Fig. 10.

Region A is the region of fast exchange where, generally, $T_1 = T_2$. In region B, the exchange contribution is becoming increasingly important, and the lines broaden. In this region,

$$\frac{1}{T_2} - \frac{1}{T_{2_F}} = X_B\left(\frac{1}{T_{2B}} + 4\pi^2\delta\nu^2\tau_B\right) \tag{34}$$

In region C, there is separation of both bound and free peaks. As the temperature is decreased, the peaks narrow as separation improves. In practice, it is rarely possible to see all three regions. For

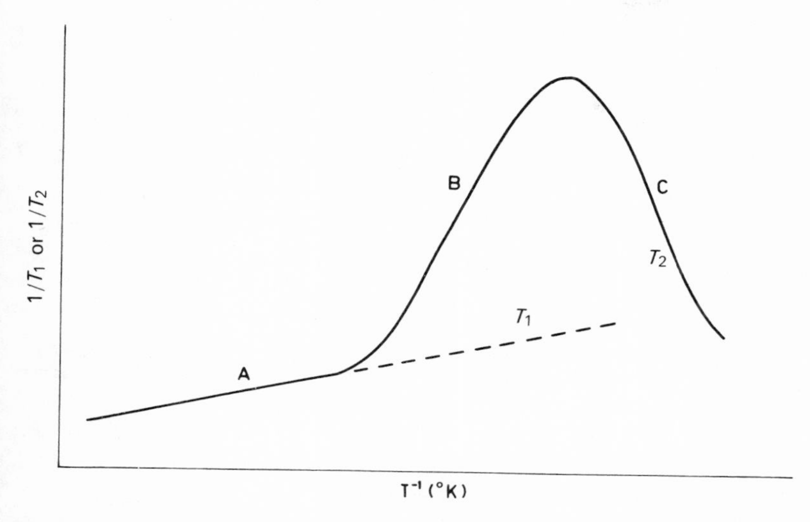

Figure 10. Schematic representation of variation of T_1, T_2 with temperature.

proton NMR in water, most measurements between 0 and 100°C lie within region A, although it has been possible to make exchange studies for some nuclei. For a complete review of the results, the reader is referred to the review of Hinton and Amis.[153] Most exchange studies in water have involved ^{17}O measurements, which will be considered in the next section. Proton linewidth studies yield exchange parameters for solvents with slightly slower exchange than water. This condition is fulfilled for most nonaqueous solvents.

(ii) ^{17}O Shifts

A small but significant amount has been published on ^{17}O shift measurements. Jackson *et al.*[154] observed separate ^{17}O signals for bound and free water in Be^{2+}, Al^{3+}, Ga^{3+}, and Mg^{2+} solutions when a paramagnetic ion was added to shift the free-water signal (Fig. 11). In the absence of a paramagnetic ion, the bound peak was hidden under the free peak. Using 11.48% enriched water, Connick and Fiat[155,156] were able to integrate the areas of bound and free water and obtain hydration numbers of 5.9, 4.2, and 6.0 for Al^{3+}, Be^{2+}, and Ga^{3+}, respectively. Using a closely related approach, Alei and Jackson[157] were able to estimate values of 5.9, 3.9, and 6.8 for Al^{3+}, Be^{2+}, and

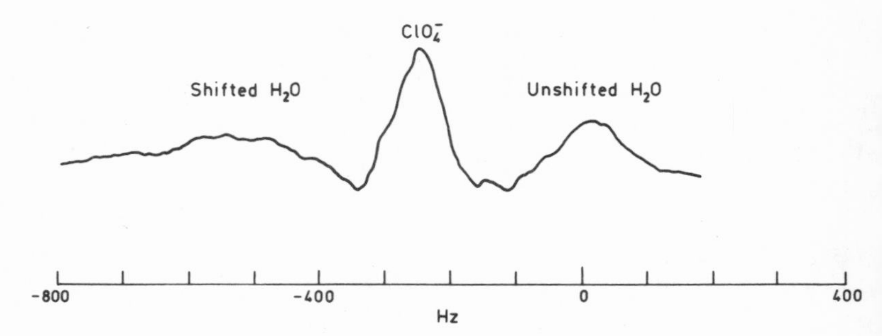

Figure 11. ^{17}O spectrum of 1.6 mol liter^{-1} aluminum perchlorate + 0.24 mol liter^{-1} cobalt(II) perchlorate in 0.5 mol liter^{-1} perchloric acid at 2.6% enrichment (after Jackson *et al.*[154]).

Cr^{3+}, respectively. In this technique, the shift caused by a paramagnetic ion with a labile solvation shell is inversely proportional to the free water in solution. If an added ion, e.g., Al^{3+}, ties up a certain amount of free water, comparison of paramagnetic ion shifts with and without Al^{3+} enables the solvation number of Al^{3+} to be obtained.

The effect of paramagnetic ions on solvent resonances has already been mentioned. As discussed for aqueous solutions, ^{17}O measurements are more useful in the study of solvent exchange between solvation shell and bulk solvent. Thus, Swift and Connick,[152] as well as developing the detailed theory, made extensive measurements on the ^{17}O line-broadening due to the effects of Mn^{2+}, Fe^{2+}, Co^{2+}, Ni^{2+}, or Cu^{2+} and derived relevant exchange parameters for these systems. Recently, the measurements on the effects of Ni^{2+} have been reexamined by Neely and Connick.[158] The earlier results have been widely reviewed, in particular by Hinton and Amis,[153] by Deverell,[159] and by Silver and Luz.[160] An interesting observation made by Jackson *et al.*[154] is that addition of ClO_4^- ions to Cr^{2+} solutions narrowed the water linewidth, which suggested replacement of water in the solvation shell by ClO_4^-. A similar result also held for MnO_4^- ions and Co^{2+}. It had previously been noted that in general, anions have no effect on linewidths. This result is somewhat surprising in view of the recent observation[161] that chloride ions have a marked effect on the proton linewidths of aqueous Cu^{2+} solutions. For the addition of SCN^- to Fe^{3+} solutions, the effect is so large that semiquantitative estimates of all six association equilibria

$$Fe(SCN)_{i-1}^{(3-i+1)+} + SCN^- \underset{}{\overset{K_i}{\rightleftharpoons}} (FeSCN)^{(3-i)+}; \quad i = 1\text{–}6$$

can be obtained.[162]

The effect of diamagnetic electrolytes on the ^{17}O shift of water has been studied by Luz and Yagil[53] and by Fister and Hertz[52] (Fig. 12). Despite inconsistencies between the two sets of data, it is readily apparent that the molar shifts bear no resemblance to those for protons. Particularly when taken in conjunction with the relatively smaller temperature dependence of the water shift for ^{17}O compared with ^{1}H, it must be concluded that a completely different shift-producing mechanism (or mechanisms) is operating. In particular, the results are not compatible with the electrostatic bond polarization

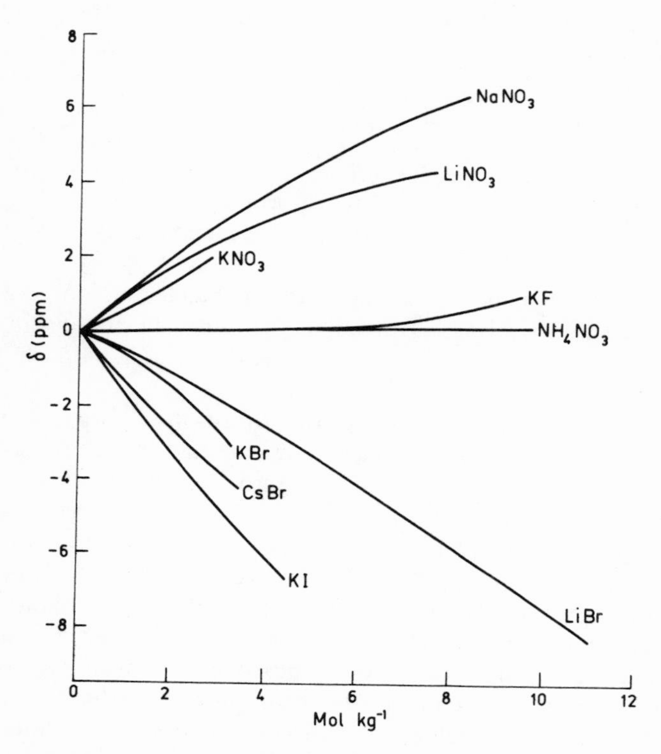

Figure 12. ^{17}O shifts of aqueous solutions of diamagnetic electrolytes (after Luz and Yagil[53]).

mechanism discussed by Buckingham.[59] Variable-temperature measurements would indeed be very useful to assess structural effects, but the variations of molar shift with ion size at room temperature do suggest that structural effects do not predominate. It is interesting that for both cations and anions, the relative order of molar shifts is the same as that noted for the shifts produced by counter ions when studying cations (e.g., ^{23}Na) and anions (e.g., ^{35}Cl). The mechanism of counter-ion dependence on cations and anions is fairly well established (see below) and is due to a repulsive overlap mechanism whereby direct ion–ion encounters restrict the free rotation of the closed-shell electrons of the ions, giving rise to a downfield shift.

As far as is known, no high resolution studies have been made on the effect of electrolytes on the ^{2}H resonance of D_2O. It would be expected, however, that behavior similar to that observed for ^{1}H would be found, although isotope effects could be important.

(iii) Solute Shifts

Although not as easy experimentally as solvent proton studies, solute NMR studies are extremely important. They are one of the few techniques that can give information concerning the environment of a single ion.

For the alkali metal, alkaline earth, and halide ions, only a single resonance line is observed at all temperatures in all solvents, so rapid exchange occurs between ions and any ion pairs in such solutions.

The first observation that closed-shell ions showed concentration-dependent chemical shifts was made by Gutowsky and McGarvey,[163] who found a shift of approximately 30 ppm between saturated and half-saturated ^{133}Cs.

Shoolery and Alder[85] later observed a shift of approximately 1.67 ppm between 4 mol liter^{-1} $K^{19}F$ and 12 mol liter^{-1} $K^{19}F$. The first comprehensive study was made by Richards *et al.*[164] on the ^{205}Tl nucleus. This nucleus is remarkable for the size of its shifts; indeed, shifts as large as these are almost impossible to measure on any but specially constructed or modified spectrometers. Richards and co-workers found that the shifts depended on both the concentration and the nature of the cation. Although there was a linear shift with concentration at high concentrations, there was pronounced curvature at low concentrations. The anomalous behavior of TlOH was ascribed to ion-pair formation.

Shortly after this, Carrington *et al.*[165] studied ^{19}F and ^{133}Cs shifts in a variety of solutions. They found for ^{19}F shifts of 0.4 mol liter^{-1} KF in the presence of added electrolytes that the shift depended on the concentration and nature of both cation *and* anion, whereas for ^{133}Cs, in agreement with the Tl data of Richards and co-workers, the shift depended principally on the nature of the counter ion. Around this time, workers in the field of solid state physics began to use NMR methods to study the short-range attractive and repulsive forces present in ionic crystals. In particular, Van Kranendonk[166] demonstrated that an idealized ionic model was unsatisfactory in explaining the chemical shifts and relaxation times of magnetic nuclei in ionic crystals. Kondo and Yamishita[167] were able to explain the relaxation of magnetic nuclei in ionic crystals in terms of the repulsive overlap of the closed-electron shells of the ions. These repulsive forces, arising mainly from the Pauli principle, tend to produce a nonspherical charge distribution around the ion, which gives rise to a more efficient quadrupolar relaxation mechanism. With account taken of second-nearest neighbors, Yamagata[168] and Ikenberry and Das[169–170] have shown that an extension of the Kondo and Yamashita approach yields reasonable values for the chemical shifts of magnetic nuclei in ionic crystals. In particular, this theory predicts the correct order of efficiency of shift-producing mechanism, whereas polarization and covalency effects give rise to the inverse order. As will be seen below, the same order of shift-producing effects is observed in solution as in crystals, which suggests a similar shift mechanism.

Deverell and Richards, in two classic papers,[171,172] studied the effect of concentration and counter ion on both the magnetically active alkali metals (^{7}Li, ^{23}Na, ^{39}K, ^{87}Rb, and ^{133}Cs) and the halide ions (^{35}Cl, ^{87}Br, ^{127}I) (Figs. 13 and 14). From the cation NMR, they noted that the downfield shift produced by counter ions was in the same order as that produced in the crystalline state. The Kondo–Yamashita approach yields the following equation for the shielding constant for an ion in a cubic crystal:

$$\delta = K\left\langle \frac{1}{R^3} \right\rangle \frac{1}{\Delta} \sum S^2 \tag{35}$$

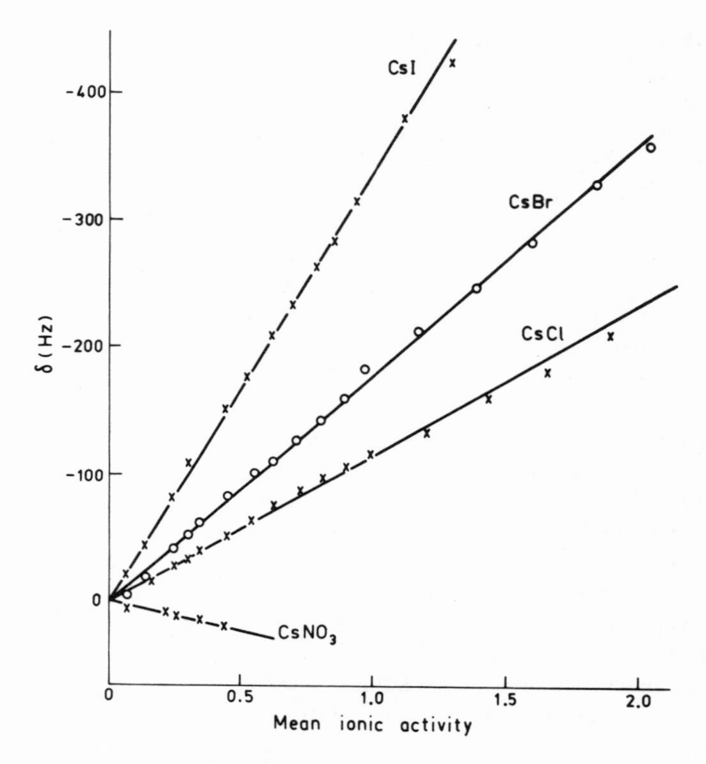

Figure 13. Variation of ^{133}Cs shift with activity of aqueous solutions of cesium salts (at 7.35 MHz) (reproduced from Deverell and Richards[171] by permission of Taylor and Francis Ltd.).

where K is a constant, $\langle 1/R^3 \rangle$ is the expectation value taken over all the outer p electrons of the central ion, and ΣS^2 is the sum of the squares of appropriate overlap integrals of the p orbitals of the central atom with s orbitals and with p orbitals in both σ and π bonding configurations of the surrounding ions. Deverell and Richards used the Kondo–Yamashita formalism and obtained for the chemical shift of an ion in solution the following:

$$(c) = K\left\langle \frac{1}{R^3} \right\rangle \frac{1}{\Delta}\left[\Lambda^{(c)}_{\text{ion--ion}} + \Lambda^{(c)}_{\text{ion--ion}} - \Lambda^{(o)}_{\text{ion--water}}\right] \qquad (36)$$

where $\Lambda^{(c)}_{\text{ion--ion}}$ and $\Lambda^{(c)}_{\text{ion--water}}$ are the appropriate sums of overlap integrals with all surrounding ions and water molecules at a solute concentration c, and $\Lambda^{(o)}_{\text{ion-water}}$ is the sum at infinite dilution. Deverell and Richards concluded from the observed linearity of shift with solute activity that the first term, $\Lambda^{(c)}_{\text{ion-ion}}$, being dependent on ion–ion collisions, is predominant. However, if each ion were to displace one water molecule on collision, then $\Lambda^{(c)}_{\text{ion-water}}$ would change also, but would do so in exactly the same was as $\Lambda^{(c)}_{\text{ion-ion}}$. Hence, it would be impossible from this treatment to separate the two possible shift-producing mechanisms. Change of solvent is known to produce large shifts in ion resonances, and clearly, ion–solvent overlap integrals are

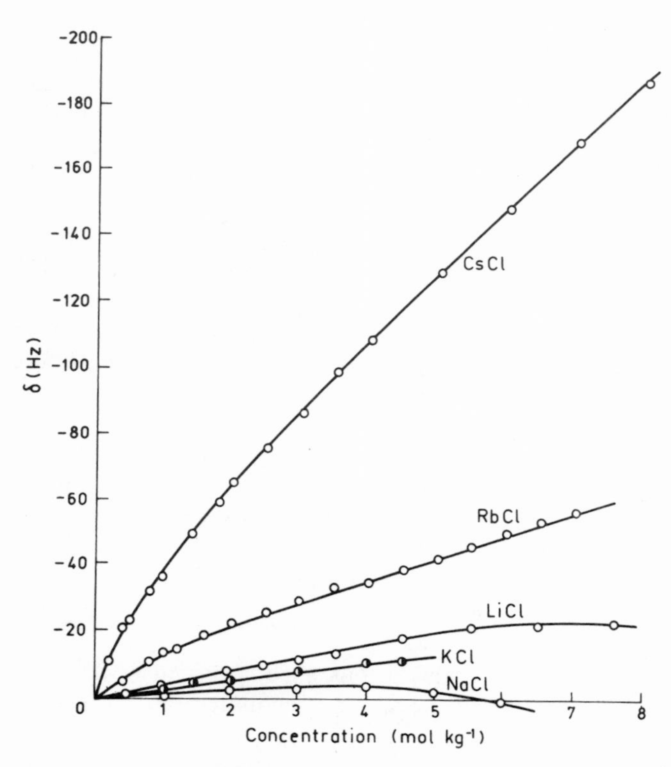

Figure 14. Variation of ^{35}Cl shift with concentration of aqueous solutions of alkali metal chlorides (at 5.63 MHz) (reproduced from Deverell and Richards,[172] by permission of Taylor and Francis Ltd.).

important, and it is more realistic to recast Eq. (36) as follows:

$$\delta(c) = K\left\langle \frac{1}{R^3} \right\rangle \frac{1}{\Delta}[\Lambda_{\text{ion–ion}} - \Lambda_{\text{ion–water}}]P \tag{37}$$

where $\Lambda_{\text{ion–water}}$ is the change of overlap when a water molecule is removed from the close vicinity of an ion, $\Lambda_{\text{ion–ion}}$ is change in overlap when an ion is placed in the hole left by the water molecule, and P is the probability that such an encounter takes place. This formalism assumes that ion–ion collisions take place by water replacement. It is probably the simplest conceivable model that will take into account loss of overlap with solvent.

Experimentally, Deverell and Richards found the shifts to be closely linear with salt activity. They have used Eq. (36) to obtain values for the bracketed overlap term at unit activity. For cation NMR, the values increased going from Li^+ to Cs^+. For LiCl and LiBr, the probability polarization mechanism was postulated. The values for ^{23}Na are much smaller than those for the larger alkali metal ions. Deverell and Richards concluded that the reason was that the structure-breaking properties of the large ions allow direct ion encounters, but for Na^+, such encounters are less likely. This conclusion must be tempered by the fact that according to the more realistic formalism given by Eq. (37), there are two possible causes for large shifts, high probability of collision or large overlap integral values. It is possible that high probability will affect the curvature of the shift vs. concentration dependence, whereas large changes in overlap integral will affect the shift magnitude.

In two more recent papers, Richards and co-workers[54,173] have examined the problems discussed above in more detail and reported very accurate low-concentration ^{133}Cs shifts using a superconducting magnet of field strength 5.02 T.

They formulate the shift as follows:

$$\delta_{\text{obs}} = P(c)[\delta_{\lambda,v} + \delta_s^r - \delta_s^{\text{dis}}] \tag{38}$$

where $P(c)$ is the probability of ionic encounter, and $\delta_{\lambda,v}$, δ_s^{dis}, and δ_s^r are the shift contributions due to ionic encounter, solvent displacement, and solvent reorientation, respectively.

They then neglect the latter two contributions. In this, as in their earlier papers, one must query their disregard of δ_s^{dis}. In order to calculate the shift, the radial dependence of the shift-producing interaction and the radial distribution are required. Thus,

$$\delta_{\text{obs}} = 4\pi \int_a^\infty n(R, c)\sigma(R)R^2\, dR \tag{39}$$

where n is the local number density of counter ions around the ion of interest and $\sigma(R)$ is the shift contribution of an ion at a distance R from the ion of interest. These authors choose a Debye–Hückel (DH) potential, but modify it by addition of an oscillating potential to take into account the particulate nature of the solvent and the tendency for solvent molecules to form well-defined shells around an ion. Thus, they take

$$n(R, c) = n^\infty(c)\exp\left(-\frac{U^{\text{osc}}(R) + e\psi_{\text{DH}}(R, c)}{kT}\right) \tag{40}$$

whereas for a Debye–Hückel continuum model

$$n_{\text{DH}}(R, c) = n^\infty(c)\exp\left(-\frac{e\psi_{\text{DH}}(R, c)}{kT}\right) \tag{41}$$

The integrals over all space of both Eq. (40) and Eq. (41) clearly are equal. As the integrating limit of both equations is increased, the integral of the first will oscillate around that of the second. At some radius R' close to the first minimum in the r.d.f., the two integrals will be equal. For distances greater than R', a continuum r.d.f. is assumed, whereas all ions closer than R' to the ion are assumed to be at the distance of closest approach, a. From the known short-range dependence of shifts deduced from repulsive overlap theory, it was assumed that beyond R' (approximately equal to the sum of ion radius and solvent molecule diameter), the contribution to the shielding was negligible. Thus,

$$\delta_{\text{obs}} = \sigma(a)4\pi \int_a^R n^\infty(c)\exp\left(\frac{-e\psi_{\text{DH}}}{kT}\right)R^2\, dR \tag{42}$$

Empirically, it was found that the shifts obey the following relationship:

$$\delta - \delta_0 = C^g \tag{43}$$

where δ_0 is the infinite dilution shift and g is a parameter, that is a constant for any given salt (and solvent). Approximate numerical integration of Eq. (42) was in accord with Eq. (43).

For aqueous solutions of Cs^+ salts, the values of g, varying slightly with anion due to changes in the distance of closest approach, in all cases agreed closely with calculated values. For nonaqueous solutions, g changes due to the different dielectric constant, affecting Ψ_{DH}. Very reasonable agreement between observed and calculated values of g was found (when the dielectric constant varied from 32 to 130, g varied from 0.5 to 0.9). In the more recent paper, Richards and co-workers[173] have made a much more detailed computing study of the integration of Eq. (42), in which it was found that g was no longer a constant: the full integral equation in fact fits the experimental data better than the empirical logarithmic equation. Analysis of the data bears out that it is indeed the bulk dielectric constant that is required in the DH potential, a fact supported by variable temperature and nonaqueous solvent studies. It is interesting that in mixed solvents of acetic acid and water and methanol and water, the data for Cs^{133} shifts cannot be fitted well except at very low concentrations, whereas the data for each individual solvent fit very well. Recent unpublished calculations[174] suggest that the sorting of solvent molecules around an ion in a mixed solvent will be salt-concentration-dependent, and will thus affect the concentration dependence of the shift (solvent sorting will be dealt with in greater depth in Section IV.3).

Much less extensive studies have also been made on ^{23}Na,[175–177] ^{39}K,[176] ^{133}Cd,[178–179] and ^{25}Mg.[180] In the ^{23}Na study, Templeman and Van Geet[175] observed that for the sodium salts studied ($NaClO_4$, NaOH, and NaCl), the observed shifts were closely linear in salt mol fraction. The ^{133}Cd shifts showed very pronounced curvature with concentration (Fig. 15), indicative of ion association, which is well established for Cd^{2+} solutions. There is no reason (apart from activity coefficient estimation) why such results should not be used for ion association constant measurement; indeed, approximate determinations have been made by this method in nonaqueous solutions (see Section IV.3).

It is interesting that for all cation resonances studied, the order of a shift-producing mechanism is always the same ($I < Br < Cl \simeq OH < H_2O < NO_3 < ClO_4$). Thus, relative to the infinite dilution resonance of water, the iodide ion produces a large downfield shift and the nitrate ion a small upfield shift.

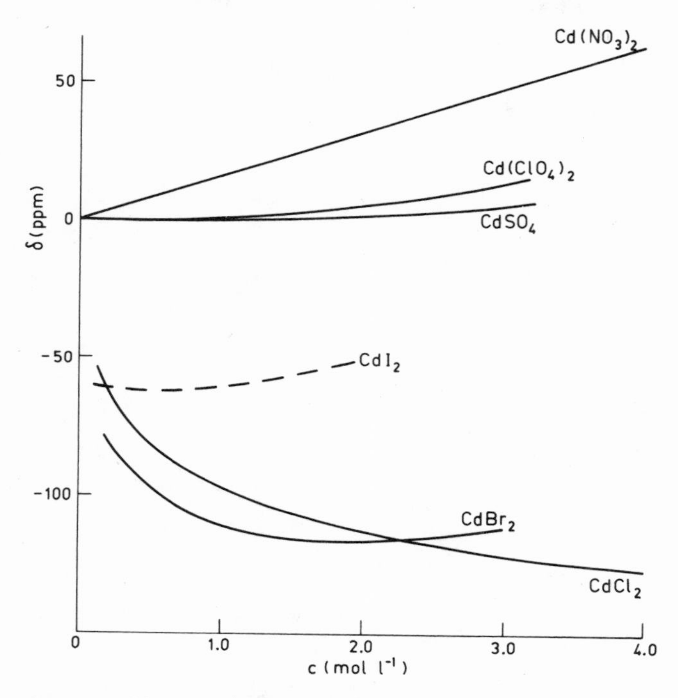

Figure 15. ^{113}Cd shifts in various cadmium salt solutions as a function of concentration (from Kostelnik and Bothner-By[179]).

Akitt[177] has suggested that the upfield shift caused by oxyanions (and tetrafluoroborate) arises because these anions do not overlap the cation orbitals, but displace water, which does have an overlap contribution. This is unlikely, since in solvents in which the infinite dilution shift is considerably upfield of water, the oxyanions give downfield shifts (see Fig. 23).[181] If Akitt's suggestion were correct, upfield shifts would be observed.

A few studies have been made of the effect on cation (^{133}Cs) and anion (^{35}Cl, ^{81}Br, and ^{127}I) resonances of varying ionic environment at constant overall concentration. Deverell and Richards[172] report the effect of varying halide composition at constant cesium concentration of 4 *m*. Similar results obtained at 2 *m* by Porthouse *et al.*[182] are shown in Fig. 16. In contrast to shifts of over 100 Hz, shifts 20 times smaller in

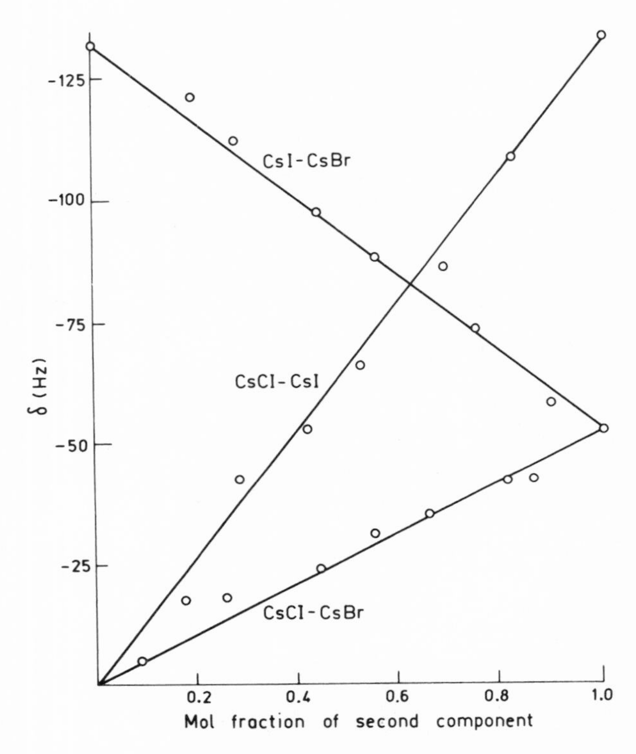

Figure 16. ^{133}Cs shifts in mixtures of cesium halides at 2 *m* (at 5 MHz) (Porthouse *et al.*[182]).

the cesium resonance are obtained on varying the cation composition (Fig. 17) at 4 *m*.[182] Related effects are observed on ^{81}Br shifts[172] (Figs. 18 and 19), where the effect of changing the cation is far greater than that of changing the anion. Both sets of experiments are good confirmation of Brønsted's principle of specific interaction of ions, that like–unlike charge interactions are dominant at moderate concentrations, a principle previously justified on thermodynamic evidence. Nevertheless, like–like interactions play an increasing role as the concentration is increased.

Some work has been reported on the concentration dependence of ^{1}H shifts of organic ions. Thus, Anderson and Symons[183] found

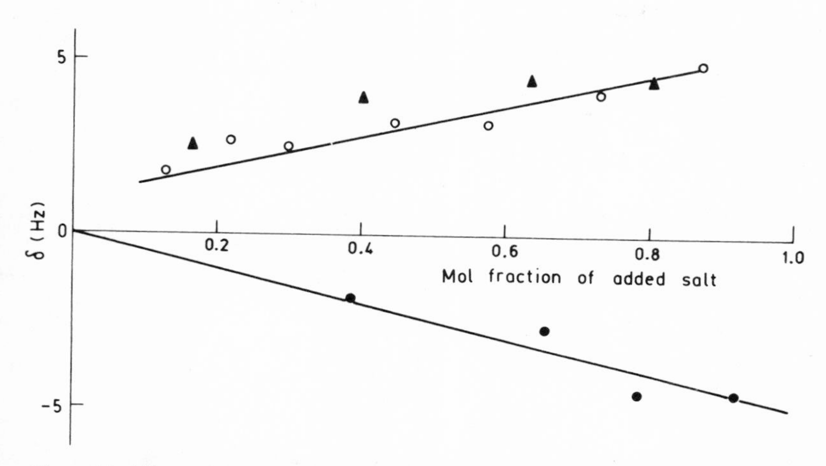

Figure 17. ^{133}Cs shifts in mixtures of alkali chlorides at 4 *m* (at 5 MHz) (Porthouse *et al.*[182]): ○ LiCl, ▲ $CaCl_2$, ● KCl.

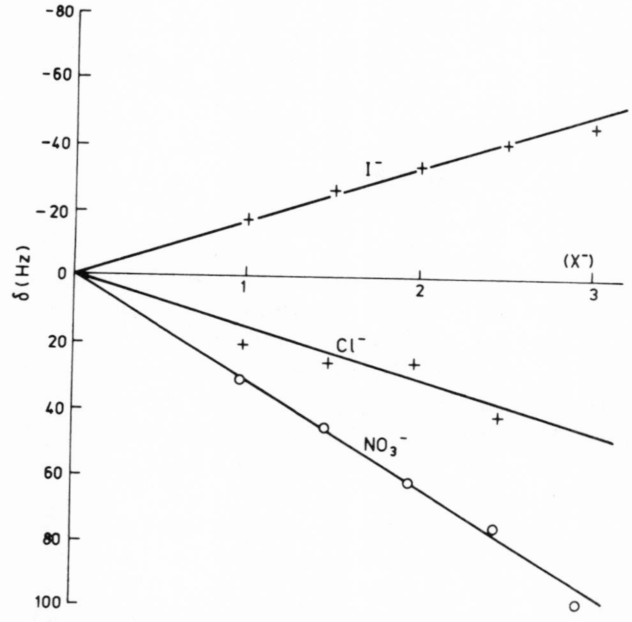

Figure 18. ^{81}Br shifts in solutions of mixed sodium salts, $Na^+ = 4\,m$ (at 14.89 MHz), $Br^- + X^- = 4\,m$ (reproduced from Deverell and Richards[172] with permission of Taylor and Francis Ltd.).

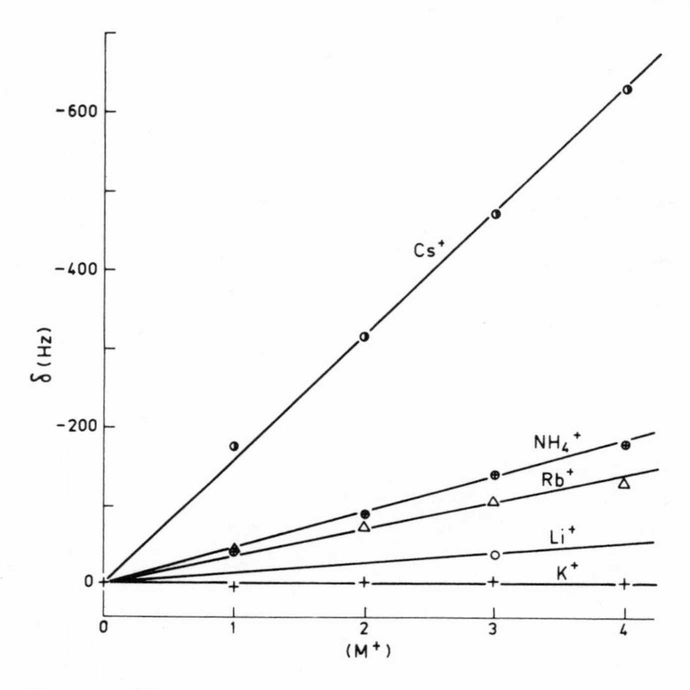

Figure 19. ^{81}Br shifts in solutions of mixed bromides. $Br^- = 4\,m$ (at 14.89 MHz) $Na^+ + M^+ = 4\,m$ (reproduced from Deverell and Richards[172] with permission of Taylor and Francis Ltd.).

both a concentration and a solvent dependence for the shifts of *N*-alkylpyridinium salts. In water, such shifts were concentration- but not anion-dependent, which suggested the presence of cation–cation interactions. In other solvents, the shifts, anion-dependent, correlate with Kosower's solvent polarity scale, and were interpreted in terms of ion-pairing (being favored in low-polarity solvents). Covington *et al.*[184] have observed that the trimethylsulphonium iodide ^{1}H resonance was concentration-dependent in both water and acetonitrile. From the concentration dependence and an estimate of activity coefficients, ion association constants were obtained that were in close accord with those estimated from conductance measurements.

A potential technique for studying interactions of multiatomic ions involves the measurement of coupling constants. To date, however, few ions have been found that show solvent- and concentration-dependent coupling. One exception is the BF_4^- ion.

Following initial work by Kuhlmann and Grant[185] and Haque and Reeves,[186] Gillespie *et al.*[187] made an extended study of both solvent and counter-ion effects for ^{11}B–^{19}F coupling. For Ag^+, NH_4^+, and NMe_4^+ salts, they found no counter-ion effects, whereas for Li and Na salts, the coupling ranges from 1.2 Hz at zero concentration to 2.9 Hz and 4.8 Hz at 8 *M*, respectively. These authors concluded that solvent-separated ion pairs were the reason for the change in coupling constant. They ruled out direct pairs, since these should cause broadening of the ^{11}B line and hence collapse of the ^{11}B–^{19}F coupling. The idea of a rigidly held solvation shell for the Li^+ ion also suggests that no direct encounters occur in $LiBF_4$ solutions.

Akitt[177] has recently made a multinuclear shift of aqueous fluoroborate solutions. He observed that anions have little affect on ^{19}F shifts and on ^{19}F–^{11}B coupling, in contrast to the behavior of cations. ^{23}Na shifts are markedly concentration-dependent (upfield) in $NaBF_4$ solutions, and are thus indicative of direct cation–anion encounters. As Akitt points out, these encounters must be very shortlived so as not to cause rapid ^{11}B quadrupolar relaxation. The ^{19}F shifts and couplings were interpreted in terms of the electrostatic polarization ideas earlier developed for proton shifts.[114]

2. Nonaqueous Solutions

(i) Solvent Shifts

Inevitably, the work done on nonaqueous solvents follows closely on that done on aqueous solutions, although the former is very much less extensive. With the exception of the hydroxylic protons of the lower alcohols, proton shifts in nonaqueous solvents are generally small, since the proton site, usually on a carbon atom, is at least two bonds away from the active site. Thus, for the alcohols, it is valid to use the CH_3 or CH_2 group as internal reference for the hydroxyl proton shifts.

(*a*) *Alcohols.* The earliest observations of separate bound and free signals were made with MeOH as a solvation in 1962 by Swinehart and Taube.[188] Using a solution containing 1 mol $Mg(ClO_4)_2$ to 17.1 mol CH_3OH and 1.4 mol water, they obtained a coordination number for MeOH of 5.0 and for water of 0.7. They also obtained some indirect evidence that there was the possibility of ion association. Luz and Meiboom[189,190] made extended studies of both Co^{2+} and Ni^{2+} in MeOH and MeOH + H_2O mixtures. For Co^{2+}, they were able to measure the hyperfine interaction at $-60°C$ and thence the ion–proton distances, which were found to be 2.8 and 3.6 A for the hydroxyl and methyl protons, respectively. Exchange parameters were obtained both in pure MeOH and in the mixtures. Matwiyoff and Taube[109] found a coordination number of 6.0 for the Mg^{2+} ion in both aqueous and methanolic solution of $Mg(ClO_4)_2$ to which acetone had been added to lower the freezing point of the solvent. Nakamura and Meiboom[191] obtained a coordination number of 6.0 for the Mg^{2+} ion in MeOH directly from low-temperature bound and free signals. By adding Co^{2+} ions, they were able to shift the free CH_3 resonance and observe the bound CH_3 group. From kinetic studies, they showed methanol exchanged as a whole molecule. The exchange process was not that of proton transfer.

Grasdalen[192,193] has measured the coordination numbers of Al^{3+} in both ethanol[192] and *n*-propanol[193] and obtained a value of 4 for both solvents, whereas for methanol and solvents of better solvating ability, e.g., H_2O, DMSO, and DMF, values of 6 were obtained. It is likely that the low values are due to polymerization or anion association or both.

^{13}C is likely to prove most valuable in the study of the solvation of ions in organic solvents. Thus, for example, Stockton and Martin[194] have observed separate peaks from bound and free solvent for both Mg^{2+} and Al^{3+} in both ethanol and *n*-propanol. The exchange parameters obtained from variable temperature studies were the same as those derived from ^1H studies, which shows that the proton exchange is in fact whole-molecule transfer. Signal-intensity measurements are difficult, since the broad-band proton decoupling produced nuclear Overhauser enhancement (NOE),[195] which can selectively enhance the amplitude of different signals. However, pulsed decoupling or no decoupling at all can remedy this problem. Stockton and Martin[194] obtained coordination numbers of 5.5 ± 1.0 and 4 ± 1 for Mg^{2+} and Al^{3+}, respectively, in ethanol. They observed the bound β carbon atoms to resonate downfield, whereas the γ carbons resonated upfield and the δ carbons slightly upfield. Such behavior is consistent with Buckingham's electrostatic theory, in which the downfield shift is due to the term in E^2 (the parallel contribution is always downfield), and the upfield shift is due to the angular term. It may indeed be possible to estimate solvation geometries from such considerations.

Following the initial molal shift measurements of the effect of ions on the hydroxyl resonance of methanol by Hammaker and Clegg,[196] Symons and co-workers,[114,197,198] made an extended study on the effect of salts on methanol. Butler and Symons[197] made shift measurements of both bound and free solvent in methanolic solution of Mg, Zn, and Al perchlorates at low temperatures. They noted that the free water shifted linearly with salt concentration in a similar manner for all three salts, which suggested that the secondary solvation of all three cations is similar, and in the authors' view, negligible, and hence the free solvent shift is due simply to the anion. In a later publication, the same authors[198] used this technique to separate cation and anion effects, and as mentioned in Section IV.1(i), this same method has been used to separate ion effects in water. They interpret both cation and anion effects in terms of hydrogen-bond enhancement or decrease, and do not take into account any electric field effect, which must be expected to be very important, particularly for ions of large charge–radius ratio. Butler and Symons, together with Davies,[114] have also noted that both the free and bound solvent resonances of Mg^{2+} exhibit identical anion dependence for NO_3^-, Br^-, and ClO_4^-. Such behavior does not hold for the acetate ion. These authors rationalized this behavior by saying that the anions interact with both bound and free solvent in an identical manner; it is only the highly basic acetate anion that prefers coordination to the bound solvent, which is made more acidic than the free solvent due to the presence of the highly charged Mg^{2+} ion.

Lindheimer and Brun[199] and Vogrin and Malinowski[200] have applied the Malinowski technique to the study of solutions of some salts in methanol to obtain solvation numbers similar to those for aqueous solutions.

(*b*) *Other Solvents.* Ionic molal shift studies, apart from those in the lower alcohols, are not common. Allred and Wendricks[201] have measured molal proton shifts at 25°C in liquid ammonia. All shifts are downfield, unlike the shifts with these ions in water, in which all shifts are upfield. The results were explained in terms of increasing polarization of the lone pair with decreasing ionic size for cations, giving rise to proton deshielding. For the anions, however, ion-pair formation reduces the effective solvation number, the formation constants being in the order $I < Br < Cl$. It is interesting that the gas-phase shift (at $-77°C$) is 1.05 ppm, whereas for water (at 0°C), it is 4.58 ppm. Structural effects should be of minor importance; the observed shifts suggest a polarization mechanism for cations.

Rode[202] has made a brief study of the effects of alkali metal (and ammonium) chlorides on anhydrous formic acid. The shifts of the carbon-bonded hydrogen are always downfield, with the cesium shift being the most downfield. The overall order is $Cs^+ > Rb^+ > Na^+ > K^+ > Li^+ > NH_4^+$, which is surprising, since the structural contribution must be paramount.

Proton NMR studies of dipolar aprotic solvents might be expected not to be fruitful, due to the distance of the protons from the active site of the solvent. However, Coetzee and Sharpe[203] have measured molar shifts of various salts in DMSO, AN, and sulpholane. To compare shifts in different solvents, it is necessary to choose the right solution composition scale; in this case, since the observed shifts are the mol-fraction-weighted average of solvation shell and free solvent, one should choose a mol fraction scale, or more conveniently, an aquamolality scale, which is defined as the number of moles of solute in 55.51 mol solvent. In water, this scale is identical to the molality scale. Even after conversion of the data of Coetzee and Sharpe[203] to a common concentration scale, it is difficult to see any order in the results. All shifts are downfield except for the tetraphenyl borates, which are strongly upfield. The aquamolal shifts for AN are much larger than those for the other solvents. They are of the same magnitude as they are in water. It is likely that they are larger due to the small size of the AN molecule, which allows close approach of an ion. Coetzee and Sharpe,[203] from temperature dependence and dilution studies, concluded that AN is not associated, whereas the other two solvents are; in particular, the protons of sulfolane shift upfield with increasing temperature and similarly DMSO protons shift upfield on dilution with CCl_4. It is likely that the AN electrolyte shifts are due to electric field effects, whereas for the other two solvents, structural effects are also likely to be important. Variable temperature and measurement of the shifts of bound and free solvent in solutions of multicharged ions, in a similar way to the work done in water, would shed further light on this problem.

The studies of paramagnetic ions have involved both low-temperature observation of separate bound and free solvent and conventional broadening techniques. Thus, Matwiyoff[204] has obtained separate signals for bound and free DMF in Co(II) solutions below $-38°C$, but was unable to observe bound Ni(II) due to rapid exchange. However, in both cases, from temperature dependence studies, exchange parameters were obtained. Vigee and Ng[205] have studied DMSO line-broadening caused by Ni(II), Co(II), Fe(III), Mn(III), and Cr(III). Similar studies have been made of Ni(II) in AN,[206–207] EtOH,[208] DMSO,[209–211] Co(II),[205,209,211] Fe(III),[212] and Mn(II)[213] in DMSO. The last-mentioned paper[213] is a particularly good example of a thorough study illustrating the wealth of information obtainable from such work. For further information on NMR shift and exchange parameters, one of the reviews[153,159,214] should be consulted.

Beech and Miller[215] have studied the proton shifts of Co(II) halides as a function of temperature in a variety of nonaqueous solvents. The differently solvated species (e.g., $CoAN_4X_2$ and $CoAN_2X_2$) have different coupling constants [see, for example, Eq. (26)], and the authors estimated equilibrium constants and enthalpies for the various solvation equilibria.

In addition to proton studies, ^{14}N resonances have been used in particular for liquid NH_3 studies, to measure the coordination number of Al(III).[216] In this case, it was necessary to add a paramagnetic ion to broaden the free solvent peak beyond detection. Studies of paramagnetic line-broadening caused by Ni(II) in the same solvent have also been made.[217] ^{17}O line-broadening caused by Ni(II) in ^{17}O-enriched DMF has also been observed.

Almost without exception, all other work has involved the study of doubly or triply charged ions and the observation of free and bound solvent, and hence the evaluation of coordination numbers, and sometimes exchange parameters.

Thus, the coordination number of Al^{3+} (mostly as the chloride) was found to be 6.0 in DMSO[218] and DMF,[219,220] and in AN,[221] in which it is very low and depends on the nature of the anion, it is 1.5 for $AlCl_3$ and 2.9 for $Al(ClO_4)_3$. The coordination numbers are independent of temperature, and it is likely that in a poorly solvating medium like AN with a very low dielectric constant, anions are likely to compete with solvent for the solvation sites. Fratiello *et al.*[222] have also obtained bound and free signals for $AlCl_3$, $BeCl_2$, $GaCl_3$, $SbCl_5$, and $TiCl_4$ in DMF, and have obtained coordination numbers and exchange parameters. In a more extended study of $SbCl_5$ in DMF, the same authors[223] noted that for the bound solvent, the aldehyde proton shifted most, which suggested coordination through the aldehyde group, not through the nitrogen, and the two methyl signals shifted different amounts. Also, the aldehyde proton–methyl proton coupling is double that of the free solvent. Above 100°C, the two methyl signals coalesce (Fig. 20) due to rapid rotation around the carbon–nitrogen single bond,[224] whereas for the bound solvent, no such coalescence occurs.

Lincoln[225] has obtained a very low value for the coordination number of Ga^{3+} in AN (<0.8), which depends on added chloride ion concentration. ^{71}Ga NMR revealed two resonances, one of which he ascribed to $Ga(AN)_6^{3+}$ and the other to $GaCl_4^-$ or Ga_2Cl_6. From the relative areas, he was able to estimate coordination numbers that agreed with the proton results. Ahmed[226] and Ahmed and Schmulbach[227] have also studied Ga^{3+} in AN. In perchlorate solutions, the coordination numbers show large concentration dependence (5.2 at 0.1 *m*, 2.9 at 0.32 *m*), whereas for chloride solutions, the value is much lower. Comparison of the exchange parameters derived from line-broadening suggests that the exchange mechanism in $GaCl_3$ is very different from that for the related halides of aluminum, boron, and antimony.

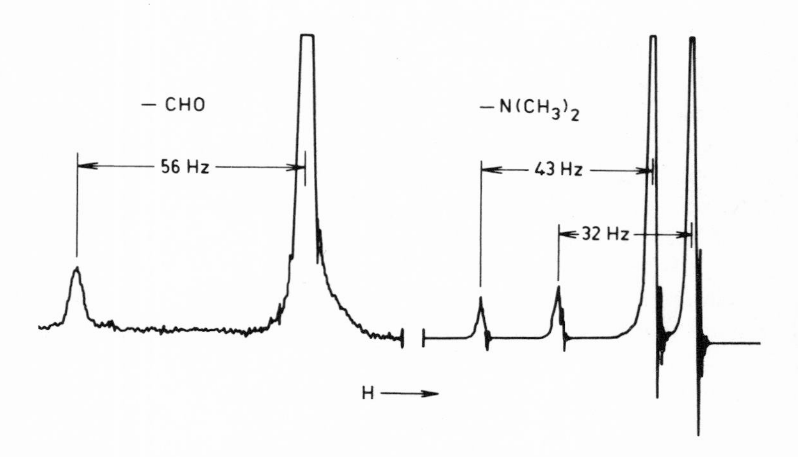

Figure 20. ^{1}H spectrum of solutions of $SbCl_5$ in DMF. The aldehyde signal is displaced downfield from the bulk DMF to a much greater extent than are the methyl signals (reproduced from Fratiello *et al.*,[223] with permission of Taylor and Francis Ltd.).

(ii) Solute Shifts

There has been made, inevitably, an artificial division between aqueous and nonaqueous systems. In fact, many of the studies of aqueous systems discussed earlier also include work on nonaqueous solvents. It is only where important differences arise between aqueous or nonaqueous systems that this work will be considered in depth. Also included in this section is some discussion of the factors that cause solute nuclei to resonate at different frequencies in different solvents, although much of the theory centers around aqueous solutions.

^7Li shifts in aqueous solution are virtually concentration- and counter-ion-independent,[228] due presumably to a low probability of direct ionic encounter—an impenetrable solvent shell. (Alternatively, the ion-pair shift may be very similar to that of the free cation.) However, Maciel *et al.*[229] have observed large solvent shifts (up to 6 ppm) and some slight concentration dependence in poorly solvating solvents. Also, in very poorly solvating solvents, such as THF, DME, or Et_2O,[230] the ^7Li resonance in the presence of the fluorenyl anion is approximately 6 ppm downfield of the free Li^+ ion, which indicates direct ion association with the cation sitting above the ring system. These findings lead one to suggest that in aqueous solution, the solvation shell is tightly bound, and as the solvent is changed to less solvating solvents, ion–ion interactions become more probable. Similar results have been obtained by Herlem and Popov,[231] who observed that for the solvents ethylamine and isopropylamine, the ^{23}Na shifts of NaI and $Na\phi_4B$ apparently did not extrapolate to the same infinite dilution value. It seems likely that down to the lowest concentration measurable (0.1 *M*), Na–I contact ion pairs are formed extensively, and presumably at some lower concentration, the dependence must change so that both shifts will extrapolate to the same infinite dilution value. It would then be perfectly feasible to use the approach of Richards and co-workers,[54] discussed in Section IV.1, to study the radial distribution function of the ions.

A linear correlation was found[231] between ^{23}Na chemical shift in various solvents and Gutmann donor number (which equals minus the enthalpy of complex formation between the solvent and $SbCl_5$). For the solvents ethylamine and isopropylamine by extrapolating the linear correlation, values of 55.5 and 57.5, respectively, were found (cf.

water ~ 33). It is difficult to see how such solvents should have such high solvating ability and yet ion-pair so extensively. Intuitively, an inverse relationship would be expected to exist between solvating ability and ion association.

Recently, Popov and co-workers[232,233] observed such a correlation. Using a superconducting magnet and an operating frequency of 60 MHz, the ^{23}Na resonance of a variety of salts in many solvents was observed down to a concentration of 0.01 mol liter^{-1}. For a very strongly solvating solvent, DMF, no concentration dependence was observed for perchlorate and tetraphenylborate solutions (Fig. 21), whereas for MeOH and EtOH, the dependence was slight. For poorly solvating solvents, ClO_4^- and ϕ_4B^- solutions were again relatively concentration-independent. For the halides I^-, Br^-, and

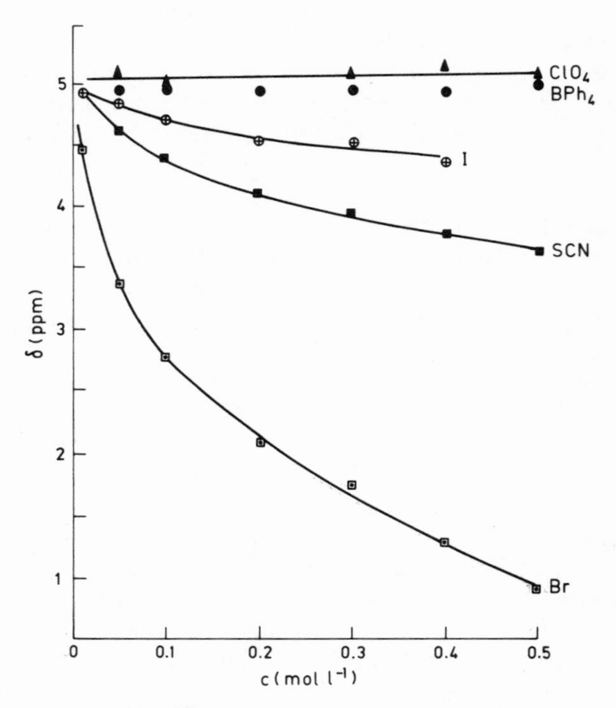

Figure 21. ^{23}Na shifts of various sodium salts in *N*, *N*-dimethyl-formamide (DMF) (reproduced from Greenberg *et al.*,[232] with permission of American Chemical Society).

SCN^-, the dependence was pronounced and highly nonlinear in all solvents. For tetramethylguanidine, the concentration dependence appears to be slight, but additionally, the lines do not extrapolate to the same point (Fig. 22), and so it must be concluded that above 0.01 mol liter^{-1} in these solutions, ion association is very extensive. For propylene carbonate, an approximate association constant of 36 ± 10 liter mol^{-1} was calculated for NaSCN.

Other ^{23}Na studies of a nature similar to those of Popov [e.g., NaBϕ_4 and $NaBH_4$ in polyethers $CH_3(OCH_3CH_2)_nOCH_3$, where $n = 1,4$ Ref. 234] suggest that ion association occurs for the borohydride, but not for the tetraphenylborate.

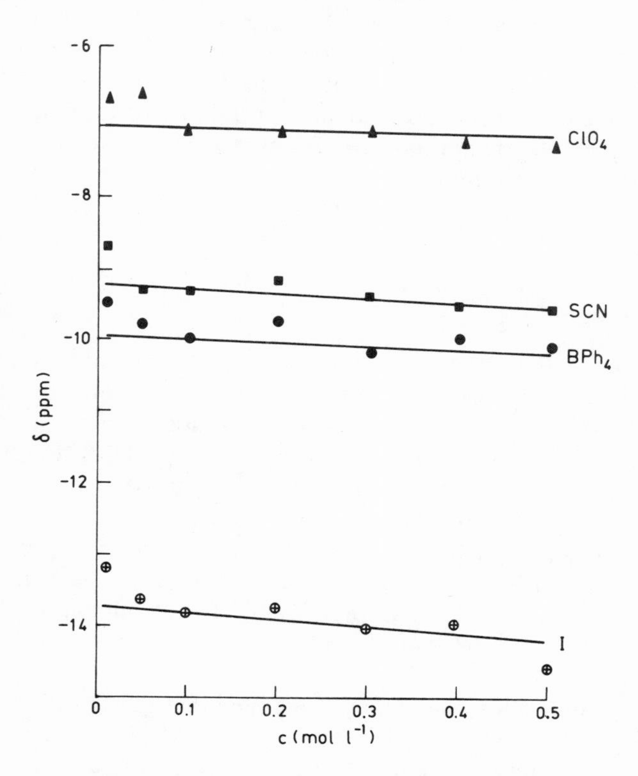

Figure 22. ^{23}Na shifts of various sodium salts in 1,1,3,3,tetramethylguanidine (reproduced from Greenberg *et al.*,[232] with permission of American Chemical Society).

The interpretation of the infinite dilution shifts of solutes, particularly for closed-shell ions, is limited by the very nature of the NMR experiment. The theoretical quantity to which the chemical shift is related is the shielding constant. According to the formalism of Ramsey, the shielding constant may be separated into two parts, a diamagnetic and a paramagnetic contribution, and it is valid to neglect changes in the diamagnetic contribution for all nuclei except ^1H (and possibly ^7Li). The paramagnetic contribution may be considered as the shift between a solvated and a gas-phase ion, thus, it is the downfield shift caused by the restricted rotation of the spherical electron distribution arising from collisions between the ion and its surrounding molecules (and ions). An alternative, but entirely equivalent, quantum mechanical interpretation is that collisions mix in nonspherical excited states to the spherical ground state of the atom. In practice, in any NMR experiment, it is necessary to make shift measurements relative to a standard. In the case of variable solvent studies, the shifts are normally recalculated to be relative to an infinite dilution aqueous solution.

Thus

$$\delta_{obs} = \sigma_p(s) - \sigma_p(aq) \tag{44}$$

where δ_{obs} is the observed shift and $\delta_p(s)$ and $\sigma_p(aq)$ are the infinite dilution paramagnetic contributions to the shift in both solvent and aqueous solutions, respectively. Various authors[168,170,235,236] have used the Kondo–Yamashita repulsive overlap model[167] to estimate the shift between free alkali or halide ions and those in ionic crystals. Simply by subtracting the calculated values and measurement of crystal shifts relative to an infinite dilution reference, it is possible to estimate $\sigma_p(aq)$. Such calculated values should be independent of the crystal counter ion. Ikenberry and Das[169] found values for ^{87}Rb of -0.88×10^{-4}, -0.63×10^{-4}, and -0.60×10^{-4} from RbCl, RbBr, and RbI, respectively.

Itoh and Yamagata[237] have suggested that the solvation effects responsible for σ_p should be the same as those that affect relaxation for quadrupolar nuclei. These authors assumed a direct relationship between σ_p and the quadrupolar coupling constant, and were thus able to estimate $\sigma_p(aq)$ for ^{127}I. Deverell[238] has considered this relationship in some detail. He concluded that

$$\frac{1}{T_1} = \frac{3}{4000} \frac{2I+3}{I^2(2I-1)} \left(\frac{e^2Q}{h}\right)\left(\frac{\Delta\sigma}{d^2}\right)^2 \tau_c \tag{45}$$

where I is the nuclear spin quantum number, e is the electronic charge, Q is the nuclear quadrupole moment, $\Delta\sigma$ is the change in shielding constant, α is the fine structure constant, and τ_c is the correlation time. From proton-derived τ_c values and experimentally observed T_1 values, Deverell estimated σ_p values.

For the cases where T_1 and chemical shift are known as a function of concentration, Deverell observed that

$$\left(\frac{1}{T_1\tau_c} - \frac{1}{(T_1\tau_c)_0}\right) \Big/ \frac{1}{(T_1\tau_c)_0} = \frac{\delta}{\delta_p(aq)} \tag{46}$$

where the subscript 0 refers to infinite dilution. Assuming that τ_c is proportional to viscosity, Deverell was able to estimate $\sigma_p(aq)$.

A related technique was used by Baker *et al.*,[239] who used the relationship between σ_p for HF and the spin–rotation interaction constant derived from gas-phase relaxation studies. From the shift between the gas-phase HF molecule and infinitely dilute aqueous F^- ion, they obtained $\sigma_p(aq)$ for the F^- ion as -169 ppm.

Deverell *et al.*[240] have also observed ^2H isotope solvent effects on solute chemical shifts in water. Assuming that such shifts are caused by a zero-point energy change affecting the excitation energy term in the paramagnetic contribution, they were able to estimate approximate values of $\sigma_p(aq)$.

It has proved possible to estimate the shielding constant of gaseous ions by optical pumping or molecular beam measurements.[241] As Deverell has observed,[238] all the techniques where they coincide give approximately consistent results. Unfortunately, except for F^-, none of the measurements is very accurate (10 ppm). Indeed, they do not come anywhere near the accuracy with which NMR shifts can be measured.

It is to be hoped that further work will include the extension of the work of Deverell[238] and the application of Eqs. (45) and (46) to non-aqueous solvents, which should provide a definitive test for the proposed relationship between quadrupole coupling and shielding constant, and thence lead to a more accurate value of $\sigma_p(aq)$.

Systematic studies of the effect of solvent on solute shifts have not been extensive, and have generally involved correlations with other solvent properties. As mentioned earlier, the shielding constant is dominated by the restriction to free rotation of the electron term. Various authors have tried to relate this idea of hindrance to the strength of the solvating ability. Thus, Bloor and Kidd[242] attempted to correlate solvent shifts for ^{23}Na ions with the pK_a values of the solvents acting as acids in water. Interestingly, they note all oxygen coordinating solvents lie to high field of water and all nitrogen coordinations to low field. This same effect was recently observed for ^{205}Tl shifts.[243] Maciel *et al.*,[229] in their ^{7}Li study, reported that the shifts correlated with solvent polarizibilities. Popov and co-workers, in a series of papers,[231–233,244,245] have studied ^{23}Na (and ^{7}Li) shifts in a variety of solvents. At the outset, it must be admitted that in certain instances, there are considerable differences between the shifts measured.[231,242,244,245] However, as already mentioned, Popov and co-workers find a very good correlation with Gutmann donor numbers. The complex so formed evidently has a large amount of covalent character, and Popov argues that the ion–solvent interaction must therefore have covalent character.

Richards and co-workers[54] studied ^{133}Cs resonances in many solvents. It is immediately obvious that there is no correlation between the ^{133}Cs and ^{23}Na shifts measured by Popov and by Bloor and Kidd. It is therefore not possible for the Gutmann donor number correlation to work for ^{133}Cs. Popov has noted that it does not work for ^{7}Li either, which does not inspire confidence in its existence. As mentioned earlier, the theory of Richards and co-workers fails to take into account solvent displacement during the direct encounter of ions. This effect is clearly important, as may be seen (Fig. 23) from the results of Covington *et al.*[181] In aqueous solution, the effect of NO_3^- on the ^{133}Cs resonance is upfield (diagmagnetic), whereas in hydrogen peroxide and water mixtures, the dependence changes through zero to downfield (paramagnetic) as the peroxide proportion increases. One must therefore treat Richards' quantitative results with some caution. However, he demonstrated that for ion–ion interactions, the most important contribution is repulsive overlap, with the contribution of a dipole induced in the observed ion caused by close encounter with a counter ion much less important except for small ions. A van der Waals contribution is of negligible importance. It must therefore be expected

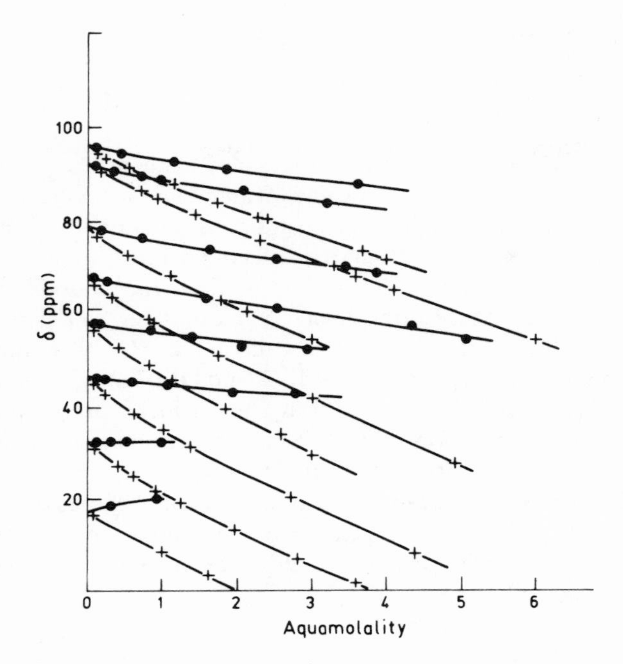

Figure 23. Concentration dependence of ^{133}Cs shifts in hydrogen peroxide + water: (+) CsCl; (●) $CsNO_3$. Mol fractions of peroxide reading from top: 0.918, 0.771, 0.503, 0.359, 0.259, 0.161, 0.075, 0 (from Covington *et al.*[181]).

that the first two contributions will be important factors affecting the solvent shifts. Similar overlap effects on the lone pairs of oxygen atoms would explain the grouping of oxygen-containing solvating solvents (and similarly for nitrogen-containing solvents), but for the ^{133}Cs shifts, it can be seen from Richards's paper[54] that the shifts range from 60 ppm for MeOH through zero for water to −60 ppm for DMSO. The nitrogen solvents appear throughout the range, although they are centered near the middle (along with many oxygen solvents). It is interesting to note that Ikenberry and Das,[169] using a simple localized electron-pair model for water, obtained reasonable values for $\sigma_p(aq)$, from the assumption of an octahedral arrangement of water molecules around a Rb^+ ion, and the use of the theory of Kondo and Yamashita.[167]

Anion NMR studies are not very extensive.[31] ^{19}F studies have proved to be difficult to perform due to the limited solubility of most fluorides in organic solvents. Carrington *et al.*[165] and Tong *et al.*[246] have measured the ^{19}F shifts of salts in water, methanol, and mixtures of these two solvents. The latter authors also made measurements in ethanol, ethylene glycol, glycerol, and formamide; they noted that the alcohols resonated to high field of aqueous solution, whereas the formamide was downfield. Richards and co-workers[247] have studied the ^{35}Cl and ^{81}Br resonance in methanol, where they found for both nuclei that the resonance is upfield of its value in water. Stengle and co-workers[248,249] have made measurements on the ^{35}Cl, ^{81}Br, and ^{127}I resonances in H_2O, MeOH, AN, DMSO, and DMF. Where such data overlap, these authors have been able to show a linear correlation with the charge-transfer-to-solvent (CTTS) uv absorption maxima for all three halides in these various solvents. It must therefore be concluded that at least for the halide ions, one of the principal causes of solvent shifts arises from the excitation energy term in the denominator of the paramagnetic term of the Ramsey shielding expression (and also in the Kondo–Yamashita approach).

3. Mixed Solvents

Work on mixed solvents has led to the gradual realization that solvation around an ion in a mixed solvent can be regarded as a competition between the solvent components for the available solvation sites; this competition is a useful measure of solvating ability.

(i) ^1H Shifts

In one of the earliest studies (mentioned earlier), Swinehart and Taube[188] observed both water and methanol in the solvation shell of Mg^{2+} from low-temperature proton NMR. In particular, they found that the ratio of water to methanol in the solvation shell was greater than in bulk solvent, which gave rise to the concept of solvent sorting. Luz and Meiboom studied low-temperature line-broadening and observed separate bound and free signals due to $^{189}Co^{2+}$ and $^{190}Ni^{2+}$ in MeOH + H_2O mixtures. Estimating the equilibrium constants for the equilibria they obtained the values:

$$K_1 \quad = \quad 44\,\mathrm{l\,mol^{-1}} \text{ at } -60^\circ\mathrm{C}$$

$$K_2^{\mathrm{trans}} = 750\ \mathrm{l\ mol^{-1}}\ \text{at}\ -60°\mathrm{C}$$

$$K_2^{\mathrm{cis}} = 240\ \mathrm{l\ mol^{-1}}\ \text{at}\ -60°\mathrm{C}$$

They also estimated the various exchange parameters, and found that the highly aquated complexes lost MeOH faster than highly methanolated ones. Craig and Richards[250] measured the proton shifts of water in water–dimethylformamide mixtures with and without added LiCl. The LiCl concentration was kept constant at mol fraction 0.1132. There were thus eight solvent molecules for every pair of ions. Without added salt, the water protons shifted upfield on addition of DMF, due presumably to breakdown of water structure. In the salt solution, the water resonance did not shift appreciably with solvent mixture composition. This suggested that all the water molecules were in proximity to the ions throughout the solvent range.

In an extensive series of papers, Fratiello and co-workers[251–256] studied solvent-sorting effects for a wide variety of both solvent mixtures and salts. In early studies in dioxan–water mixtures, they found[251] no evidence for solvation by the organic component. The water proton shifts in 50% vol/vol dioxan were, however, upfield for alkali halides, which suggested that structural changes were occurring in solution. For multicharged ions, however, the shifts suggested that an electric field effect was operating. In pyridine–water mixtures, however,[252] the protons of both components shift, the pyridine peak downfield, whereas for the alkali ions, upfield water shifts were produced. Paramagnetic line-broadening suggests coordination through nitrogen, since the γ protons broaden least, and as with the other ions, both components solvate the ions.

In later work on line-broadening by paramagnetic ions in THF–water mixtures, Fratiello and Miller[253] found no evidence for solvation by THF. The same authors[254] also studied the effects of ions on 10:1 aqueous mixtures of acetone, DMF, DMSO, dioxan, EtOH, MeOH, or NMF. They argued that if the second component simply acts as a diluent and does not solvate the ion, then the observed water shift should be inversely proportional to the concentration of water. This may be seen as follows:

Assuming that bound and free water are in rapid equilibrium, then

$$\delta_{\mathrm{obs}} = X_B\delta_B + X_F\delta_F \tag{47}$$

which rearranges to

$$\delta_{\text{obs}} - \delta_F = X_B(\delta_B - \delta_F) \tag{48}$$

where δ_{obs} is the observed shift, and X_B and X_F are the mol fractions of bound and free water of chemical shifts δ_B and δ_F, respectively. Assuming a constant solvation number n gives

$$X_B = \frac{nm}{[H_2O]} \tag{49}$$

where m is the concentration of salt and $[H_2O]$ is the concentration of water (in the same units). Clearly, when comparing solutions of the same solute concentration, $\delta_{\text{obs}} \propto 1/[H_2O]$. The authors argued that their early work gives sufficient evidence that dioxan acts solely as diluent. Using the data in pure water, they were able to construct graphs of δ_{obs} vs. $1/[H_2O]$ for various solutes. They found that except for dioxan and acetone, all other solvents fall below the line, which suggests that there is competition between both solvent components. For multicharged ions, they were able to confirm this (for DMSO, DMF, and NMF) by observation of both bound and free organic component. Co^{2+} line-broadening measurements gave similar results. Although it is not discussed by Fratiello, this technique may be a method of estimating changes in n (i.e., solvent sorting), although there may be complications from structuring effects.

Similar studies[176] with $AlCl_3$, $TiCl_4$, and $CoCl_2$ in aqueous organic mixtures have led to the following order for (sorting) solvating ability:

$$\text{DMSO} > \text{alcohols} > \text{amides} > \text{TMU} \simeq \text{THF} > \text{acetone} > \text{AN} > \text{dioxan}$$

For aluminum salt solutions, it is possible to observe separate bound and free organic solvent molecules at or above room temperatures. Fratiello and co-workers[256] have observed (Fig. 24) separate bound and free DMF and DMSO in aqueous solution, but in the restricted solvent mixture range, no bound dioxan, THF, or TMU was observed. By supercooling the aqueous DMSO solutions to −35°C, separate bound and free water signals could be observed. It is interesting that over this temperature range, the relative proportions of DMSO/water in the solvation did not appear to change, since the sum

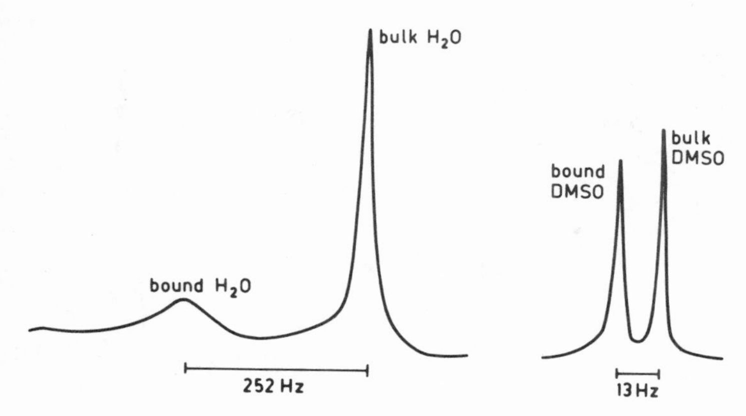

Figure 24. ^1H spectrum of 2 mol liter^{-1} $AlCl_3$ in 10:1 mixture of H_2O:DMSO (after Fratiello *et al.*[256]).

of the two component solvation numbers remained at 6.0 ± 0.3. Their results suggested slight preferential solvation by DMSO, but this was not borne out by the extensive study of this system by Ölander *et al.*,[257] who estimated approximate values for the apparent equilibrium constants between differently solvated Al^{3+}.

This work has since been extended by several authors to include several new solvent systems. Thus, Crea and Lincoln[258] studied mixtures of TMP + H_2O to which acetone had been added to lower the freezing point so that they could study the relative solvating of TMP to Al^{3+}. Delpuech and co-workers[259–262] have made extensive studies of aqueous mixtures of organic solvents, particularly those containing phosphorus. As well as using ^1H NMR[259–261] to study the apparent equilibrium constants between differently solvated Al^{3+} (and Be^{2+}). Delpuech's group have used[261,262] ^{31}P and, with the advent of Fourier-transform NMR, ^{13}C as well.[262] It is interesting that in water-poor solutions, the Al^{3+} salts were not very soluble, and it was convenient to compare the bound peak with the ^{13}C–^1H coupling satellites of the free peak, rather than with the free peak itself.[261] It has also proved possible to observe separate ^{13}C resonances of the carbonyl groups of acetamide (A) in aqueous solution for the species $[Al(H_2O)_5A]^{3+}$ and $[Al(H_2O)_4A_2]^{3+}$ at room temperature[262] (Fig. 25).

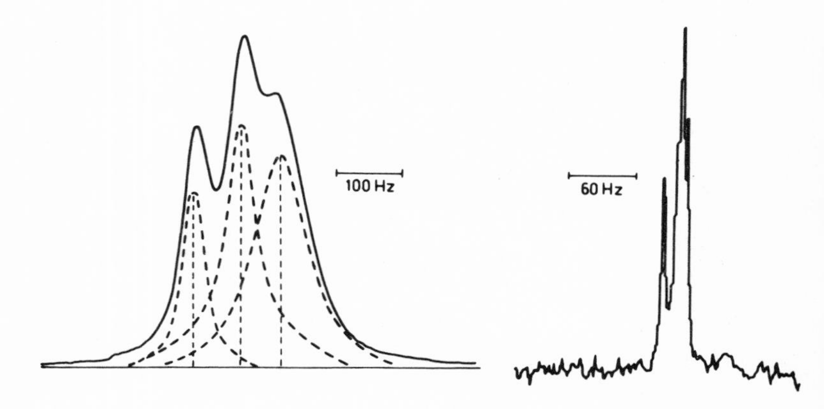

Figure 25. ^{27}Al and ^{13}C spectra of an aqueous solution of aluminum perchlorate in acetamide (*A*) at 22.63 MHz. ^{27}Al (left): three signals correspond to $Al(H_2O)_6^{3+}$, $AlA(H_2O)_5^{3+}$, and $AlA_2(H_2O)_4^{3+}$; ^{13}C (right): three signals correspond to A, $AlA(H_2O)_5^{3+}$, and $AlA_2(H_2O)_4^{3+}$ (reproduced from Canet *et al.*,[266] with permission of Academic Press).

Green and Sheppard[263] and Toma *et al.*[264] have demonstrated that it is possible to obtain separate ^1H signals for Mg^{2+}–water complexes in dilute aqueous acetone solutions at 185°K, i.e., at the temperature lower than that used by Schuster and Fratiello.[107] When the water concentration becomes less than that required to solvate the ions completely, i.e., ratio of water to Mg^{2+} less than 6, signals corresponding to $Mg^{2+}(H_2O)_i$ were observed for i ranging from 1 to 6. This is shown derived from later work[265] in Fig. 26. Bound acetone signals were also observed.[263] From the areas under the resolved peaks, Covington and Covington[265] obtained equilibrium constants for the formation of solvates and showed that the solvate distribution is statistical despite the strong preference of Mg^{2+} for water over acetone (Fig. 27). Using a 250-MHz instrument, Toma *et al.* have shown the presence of isomeric dihydrates (Fig. 28). Delpuech and coworkers[266,267] have been able to separate ^{27}Al signals for all seven of the species $[Al(H_2O)_{6-i}A_i]^{3+}$, where $i = 0$ to 6 and where A is dimethylmethylphosphonate, dimethylphosphite, or hexamethylphosphoramide. They found that shifts were additive on replacing a water molecule by an organic molecule, a condition that will clearly be important for studies of ions where no separate signal can be observed.

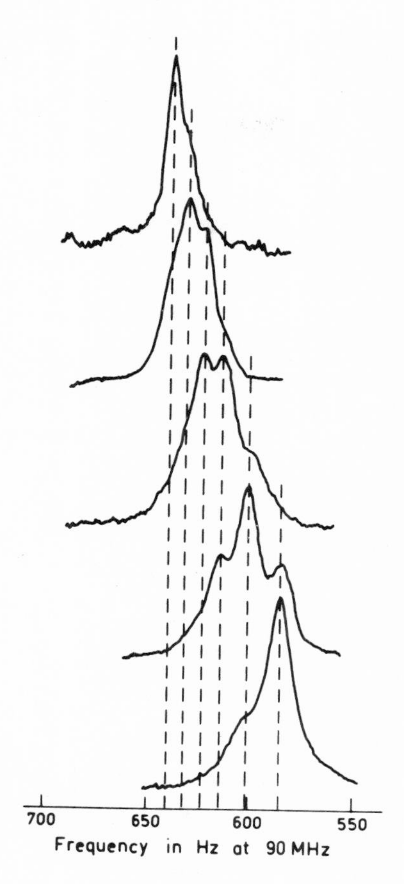

Figure 26. ^{1}H signals from bound water protons in magnesium perchlorate–acetate–water solutions at 185°K and varying composition (Covington and Covington[265]).

Gudlin and Schneider,[268] however, studied the ^{27}Al resonance of aluminum salts in DMSO + DMF mixtures and observed separate resonances for differently solvating ions at room temperature, also observing separate lines for different geometrical isomers (Fig. 29). They found that the shifts were not directly additive, but could be adequately explained on an additive basis if nearest-neighbor solvent molecules were taken into account.

Quantitative measurement of apparent exchange equilibria have also been made by different techniques by Grunwald and co-workers[269] and by Stockton and Martin.[270] The former group studied the shift when small quantities of water were added to propylene

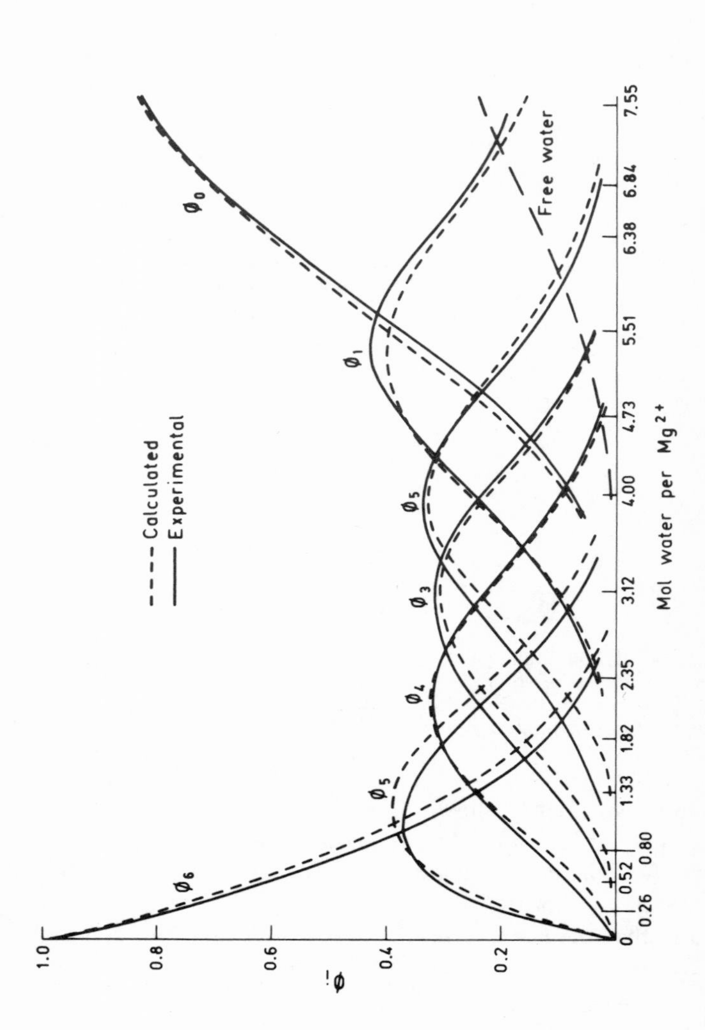

Figure 27. Distribution of solvate species in the system magnesium perchlorate–acetone–water solutions at low water concentration. (Covington and Covington[265]).

carbonate solutions of electrolytes. From these measurements, they were able to estimate equilibrium constant for the process

$$X^{\pm} + H_2O \overset{K_1}{\rightleftharpoons} X^{\pm}(H_2O) \tag{50}$$

They were able to find only a fair correlation between $\log K_1$ and ΔG_t^0, the free energy of transfer of ions from water to propylene carbonate. The behavior of Cl^- was highly anomalous.

Stockton and Martin,[270] however, observed the proton shifts of water and alcohols in dilute solution in AN when salt was added. Using a Benesi–Hildebrand type of treatment, they also estimated apparent equilibria for the process

$$X^{\pm} + ROH \overset{K_1}{\rightleftharpoons} X^{\pm}(ROH) \tag{51}$$

They interpreted the proton shifts in terms of an electric field effect and the log K_1 values in terms of cation–permanent dipole and –induced dipole interactions.

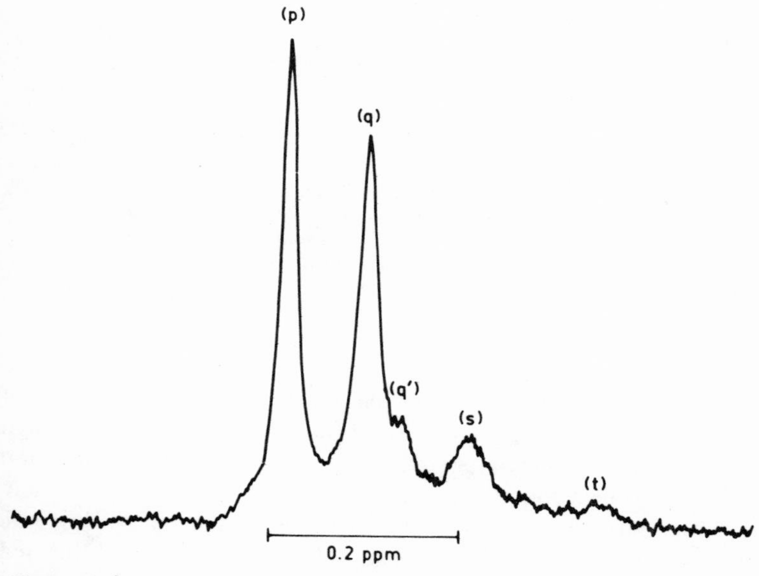

Figure 28. ^{1}H spectrum at 250 MHz of solvate species in magnesium perchlorate–acetone–water solution at 188°K. (p), (q), and (s) are assigned to the mono-, di-, and trihydrates; (t) corresponds to the tetrahydrate, and (q') to the second isomer of the dihydrate (reproduced from Toma *et al.*,[264] with permission of American Chemical Society).

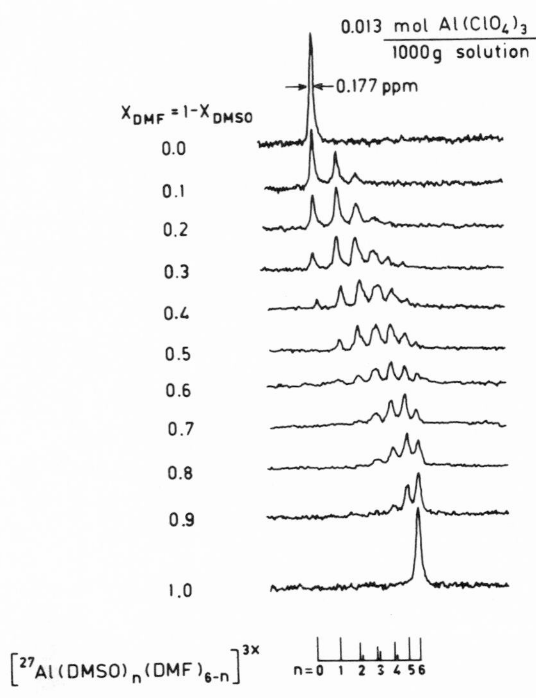

Figure 29. ^{27}Al spectra of 0.013 mol liter^{-1} aluminum perchlorate solution in mixtures of DMSO + DMF in nitromethane at ambient temperature (reproduced from Gudlin and Schneider[268] with permission of Academic Press).

The use of proton NMR in mixed solvents has done much to increase our understanding of the relative solvating ability of different solvents in a systematic way in terms of competitive process, although there are difficulties with regard to the separation of cation and anion effects, a problem common to all spectroscopic studies of the solvent.

No such problems arise when a spectroscopic property of a solute species is studied. An area in which mixed solvents have been little used and could probably be used successfully is to modify, and, it is to be hoped, nullify, the effect of extensive water structural effects on chemical shifts. Structural effects play a very large part in aqueous proton shifts, as discussed earlier, whereas nonaqueous shifts are only little affected by structural effects. Despite the complexity of a ternary system as opposed to a binary system, disruption of solvent structure by an organic solvent could lead to simplification of interpretation.

(ii) Solute Shifts

Carrington and co-workers[165] studied the ^{19}F resonance of KF in a variety of organic aqueous solvent mixtures and observed for MeOH a linear change in shift with solvent mol fraction. The results for ethylene glycol were curved, whereas the shifts in solvent mixtures of low dielectric constant (AN, dioxan, isopropanol) in water-rich mixtures only, due to solubility problems, showed no solvent dependence. These authors concluded that for methanol–water mixtures, the immediate environment of the F^- ion was effectively the same as bulk solvent, whereas for ethylene glycol, the curvature suggested preference for the glycol. The results for low dielectric solvents suggest that the F^- ion retains an aqueous environment. These qualitative ideas were initially developed by Smith and Symons[271] for CTTS uv shifts, and have since been used extensively. Thus, Bloor and Kidd[242] found by ^{23}Na NMR that the sodium ion was strongly preferentially solvated by water in H_2O–AN mixtures, but by ethylene diamine (en) in H_2O–en mixtures. Stengle and co-workers[272,273] put such ideas on a more quantitative basis when they studied ^{57}Co and ^{1}H line-broadening of $Co(acac)_3^{3+}$ and $Cr(acac)_3^{3+}$ (where acac is acetylacetone), respectively, in a variety of organic solvent mixtures of $CHCl_3$. They defined the "isosolvation point" as the mol fraction at which the solvation shell contained equal numbers of both solvent molecules. They attempted to relate this point and the consequent shift–composition plots to thermodynamic ideas of solvent sorting. In further studies of preferential solvation, Stengle and co-workers[246,248,249] studied the halide resonances (^{19}F, ^{35}Cl, ^{81}Br, and ^{127}I) in a variety of aqueous organic mixtures. The ^{19}F results[246] agree very closely with those of Carrington *et al.*,[165] while the results for other nuclei suggest very strong preferential solvation of the halides for water over MeCN, but not so for DMSO.

These qualitative or semiqualitative ideas of preferential solvation were placed on a sounder thermodynamic basi. in a series of papers by Covington and co-workers.[181,274–276] Experimental results for infinite dilution solute ion shifts of alkali metal ions and halides in hydrogen peroxide–water mixtures (see Fig. 23) led to qualitative conclusions that Rb^+, Cs^+, and F^- were preferentially solvated by peroxide and Li^+ by water (Figs. 30 and 31). By treating the overall solvation process

$$X(H_2O)_n + nH_2O_2 \rightleftharpoons X(H_2O_2)_n + nH_2O \tag{52}$$

as series of successive equilibria, written for example for $n = 4$, using the abbreviations $W = H_2O$ and $P = H_2O_2$ for a solution containing 1 mol of X, w mol of W, and p mol of P

$$\begin{aligned} XW_4 + pP + (w-4)W \overset{K_1}{\rightleftharpoons} XW_3P + (p-1)P + (w-3)W \overset{K_2}{\rightleftharpoons} \\ \overset{K_2}{\rightleftharpoons} XW_2P_2 + (p-2)p + (w-2)W \overset{K_3}{\rightleftharpoons} \\ \overset{K_3}{\rightleftharpoons} XWP_3 + (p-3)P + (w-1)W \overset{K_4}{\rightleftharpoons} \\ \overset{K_4}{\rightleftharpoons} XP_4 + (p-4)P + wW \end{aligned} \tag{53}$$

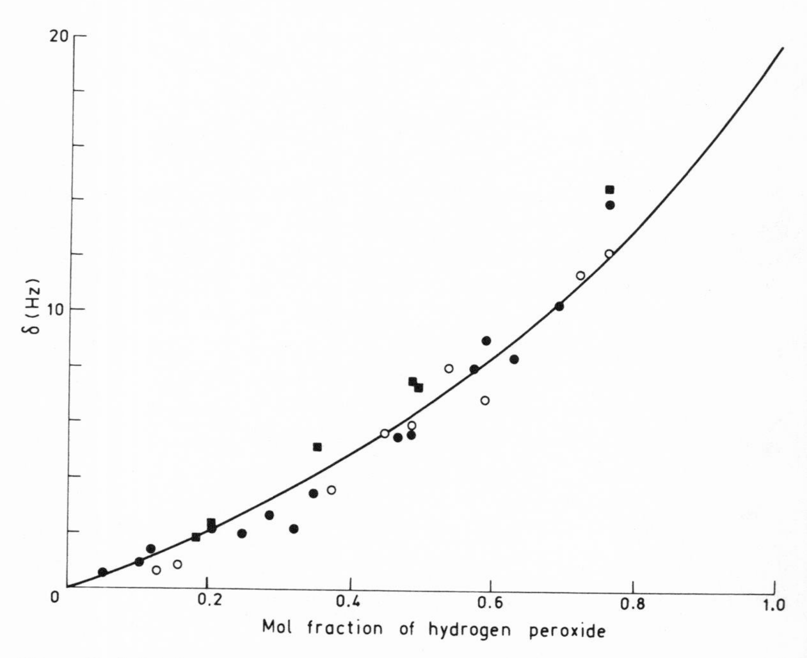

Figure 30. ^{7}Li shifts for lithium chloride (●), sulfate (■), and nitrate (○) in hydrogen peroxide–water mixtures at one aquamolal concentration (at 36.43 MHz) (from Covington *et al.*[181]).

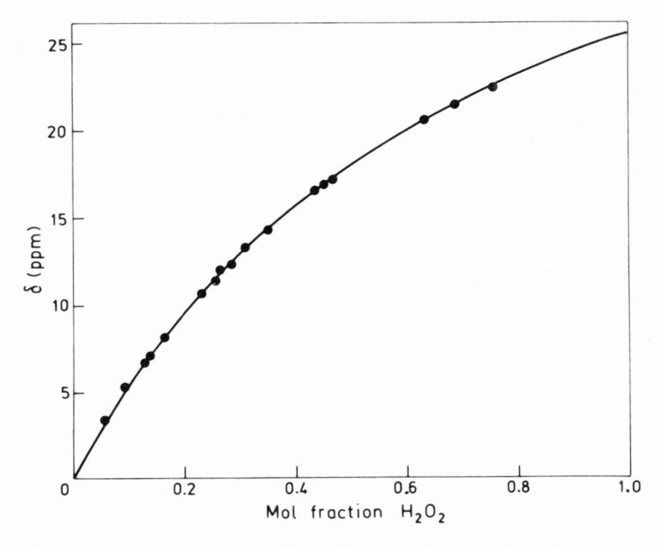

Figure 31. Infinite dilution ^{19}F shifts in hydrogen peroxide–water mixtures (from Covington *et al.*[181]).

the results could be put on a quantitative basis. The individual equilibrium constants were assumed to be related statistically, viz.,

$$K_1 = 4K^{1/4}, \quad K_2 = \tfrac{3}{2}K^{1/4}, \quad K_3 = \tfrac{2}{3}K^{1/4}, \quad K_4 = \tfrac{1}{4}K^{1/4} \qquad (54)$$

where $K^{1/4} = (K_1K_2K_3K_4)^{1/4}$.

Using for the observed shift

$$\delta_{\text{obs}} = \sum_i \phi_i \delta_i \qquad (55)$$

where δ_i is the intrinsic shift of the ith solvated species $XW_{i-n}P_i$, assumed to be proportional to the amount of P it contains (iP), and ϕ_i is the mol fraction of the ith species, the relationship

$$\frac{\delta}{\delta_P} = \frac{K^{1/4}(X_P/X_W)}{1 + K^{1/4}(X_P/X_W)} \qquad (56)$$

was obtained, where δ_P refers to the infinite dilution shift of the nucleus

X in pure P relative to its value in W. It is not necessary to assume $n = 4$, for the quantity determinable is $K^{1/n}$ or the free energy of preferential solvation $\Delta G_{ps}/n$.

Values obtained were related to the familiar free energy of transfer for alkali halides between H_2O and H_2O_2 obtained from measurements with ion-selective electrodes.[277] For this isodielectric mixture, the two quantities should be the same, and were found to be so experimentally to within 1 kJ mol^{-1}.

In a second paper, a rigorous thermodynamic theory of preferential solvation was presented based on a treatment sketched by Grunwald *et al.*[278] It was demonstrated how for mixed solvents, the free energy of preferential solvation for ions derived spectroscopically could be related, if a value of n is assumed, to the thermodynamically derived free energy of salts by including electrostatic contributions.

These considerations are also appropriate to the proton work of Stockton and Martin[270] mentioned in the preceding section. Their apparent equilibrium constant K' for Eq. (50) can be related to the free energy of preferential solvation. Instead of determining the effect of adding salt on the proton shift of a small fixed concentration of water in acetonitrile or propylene carbonate, Grunwald and co-workers kept the salt concentration constant and varied the water concentration in propylene carbonate. Analysis also yields an apparent equilibrium constant. Both techniques suffer from the problem of separating anion and cation effects.

Using a method suggested by Van Geet,[279] it is possible to determine the value of n for small ions. The method is based on the competition by water for the solvation sphere of Na^+ in THF, and is measured by ^{23}Na shifts. The salt used must contain a large anion, which is assumed to be unhydrated during titration of the THF solution with water. For sodium tetrephenylborate, $n = 3.0$ or 3.7, depending on the initial concentration of solution. In an earlier study, Wong *et al.*[280] measured the ^1H acetone methyl shifts for lithium perchlorate solutions in acetone–nitromethane as a function of acetone–$LiClO_4$ ratio. From a break at a ratio of 4.3, they concluded that Li^+ is solvated by four molecules of acetone.

Covington and co-workers[276,277] have extended their thermodynamic analysis in an attempt to account for the asymmetrical variation of measured shift with cosolvent mol fraction observed in some experiments reported in the literature. Nonstatistical distribution of equilibrium constants K_i can lead to this effect, as can change in solvation number. In the latter situation, only a halving or a doubling of solvation number could be considered because of mathematical complexity. The treatment is analogous to monodentate vs. bidentate ligand formation in complex ion chemistry. Some amendments to the original paper[276] are given in a review of this work on preferential solvation.[281]

Such treatments lead to an explanation of even the sigmoid shift curves reported in the literature. For example, Stengle *et al.*[249] found a sigmoid curve for the ^{127}I shifts with mol fraction in DMSO–H_2O mixtures. It is tempting to conclude that this means preferential solvation of iodide by DMSO in dilute aqueous DMSO and by water in dilute solution of water in DMSO, strange though this might be. Covington and Thain[276] concluded that it indicated solvation of iodide by DMSO over the whole concentration range. Similar curious behavior of DMSO–H_2O mixtures was noticed by Clausen *et al.*[282] in ^{1}H shift studies of silver nitrate solutions. They concluded that silver ions were hydrated at low DMSO mol fractions (<0.2) and above 0.2 solvated by DMSO. A similar conclusion regarding Ni^{2+} preferential solvation was reached by Frankel *et al.*[283] from paramagnetic line-broadening of the solvent ^{1}H NMR signals in DMSO–H_2O mixtures.

It is possible that coupling constants between different nuclei in polyatomic ions may be solvent-dependent. Thus, Gillespie *et al.*[187] have studied the solvent dependence of the ^{11}B–^{19}F coupling in the BF_4^- ion as a function of solvent composition (Fig. 32). It is relatively straightforward to estimate the isosolvation point of the ion in a manner analogous to that already discussed.

Recently, there has been much interest in large polycyclic ethers due to their ability to complex alkali metal ions that remain in the cavity formed by the ring.[284] The association constant depends markedly on ionic size, the implication being that if the ion is too large, it will not fit; if too small, it will not remain complexed. Thus, Ceraso and Dye[285] were able to observe both bound and free ^{23}Na peaks below 50°C or NaCl–(I) in ethylene diamine. They calculated exchange rates that agree with those obtained by proton NMR.[286] Similarly, Shchori *et al.*[287] used ^{23}Na to study the kinetics and equilibria for

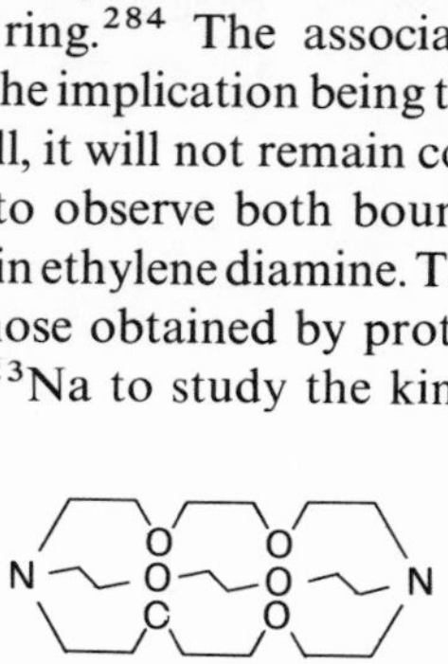

I

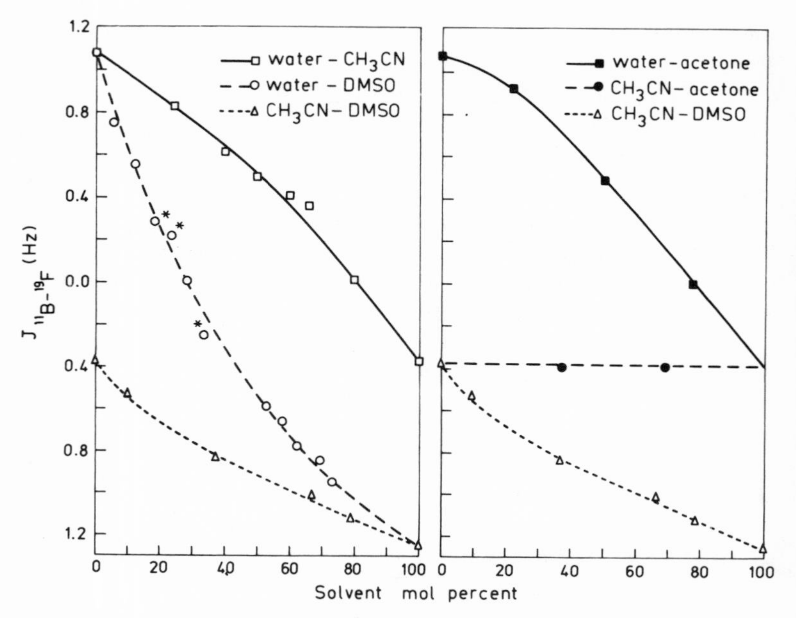

Figure 32. Variation of the coupling constant $J(^{11}B–^{19}F)$ with solvent composition in solutions of silver tetrafluoroborate. *No splitting was observed, and J was estimated from the width of a single peak (reproduced from Gillespie *et al.*,[187] with permission of the National Research Council of Canada).

complexing of the sodium ion (II) or (III) in a variety of solvents. They found that the rate was independent of solvent, which suggested that the rate-determining step was decomplexation. The effect of substituents on the ether supports a purely electrostatic complexing interaction.

II III

V. CONCLUDING REMARKS

In this review, we have tried to give an overall view of the application of NMR to the study of electrolyte solutions. The techniques used and the conclusions reached are so multifarious that a brief summary would be fatuous. However, it is hoped that the review will prove sufficiently critical to help workers in planning their researches.

REFERENCES

[1] J. T. Arnold and M. E. Packard, *J. Chem. Phys.* **19** (1951), 507.
[2] J. A. Pople, W. G. Schneider, and H. J. Bernstein, *High-resolution Nuclear Magnetic Resonance*, McGraw-Hill, New York, 1959.
[3] J. W. Emsley, J. Feeney, and L. H. Sutcliffe, *High Resolution Nuclear Magnetic Resonance Spectroscopy*, Pergamon, Oxford, 1965.
[4] A. Abragam, *The Principles of Nuclear Magnetism*, Oxford University Press, Oxford, 1961.
[5] J. W. Emsley, J. Feeney, and L. H. Sutcliffe, eds., *Progress in Nuclear Magnetic Resonance Spectroscopy*, Pergamon, Oxford, 1966, Vol. 1.
[6] E. F. Mooney, ed., *Annual Review of NMR Spectroscopy* Academic Press, New York, 1968, Vol. 1.
[7] J. S. Waugh, Ed., *Advances in Magnetic Resonance*, Academic, New York, 1965, Vol. 1.
[8] R. K. Harris, Ed. *Nuclear Magnetic Resonance*, Chemical Society London, 1971, Vol. 1.
[9] P. Laszlo, in Ref. 5, (1967), Vol. 3, p. 231.
[10] H. G. Hertz, in Ref. 5, (1967), Vol. 3, p. 159.
[11] E. von Goldammer, in *Modern Aspects of Electrochemistry*, Ed. by J. O'M. Bockris, and B. E. Conway, Plenum Press, New York, 1975, Vol. 10, p. 1.
[12] Ref. 3, p. 589.
[13] J. D. Halliday, H. D. W. Hill, and R. E. Richards, *J. Sci. Instr., Ser. 2* **2** (1969) 29
[14] Varian Associates, Palo Alto, California 94303.
[15] T. C. Farrar and E. D. Becker, *Pulse and Fourier Transform NMR*, Academic Press, New York, 1971.
[16] Ref. 2, p. 81.
[17] C. A. Reilly, H. M. McConnell, and R. G. Meisenheimer, *Phys. Rev.* **98** (1955) 264A.
[18] R. F. Spanier, T. Vladimiroff, and E. R. Malinowski, *J. Chem. Phys.* **45** (1966) 4355.
[19] K. Hatada, V. Terawaki, and H. Okuda, *Bull. Chem. Soc. Jpn.* **45** (1972) 3720.
[20] Ref. 2, p. 18.
[21] M. R. Bacon and G. E. Maciel, *J. Am. Chem. Soc.* **95** /(1973) 2413.
[22] A. K. Covington, M. L. Hassall, and I. R. Lantzke, *J. Chem. Soc. Faraday Trans. 2* **68** (1972) 1352.
[23] J. W. Akitt, *J. Chem Soc. Dalton Trans.* (1973) 42.
[24] R. N. Butler, and M. C. R. Symons, *Trans. Faraday Soc.* **65** (1969) 945.
[25] R. Freeman, R. P. H. Gasser, R. E. Richards, and D. H. Wheeler, *Mol. Phys.* **2** (1959) 75.
[26] Ref. 4, p. 4.
[27] Ref. 2, p. 199.

[28]Ref. 3, p. 21.
[29]G. C. Levy and G. L. Nelson, *Carbon-13 Nuclear Magnetic Resonance for Organic Chemists*, Wiley-Interscience, New York, 1972, p. 187.
[30]Ref. 2, p. 214.
[31]C. Hall, *Q. Rev. Chem. Soc.* **25** (1971) 87.
[32]D. Eisenberg and W. Kauzmann, *The Structure and Properties of Water*, Oxford University Press, Oxford, 1969.
[33]F. Franks, ed., *Water, A Comprehensive Treatise*, Vol. 1, *The Physics and Physical Chemistry of Water*, Plenum Press, New York, 1972.
[34]J. Morgan and B. E. Warren, *J. Chem. Phys.* **6** (1938) 666.
[35]J. D. Bernal and R. H. Fowler, *J. Chem. Phys.* **1** (1933) 515.
[36]H. S. Frank and W. Y. Wen, *Discuss. Faraday Soc.* **24** (1957) 133.
[37]C. A. Coulson, *Valence*, Oxford University Press, Oxford, (1961), p. 304.
[38]A. Rahman and F. H. Stillinger, *J. Chem. Phys.* **55** (1971) 3336.
[39]O. Ya. Samoilov, *Structure of Aqueous Electrolyte Solutions and the Hydration of Ions*, Consultants Bureau, New York, 1965.
[40]J. A. Pople, *Proc. R. Soc. London Ser. A* **202** (1950) 323.
[41]A. Ben-Naim and F. H. Stillinger, in *Water and Aqueous Solutions*, Ed. by R. A. Horne, Wiley-Interscience, New York, 1972, p. 295.
[42]W. G. Schneider, H. J. Bernstein, and J. A. Pople, *J. Chem. Phys.* **28** (1958) 601.
[43]E. R. Malinowski, P. S. Knapp, and B. Feuer, *J. Chem. Phys.* **45** (1966) 4274.
[44]E. R. Malinowski, P. S. Knapp, and B. Feuer, *J. Chem. Phys.* **47** (1967) 347.
[45]J. C. Hindman, *J. Chem. Phys.* **44** (1966) 4582.
[46]C. A. Angell, J. Shuppert, and J. C. Tucker, *J. Phys. Chem.* **77** (1973) 3092.
[47]J. C. Hindman and A. Svirmickas, *J. Phys. Chem.* **77** (1973) 2487.
[48]For example: Y. Lee and J. Jones, *J. Chem. Phys.* **57** (1972) 4233.
[49]For example: D. E. O'Reilly, E. M. Peterson, and C. E. Scheie, *Chem. Phys. Lett.* **23** (1973) 69.
[50]E. U. Franck and K. Roth, *Discuss. Faraday Soc.* **43** (1967) 108.
[51]A. E. Florin and M. Alei, *J. Chem. Phys.* **47** (1967) 4268.
[52]F. Fister and H. G. Hertz, *Ber. Bunsenges. Phys. Chem.* **71** (1967) 1032.
[53]Z. Luz and G. Yagil, *J. Phys. Chem.* **70** (1966) 554.
[54]J. D. Halliday, R. E. Richards, and R. E. Sharpe, *Proc. R. Soc. London Ser. A* **313** (1969) 45.
[55]Ref. 15, p. 95.
[56]J. W. Akitt and T. H. Lilley, *J. Chem. Soc. Chem. Commun.* (1967) 324.
[57]J. Clifford, *J. Chem. Soc. Chem. Commun.* (1967) 880.
[58]T. W. Marshall and J. A. Pople, *Mol. Phys.* **1** (1958) 199.
[59]A. D. Buckingham, *Can. J. Chem.* **38** (1960) 300.
[60]J. B. Hasted, *Prog. Dielectr.* **3** (1961) 101.
[61]A. D. Buckingham, T. Schaefer, and W. G. Schneider, *J. Chem. Phys.* **32** (1960) 1227.
[62]N. Muller, *J. Chem. Phys.* **43** (1965) 2555.
[63]G. Nemethy and H. A. Scheraga, *J. Chem. Phys.* **36** (1962) 3382.
[64]R. P. Marchi and H. Eyring, *J. Phys. Chem.* **68** (1964) 221.
[65]C. M. Davis, Jr., and T. A. Litovitz, *J. Chem. Phys.* **42** (1965) 2563.
[66]H. H. Ruterjans and H. A. Scheraga, *J. Chem. Phys.* **45** (1966) 3296.
[67]D. E. O'Reilly, *J. Chem. Phys.* **61** (1974) 1592.
[68]S. D. Christian, A. A. Taha, and B. W. Gash, *Q. Rev. Chem. Soc.* **24** (1970) 20.
[69]H. S. Frank, *Discuss. Faraday Soc.* **43** (1967) 147.
[70]C. K. Rosenbaum and J. H. Walton, *J. Am. Chem. Soc.* **52** (1930) 3568.
[71]D. N. Glew, H. D. Mak, and N. S. Rath, *Can. J. Chem.* **45** (1967) 3059.
[72]T. F. Lin, S. D. Christian, and H. E. Affsprung, *J. Phys. Chem.* **69** (1965) 2980.

[73]C. Jolicoeur and A. Cabana, *Can. J. Chem.* **46** (1968) 567.
[74]S. Nishimura, C. H. Ke, and N. C. Li, *J. Am. Chem. Soc.* **90** (1968) 234 and references cited therein.
[75]G. Mavel, *C. R. Acad. Sci. Ser. C* **248** (1959) 1505.
[76]J. R. Holmes, D. Kivelson, and W. C. Drinkard, *J. Am. Chem. Soc.* **84** (1962) 4677.
[77]B. Kingston and M. C. R. Symons, *J. Chem. Soc. Faraday Trans. 2* **69** (1973) 978.
[78]W. Y. Wen and H. G. Hertz, *J. Solution Chem* **1** (1972) 17.
[79]N. Laiken and G. Nemethy, *J. Phys. Chem.* **74** (1970) 3501 and references cited therein.
[80]C. V. Krishnan and H. L. Friedman, *J. Phys. Chem.* **75** (1971) 3598.
[81]M. J. Tait, A. Suggett, F. Franks, S. Ablett, and P. A. Quickenden, *J. Solution Chem.* **1** (1972) 131.
[82]J. Oakes, *J. Chem. Soc. Faraday Trans. 2* **69** (1973) 1321.
[83]K. E. Newman, unpublished calculations.
[84]F. Franks and D. J. G. Ives, *Q. Rev. Chem. Soc.* **20** (1966) 1.
[85]J. N. Shoolery and B. Alder, *J. Chem. Phys.* **23** (1955) 805.
[86]H. G. Hertz and W. Spalthoff, *Z. Elektrochem.* **63** (1959) 1096.
[87]P. B. Fabricand and S. Goldberg, *J. Chem. Phys.* **34** (1961) 1624.
[88]J. C. Hindman, *J. Chem. Phys.* **36** (1962) 1000.
[89]M. S. Bergqvist and E. Forslind, *Acta Chem. Scand.* **16** (1962) 2069.
[90]R. E. Glick, W. E. Stewart, and K. C. Tewari, *J. Chem. Phys.* **45** (1966) 4049.
[91]K. A. Hartman, *J. Phys. Chem.* **70** (1966) 270.
[92]A. LoSurdo and H. E. Wirth, *J. Phys. Chem.* **76** (1972) 130.
[93]E. T. Sare, C. T. Moynihan, and C. A. Angell, *J. Phys. Chem.* **77** (1973) 1869.
[94]H. S. Frank and M. W. Evans, *J. Chem. Phys.* **13** (1945) 407.
[95]G. Licheri, G. Piccaluga, and G. Pinna, *Gazz. Chim. Ital.* **102** (1972) 147.
[96]P. P. S. Saluja, in *International Review of Science, Physical Chemical Society Series 2*, Vol. 6, *Electrochemistry* (Ed. by J. O'M. Bockris), Butterworth, London, 1976.
[97]E. R. Malinowski and P. S. Knapp, *J. Chem. Phys.* **48** (1968) 4989.
[98]P. S. Knapp, R. A. Waite, and E. R. Malinowski, *J. Chem. Phys.* **48** (1968) 5459.
[99]A. K. Covington and T. H. Lilley, *Spec. Per. Rep. Electrochem.* **1** (1970) 1.
[100]R. W. Creekmore and C. N. Reilley, *J. Phys. Chem.* **73** (1969) 1563.
[101]F. J. Vogrin, P. S. Knapp, W. L. Flint, A. Anton, G. Highberger, and E. R. Malinowski, *J. Chem. Phys.* **54** (1971) 178.
[102]T. J. Swift and W. G. Sayre, *J. Chem. Phys.* **44** (1966) 3567.
[103]T. J. Swift, O. G. Fritz, and T. A. Stephenson, *J. Chem. Phys.* **46** (1967) 406.
[104]S. Meiboom, *J. Chem. Phys.* **47** (1967) 410.
[105]R. E. Schuster and A. Fratiello, *J. Chem. Phys.* **47** (1967) 1554.
[106]D.-W. Wong and E. Grunwald, *J. Am. Chem. Soc.* **91** (1969) 2413.
[107]R. E. Schuster and A. Fratiello, *J. Chem. Phys.* **47** (1967) 1554.
[108]A. Fratiello, R. E. Lee, V. M. Nishida, and R. E. Schuster, *J. Chem. Phys.* **48** (1968) 3705.
[109]N. A. Matwiyoff and H. Taube, *J. Am. Chem. Soc.* **90** (1968) 2796.
[110]A. Fratiello, V. Kubo, R. E. Lee, and R. E. Schuster, *J. Phys. Chem.* **74** (1970) 3726.
[111]A. Fratiello, S. Peak, R. C. Schuster, and J. Davies, *J. Phys. Chem.* **74** (1970) 3730.
[112]A. Fratiello, V. Kubo, and G. A. Vidulich, *Inorg. Chem.* **12** (1973) 2066.
[113]J. Davies, S. Ormonroyd, and M. C. R. Symons, *Trans. Faraday Soc.* **67** (1971) 3465.
[114]R. N. Butler, J. Davies, and M. C. R. Symons, *Trans. Faraday Soc.* **66** (1970) 2426.
[115]J. W. Akitt, *J. Chem. Soc. A* (1971) 2865.
[116]J. W. Akitt, *J. Chem. Soc. Dalton Trans.* (1973) 42.
[117]K. E. Newman, unpublished results.
[118]J. W. Akitt, *J. Chem. Soc. Dalton Trans.* (1973) 1177.
[119]J. W. Akitt, *J. Chem. Soc. Dalton Trans.* (1973) 1446.

[120]J. W. Akitt, *J. Chem. Soc. Dalton Trans.* (1974) 175.
[121]H. S. Gutowsky and A. Saika, *J. Chem. Phys.* **21** (1953) 1688.
[122]G. C. Hood, O. Redlich, and C. A. Reilly, *J. Chem. Phys.* **22** (1954) 2067.
[123]G. C. Hood, O. Redlich, and C. A. Reilly, *J. Chem. Phys.* **23** (1955) 2229.
[124]G. C. Hood, A. C. Jones, and C. A. Reilly, *J. Phys. Chem.* **63** (1959) 101.
[125]O. Redlich and G. C. Hood, *Discuss. Faraday Soc.* **24** (1957) 87.
[126]G. C. Hood and C. A. Reilly, *J. Chem. Phys.* **27** (1957) 1126.
[127]J. W. Akitt, A. K. Covington, J. G. Freeman, and T. H. Lilley, *Trans. Faraday Soc.* **65** (1969) 2701.
[128]J. W. Akitt, *J. Chem. Soc. Dalton Trans.* (1973) 49.
[129]R. W. Duerst, *J. Chem. Phys.* **48** (1968) 2275.
[130]A. Merbach, *J. Chem. Phys.* **46** (1967) 3450.
[131]O. Redlich, R. W. Duerst, and A. Merbach, *J. Chem. Phys.* **49** (1968) 2986.
[132]A. K. Covington, J. G. Freeman, and T. H. Lilley, *Trans. Faraday Soc.* **65** (1969) 3136.
[133]A. K. Covington, J. G. Freeman, and T. H. Lilley, *J. Phys. Chem.* **74** (1970) 3773.
[134]A. K. Covington and R. Thompson, *J. Solution Chem.* **3** (1974) 603.
[135]G. Mavel, *Mem. Poudres Annexe* **43** (1961) No. 3572; *Chem. Abstr.* **56** (1962) 8197h.
[136]C. S. Handloser, M. R. Chakrabarty, and M. W. Mosher, *J. Chem. Educ.* **50** (1973) 510.
[137]Y. Masuda and T. Kanda, *J. Phys. Soc. Jpn.* **9** (1954) 82.
[138]Y. Masuda and T. Kanda, *J. Phys. Soc. Jpn.* **8** (1953) 432.
[139]A. K. Covington and K. E. Newman, *J. Inorg. Nucl. Chem.* **35** (1973) 3257.
[140]A. R. Quirt, J. R. Lyerla, I. R. Peat, J. S. Cohen, W. F. Reynolds, and M. H. Freedman, *J. Am. Chem. Soc.* **96** (1974) 570.
[141]M. D. Zeidler, in *Water, a Comprehensive Treatise*, Ed. by F. Franks, Plenum Press, New York, 1973, Vol. 2, p. 529.
[142]H. G. Hertz and W. Spalthoff, *Z. Elektrochem.* **63** (1959) 1096.
[143]H. M. Marciaoq-Rousselot, A. de Trobriand, and M. Lucas, *J. Phys. Chem.* **76** (1972) 1455.
[144]J. Davies, S. Ormonroyd, and M. C. R. Symons, *J. Chem. Soc. Faraday Trans. 2* **68** (1972) 686.
[145]J. W. Akitt, *J. Chem. Soc. Dalton Trans.* (1973) 42.
[146]S. Lindenbaum and A. S. Levine, *J. Solution Chem.* **3** (1974) 261.
[147]P. S. Ramanathan, C. V. Krishnan, and H. L. Friedman, *J. Solution Chem.* **1** (1972) 237.
[148]N. Bloembergen, *J. Chem. Phys.* **27** (1957) 595.
[149]Z. Luz and R. G. Shulman, *J. Chem. Phys.* **43** (1965) 3750.
[150]A. Tzalmona and E. Loewenthal, *J. Chem. Phys.* **61** (1974) 2637.
[151]A. M. Chmelnick and D. Fiat, *J. Magn. Reson.* **8** (1972) 325.
[152]T. J. Swift and R. E. Connick, *J. Chem. Phys.* **37** (1962) 307.
[153]J. F. Hinton and E. S. Amis, *Chem. Rev.* **67** (1967) 367.
[154]J. A. Jackson, J. F. Lemons, and H. Taube, *J. Chem. Phys.* **32** (1960) 553.
[155]R. E. Connick and D. Fiat, *J. Chem. Phys.* **39** (1963) 1349.
[156]R. E. Connick and D. Fiat, *J. Am. Chem. Soc.* **88** (1966) 4754.
[157]M. Alei and J. A. Jackson, *J. Chem. Phys.* **41** (1964) 3402.
[158]J. W. Neely and R. E. Connick, *J. Am. Chem. Soc.* **74** (1972) 3419.
[159]C. Deverell, in Ref. 5, 1967, Vol. 4, p. 235.
[160]B. L. Silver and Z. Luz, *Q. Rev. Chem. Soc.* **4** (1967) 458.
[161]K. E. Newman, unpublished data.
[162]K. E. Newman and S. D. Pask, to be published.
[163]H. S. Gutowsky and B. McGarvey, *Phys. Rev.* **28** (1963) 727.
[164]R. Freeman, R. Gasser, R. E. Richards, and D. H. H. Wheeler, *Mol. Phys.* **2** (1959) 75.
[165]A. Carrington, F. Dravniecks, and M. C. R. Symons, *Mol. Phys.* **3** (1960) 174.

[166]J. Van Kranendonk, *Physica* **20** (1954) 781.
[167]J. Kondo and J. Yamishita, *J. Phys. Chem. Solids* **10** (1959) 245.
[168]Y. Yamagata, *J. Phys. Soc. Jpn* **19** (1964) 10.
[169]D. Ikenberry and T. P. Das, *J. Chem. Phys.* **43** (1965) 2199.
[170]D. Ikenberry and T. P. Das, *Phys. Rev.* **138A** (1965) 822.
[171]C. Deverell and R. E. Richards, *Mol. Phys.* **10** (1966) 551.
[172]C. Deverell and R. E. Richards, *Mol. Phys.* **16** (1969) 421.
[173]C. Hall, R. E. Richards, and R. R. Sharpe, *Proc. R. Soc. London Ser. A* **337** (1974) 297.
[174]K. E. Newman, unpublished results.
[175]G. J. Templeman and A. L. Van Geet, *J. Am. Chem. Soc.* **94** (1972) 5578.
[176]E. G. Bloor and R. G. Kidd, *Can. J. Chem.* **50** (1972) 3926.
[177]J. W. Akitt, *J. Chem. Soc. Faraday Trans. 1* **71** (1975) 1557.
[178]G. E. Maciel and M. Borzo, *J. Chem. Soc. Chem. Commun.* (1973) 394.
[179]R. J. Kostelnik and A. A. Bothner-By *J. Magn. Reson.* **14** (1974) 141.
[180]L. Simeral and G. E. Maciel, *J. Phys. Chem.* **80** (1976) 552.
[181]A. K. Covington, T. H. Lilley, K. E. Newman, and G. A. Porthouse, *J. Chem. Soc. Faraday Trans. 1* **69** (1973) 963.
[182]G. A. Porthouse, T. H. Lilley, and A. K. Covington, unpublished data.
[183]R. G. Anderson and M. C. R. Symons, *Trans. Faraday Soc.* **65** (1969) 2537.
[184]A. K. Covington, M. L. Hassall, and I. R. Lantzke, *J. Chem. Soc. Faraday Trans. 2* **68** (1972) 1352.
[185]K. Kuhlmann and D. M. Grant, *J. Phys. Chem.* **68** (1964) 3208.
[186]R. Haque and L. W. Reeves, *J. Phys. Chem.* **70** (1966) 2753.
[187]R. J. Gillespie, J. S. Hartman, and M. Parekh, *Can. J. Chem.* **46** (1968) 1601.
[188]J. H. Swinehart and H. Taube, *J. Chem. Phys.* **37** (1962) 1579.
[189]Z. Luz and S. Meiboom, *J. Chem. Phys.* **40** (1964) 1058.
[190]Z. Luz and S. Meiboom, *J. Chem. Phys.* **40** (1964) 1066.
[191]S. Nakamura and S. Meiboom, *J. Am. Chem. Soc.* **89** (1967) 1765.
[192]H. Grasdalen, *J. Magn. Reson.* **5** (1971) 84.
[193]H. Grasdalen, *J. Magn. Reson.* **6** (1972) 336.
[194]G. W. Stockton and J. S. Martin, *Can. J. Chem.* **52** (1974) 744.
[195]G. C. Levy and G. L. Nelson, *Carbon-13 Nuclear Magnetic Resonance for Organic Chemists*, Wiley-Interscience, New York, 1972, p. 8.
[196]R. M. Hammaker and R. M. Clegg, *J. Mol. Spectrosc.* **22** (1967) 109.
[197]R. N. Butler and M. C. R. Symons, *Trans. Faraday Soc.* **65** (1969) 945.
[198]R. N. Butler and M. C. R. Symons, *Trans. Faraday Soc.* **65** (1969) 2559.
[199]A. Lindheimer and B. Brun, *J. Chem. Phys.* **69** (1972) 1462.
[200]F. Vogrin and E. R. Malinowski, *J. Am. Chem. Soc.* **97** (1975) 4876.
[201]A. L. Allred and R. N. Wendricks, *J. Chem Soc. A.* (1966) 778.
[202]B. M. Rode, *Z. Anorg. Chem.* **339** (1973) 239.
[203]J. F. Coetzee and W. R. Sharpe, *J. Solution Chem.* **1** (1972) 77.
[204]N. A. Matwiyoff, *Inorg. Chem.* **5** (1966) 788.
[205]G. S. Vigee and P. Ng, *J. Inorg. Nucl. Chem.* **33** (1971) 2477.
[206]N. A. Matwiyoff and S. V. Hooker, *Inorg. Chem.* **6** (1967) 1127.
[207]J. S. Baliec, Jr., C. H. Langford, and T. R. Stengle, *Inorg. Chem.* **5** (1966) 1362.
[208]F. W. Breivogel, *J. Phys. Chem.* **73** (1969) 4203.
[209]N. S. Angerman and R. B. Jordan, *Inorg. Chem.* **8** (1969) 2579.
[210]S. Blackstaffe and R. A. Dwek, *Mol. Phys.* **15** (1968) 279.
[211]S. Thomas and W. L. Reynolds, *J. Chem. Phys.* **46** (1967) 4164.
[212]C. H. Langford and F. M. Chung, *J. Am. Chem. Soc.* **90** (1968) 4485.
[213]J. C. Boubel and J.-J. Delpuech, *Adv. Mol. Relaxation Processes* **7** (1975) 209.
[214]J. F. Hinton and E. S. Amis, *Chem. Rev.* **71** (1971) 627.

[215]G. Beech and K. Miller, *J. Chem. Soc. Dalton Trans* (1972) 801.
[216]H. H. Glaeser, H. W. Dodgen, and J. P. Hunt, *J. Am. Chem. Soc.* **89** (1967) 3065.
[217]J. P. Hunt, H. W. Dodgen, and F. Kleinberg, *Inorg. Chem.* **2** (1963) 478.
[218]S. Thomas and W. L. Reynolds, *J. Chem. Phys.* **44** (1966) 3148.
[219]W. G. Movins and N. A. Matwiyoff, *Inorg. Chem.* **6** (1967) 847.
[220]A. Fratiello and R. Schuster, *J. Phys. Chem.* **71** (1967) 1948.
[221]J. F. O'Brien and M. Alei, *J. Phys. Chem.* **74** (1970) 743.
[222]A. Fratiello, D. P. Miller, and R. Schuster, *Mol. Phys.* **12** (1967) 111.
[223]A. Fratiello, D. P. Miller, and R. Schuster, *Mol. Phys.* **11** (1966) 597.
[224]Ref. 3, p. 553.
[225]S. F. Lincoln, *Aust. J. Chem.* **25** (1972) 2705.
[226]I. Y. Ahmed, *J. Magn. Reson.* **7** (1972) 196.
[227]I. Y. Ahmed and C. D. Schmulbach, *Inorg. Chem.* **11** (1972) 228.
[228]J. W. Akitt and A. J. Downes, *Spec. Publ. Chem. Soc.* **22** (1967) 199.
[229]G. E. Maciel, J. K. Hancock, L. F. Lafferty, P. A. Mueller, and W. K. Musker, *Inorg. Chem.* **5** (1966) 554.
[230]R. H. Sox, H. W. Terry, and L. W. Harrison, *J. Am. Chem. Soc.* **93** (1971) 3297.
[231]M. Herlem and A. I. Popov, *J. Am. Chem. Soc.* **94** (1972) 1431.
[232]M. S. Greenberg, R. L. Bodner, and A. I. Popov, *J. Phys. Chem.* **77** (1973) 2449.
[233]M. S. Greenberg, P. M. Wild, and A. I. Popov, *Spectrochim. Acta* **29A** (1973) 1927.
[234]G. W. Canters, *J. Am. Chem. Soc.* **94** (1972) 5230.
[235]D. W. Hafemeister and W. H. Flygare, *J. Chem. Phys.* **44** (1966) 3584.
[236]D. Ikenberry and T. P. Das, *J. Chem. Phys.* **45** (1966) 1361.
[237]J. Itoh and Y. Yamagata, *J. Phys. Soc. Jpn.* **13** (1958) 1182.
[238]C. Deverell, *Mol. Phys.* **16** (1969) 491.
[239]M. R. Baker, C. H. Anderson, and N. F. Ramsey, *Phys. Rev. Sect. A.* **133** (1964) 1533.
[240]C. Deverell, K. Schaumberg, and H. J. Bernstein, *J. Chem. Phys.* **49** (1968) 1276.
[241]O. Lutz, *Phys. Lett. A* **25** (1967) 440.
[242]E. G. Bloor and R. G. Kidd, *Can. J. Chem.* **46** (1968) 3425.
[243]J. J. Dechter and J. I. Zink, *J. Am. Chem. Soc.* **97** (1975) 2937.
[244]R. H. Erlich, E. Roach, and A. I. Popov, *J. Am. Chem. Soc.* **92** (1970) 4989.
[245]R. H. Erlich and A. I. Popov, *J. Am. Chem. Soc.* **93** (1971) 5620.
[246]J. P. K. Tong, C. H. Langford, and T. R. Stengle, *Can. J. Chem.* **52** (1974) 1721.
[247]C. Hall, G. L. Haller, and R. E. Richards, *Mol. Phys.* **16** (1969) 377.
[248]C. H. Langford and T. R. Stengle, *J. Am. Chem. Soc.* **91** (1969) 4014.
[249]T. R. Stengle, Y. C. E. Pan, and C. H. Langford, *J. Am. Chem. Soc.* **94** (1972) 9037.
[250]R. A. Craig and R. E. Richards, *Trans. Faraday Soc.* **59** (1963), 1972.
[251]A. Fratiello and D. C. Douglass, *J. Chem. Phys.* **39** (1963) 2017.
[252]A. Fratiello and E. G. Christie, *Trans. Faraday Soc.* **61** (1965) 306.
[253]A. Fratiello and D. P. Miller, *J. Chem. Phys.* **43** (1965) 796.
[254]A. Fratiello and D. P. Miller, *Mol. Phys.* **11** (1966) 37.
[255]A. Fratiello, R. E. Lee, D. P. Miller, and V. M. Nishida, *Mol. Phys.* **13** (1967) 349.
[256]A. Fratiello, R. E. Lee, V. M. Nishida, and R. Schuster, *J. Chem. Phys.* **47** (1967) 4951.
[257]D. P. Ölander, R. S. Marianeilli, and R. C. Larson, *Anal. Chem.* **41** (1969) 1097.
[258]J. Crea and S. F. Lincoln, *Inorg. Chem.* **11** (1972) 1131.
[259]C. Beguin, J. J. Delpuech, and A. Peguy, *Mol. Phys.* **17** (1969) 317.
[260]J. J. Delpuech, A. Peguy, and M. R. Khaddar, *J. Electroanal. Chem.* **29** (1971) 31.
[261]J. J. Delpuech, A. Peguy, and M. R. Khadder, *J. Magn. Reson.* **6** (1972) 325.
[262]D. Canet, J. J. Delpuech, M. R. Khadder, and P. Rubini, *J. Magn. Reson.* **15** (1974) 325.
[263]R. D. Green and N. Sheppard, *J. Chem. Soc. Faraday Trans. 2* **5** (1972) 821.
[264]F. Toma, M. Villemin, and J. M. Thiery, *J. Phys. Chem.* **77** (1973) 1294.

[265]A. D. Covington and A. K. Covington, *J. Chem. Soc. Faraday Trans. 1* **71** (1975) 831.
[266]D. Canet, J. J. Delpuech, M. R. Khaddar, and P. Rubini, *J. Magn. Reson.* **9** (1973) 329.
[267]J. J. Delpuech, M. R. Khaddar, and P. Rubini, *J. Chem. Soc. Chem. Commun.* (1974) 154.
[268]D. Gudlin and H. Schneider, *J. Magn. Reson.* **16** (1974) 364.
[269]D. R. Cogley, J. N. Butler, and E. Grunwald, *J. Phys. Chem.* **75** (1971) 1477.
[270]G. W. Stockton and J. S. Martin, *J. Am. Chem. Soc.* **94** (1972) 6921.
[271]M. Smith and M. C. R. Symons, *Trans. Faraday Soc.* **54** (1958) 339, 346.
[272]L. S. Frankel, T. R. Stengle, and C. H. Langford, *Chem. Commun.* (1965) 393.
[273]L. S. Frankel, C. H. Langford, and T. R. Stengle, *J. Phys. Chem.* **74** (1970) 1376.
[274]A. K. Covington, K. E. Newman, and T. H. Lilley, *J. Chem. Soc. Faraday Trans. 1* **69** (1973) 973.
[275]A. K. Covington, I. R. Lantzke, and J. M. Thain, *J. Chem. Soc. Faraday Trans. 1* **70** (1974), 1869.
[276]A. K. Covington and J. M. Thain, *J. Chem. Soc. Faraday Trans. 1* **70** (1974) 1879.
[277]A. K. Covington and J. M. Thain, *J. Chem. Soc. Faraday Trans. 1* **71** (1975) 78.
[278]E. Grunwald, G. Baughman, and G. Kohnstam, *J. Am. Chem. Soc.* **74** (1970) 1376.
[279]A. L. Van Geet, *J. Am. Chem. Soc.* **94** (1972) 5583.
[280]M. K. Wong, W. J. McKinney, and A. I. Popov, *J. Phys. Chem.* **75** (1971) 56.
[281]A. K. Covington and K. E. Newman, *Thermodynamic behavior of electrolytes in mixed solvents*, William F. Furter, ed., *Advances in Chemistry Series*, No. 155 (1976) p. 153.
[282]A. Clausen, A. A. El-Harakamy, and H. Schneider, *Ber. Bunsenges. Phys. Chem.* **77** (1973) 994.
[283]L. S. Frankel, T. R. Stengle, and C. H. Langford, *Can. J. Chem.* **46** (1968) 3183.
[284]A. K. Covington, *Specialist Periodical Reports, Electrochemistry, Vol. 3*, Chemical Society London, 1973, p. 1.
[285]J. M. Ceraso and J. L. Dye, *J. Am. Chem. Soc.* **95** (1973) 4432.
[286]J. M. Lehn, J. P. Sauvage, and B. Dietrich, *J. Am. Chem. Soc.* **92** (1970) 2916.
[287]E. Shchori, J. Jagur-Grodzinski, and M. Shporer, *J. Am. Chem. Soc.* **95** (1973) 3842.

3

Solvent Dipoles at the Electrode–Solution Interface

M. Ahsan Habib*

School of Physical Sciences, The Flinders University of South Australia, Adelaide

I. INTRODUCTION

Although solvent molecules constitute an essential component of the electrode–solution interface, recognition of the effects of their presence on the double-layer phenomenon is usually overlooked and often neglected. Lange and Miscenko[1] were the first to recognize some of these roles of solvent dipoles when they divided the total potential difference across the interface into two parts, one due to charges and the other to dipoles. Bockris and Potter[2] introduced consideration of the effect of variable dipole potential on electrode kinetics. However, in recent years, there has been an increasing interest in the detailed role of solvent dipoles, particularly water, in the properties of the electrode–solution interface.[3–10] Müller[11] and Reeves[12] have reviewed some of these works.

In this chapter, we have attempted to discuss the more recent advances, both experimental and theoretical, toward understanding of the role of interfacial solvent in some interfacial phenomena to arrive at a conclusion concerning the influence of solvent dipoles on some specific properties of double layers.

*Present address: Institut für Physikalische Chemie, Freie Universität Berlin, West Berlin, West Germany.

II. EXPERIMENTAL INFORMATION ON SOLVENT PROPERTIES

1. Solvent Excess Entropy

In comparison with the large amount of work carried out on the adsorption of organic compounds[13] and ions,[14,15] little investigation of solvent properties has been done. However, important information concerning solvent dipoles has been derived by measuring the thermodynamic quantity Γ_s, the surface excess entropy of the interface, defined by

$$\Gamma_s = -\left(\frac{\partial \gamma}{\partial T}\right)_{P,\mu,E_\pm} \tag{1}$$

where γ, T, and P are the interfacial tension, temperature, and pressure, respectively, μ is the chemical potential of the electrolyte; and $E_\pm$ is the potential of the working electrode with respect to an electrode reversible with respect to an ion of sign opposite to that under consideration. Experimental determination of the temperature and pressure derivative of the interfacial tension for a system of fixed composition in contact with a reference electrode reversible to one of the ions present requires the simultaneous variation of the chemical potential and the reference electrode potential with temperature or pressure. An alternative indirect method based on the measurement of the interfacial capacity for the determination of Γ_s as a function of electrode charge may be found convenient.[8,12] Thermodynamic derivations of Γ_s may be obtained in detail in Refs. 8, 12, and 16. Only two sets of experimental measurement of Γ_s have been reported so far; one by Hills and Payne[16] for solutions of NaF, NaCl, and $NaClO_3$; the other by Harrison,[17] Schiffrin,[18] and Hills and Hsieh[19] for solutions of KF, KCl, and KNO_3. These surface excess entropies contain within them two contributions: one due to ions and the other due to water dipoles [cf. Eq. (4)]. The contribution due only to solvent dipoles to Γ_s may now be represented by[12,18,19]

$$S^* = \Gamma_s - (\Gamma_+ \bar{S}_+ + \Gamma_- \bar{S}_-) \tag{2}$$

where Γ_+ and Γ_- are the ionic surface excesses and $\bar{S}_+$ and $\bar{S}_-$ are the partial molar ionic entropies. Partial molar entropies of the ions may be evaluated experimentally.[20–22] It was thus shown by Reeves[12] and Hills and Hsieh[19] to be possible to derive the solvent excess entropy

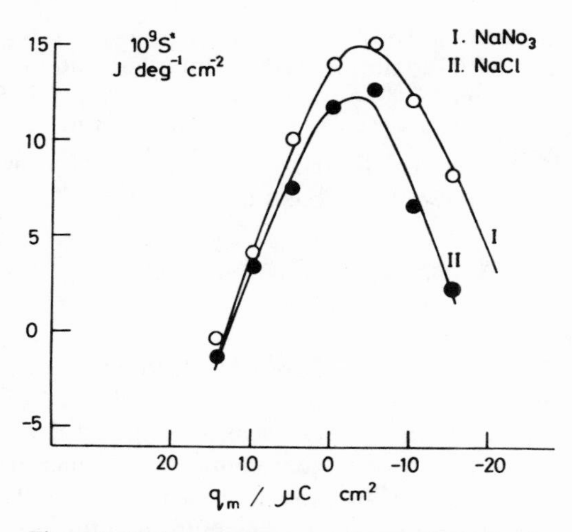

Figure 1. Surface excess entropy due to solvent dipoles, derived experimentally, as a function of electrode charge.[12]

experimentally as a function of electrode charge. These workers[12,19] found that it follows a near-parabolic course, having a maximum at a potential negative to the potential of zero charge, as shown in Fig. 1. It is clear that the orientation of the highly polar water molecules will be strongly influenced by the electric field normal to the interface, and the entropy will decrease on either side of the zero preferred orientation. This point of maximum entropy or disorder occurs, according to the results of Reeves,[12] at $q_m = -6\,\mu\text{C cm}^{-2}$, and according to the results of Hills and Hsieh,[19] at $q_m = -4\,\mu\text{C cm}^{-2}$. Thus, the minimum orientation of water molecules, as evident from these experimental results, occurs at a value of surface charge density of -4 to $-6\,\mu\text{C cm}^{-2}$.

2. Entropy of Formation of the Double Layer

Recently, Harrison *et al.*[8] measured the entropy of formation of the mercury–solution interface, given by

$$\Delta S^{m\text{–soln}} = S^{\sigma} - m^{\sigma}_{\text{Hg}} S^{\text{o}}_{\text{Hg}} - m^{\sigma}_{\text{H}_2\text{O}} S_{\text{H}_2\text{O}} - m^{\sigma}_{+} \bar{S}_{+} - m^{\sigma}_{-} \bar{S}_{-} \qquad (3)$$

where S^σ is the total entropy per unit area of the interface, m_i^σ is the number of moles of the component i in the interface, and $\bar{S}$ is the partial molar entropy of the species concerned. $\Delta S^{m\text{—soln}}$ is not the surface excess entropy Γ_s given by Eq. (1). The essential difference between $\Delta S^{m\text{—soln}}$ and Γ_s may be visualized [by comparing Eqs. (3) and (4)] when Γ_s is defined as follows on the basis of Guggenheim's approach:

$$\Gamma_s = S^\sigma - m_{\mathrm{Hg}}^\sigma S_{\mathrm{Hg}}^{\mathrm{o}} - m_{\mathrm{H_2O}}^\sigma \bar{S}_{\mathrm{H_2O}} - \sum m_{\mathrm{H_2O}}^\sigma \frac{m_i^\alpha}{m_{\mathrm{H_2O}}^\alpha} \bar{S}_i \tag{4}$$

where α represents the bulk phases.

The surface excess entropy Γ_s represents the sum of the entropies of those components in excess and the excess entropy of the other reference components that arises because of the stresses and other forces in the interphase. $\Delta S^{m\text{—soln}}$ represents the difference in entropy of the components when they are present at the interface and in the bulk of the adjoining phases; i.e., the physical interpretation of $\Delta S^{m\text{—soln}}$ is the difference in entropy of the components of the interface when they are forming part of it and when they are present in the bulk.

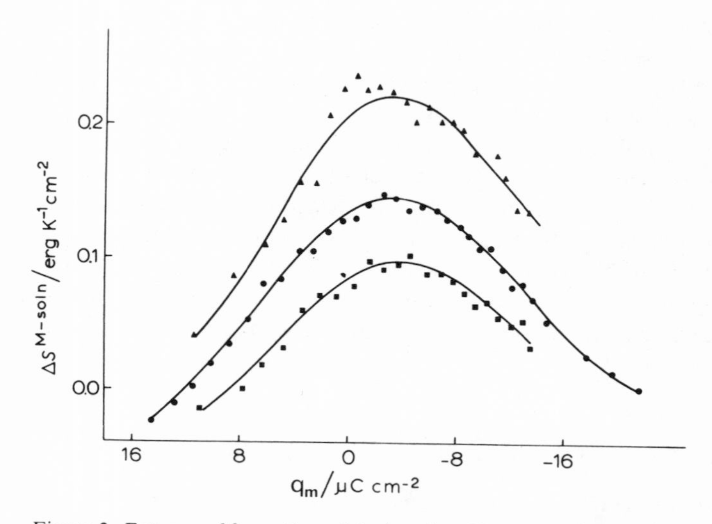

Figure 2. Entropy of formation of the interface Hg–0.1 M NaF solution at different temperatures and surface charge densities: (▲) 0, (●) 25, and (■) 45°C[8].

The entropy of formation of a Hg–solution interface as measured by Harrison *et al.*[8] varies with the charge density on the metal, as shown in Fig. 2. The maximum in $\Delta S^{m-\mathrm{soln}}$ occurs at charges -4 to $-6\ \mu\mathrm{C\ cm^{-2}}$, and this signifies that the maximum randomness in the interphase occurs at a potential negative to the potential of zero charge. Harrison *et al.*[8] attributed the observed entropy variations wholly to the change of the orientation of the water dipoles due to the strong electric field in the inner region of the double layer on the basis of the following argument:

The behavior of the metal surface is not symmetrical with respect to the charge; positive charge on the metal is present as discrete ions, and it would be expected that anions would interact more strongly with them than cations would with a smeared-out electron cloud. Thus, a strong positive polarization would produce a roughening of the Hg surface due to strong interaction of the anions with it, and hence increase the amount of Hg present at the interface. Also, the formation of adatoms on the surface would increase the entropy of the Hg atoms at the interface.[24] If this were the explanation of the observed variations of entropy, a positive contribution of the Hg to the entropy of formation should be expected when the charge on the metal is positive. This possibility is at variance with the observed decrease of the entropy of formation in the anodic branch, and the entropy variations observed must therefore be attributed mainly to the solvent contribution. The charge dependence of the entropy of formation can be accounted for in two ways: (1) the amount of water present in the inner layer has a minimum value when the charge on the metal is -4 to $-6\ \mu\mathrm{C\ cm^{-2}}$ and increases on polarization, the entropy of water at the interface remaining fairly constant at different values of q_m and smaller than that in the bulk of the solution; and (2) the amount of water is fairly constant, and all the entropy variations must be attributed to changes in the entropy of the adsorbed water molecules due to changes in the orientation of the water dipoles by the electric field (the diffuse layer correction to the total entropy is found to be small).

Hypothesis (1) is not plausible, since it leads to conclusions that are at variance with surface excess results, as shown below. If the number of water molecules at the inner region[23] is taken as $10^{15}\ \mathrm{cm^{-2}}$, the entropy of adsorption of water at this potential is $-6.3\ \mathrm{J\ K^{-1}\ mol^{-1}}$. To account for the charge dependence of $\Delta S^{m-\mathrm{soln}}$, the variation of the amount of water in the inner region from, say, -4

to $-20\,\mu\text{C cm}^{-2}$ should be equivalent to 5×10^{14} molecules. This result is in contradiction to cationic surface excess results in concentrated solutions of electrolytes, in which it has been found that the possible variations in the value[23] of the water concentration in the interphase is not very large,* certainly not of the order of 50%, as the argument set forth above should require.

The observed entropy variation must thus be wholly attributed to the change of the orientation of the water dipoles.

The position of maximum entropy or minimum orientation found by Harrison *et al.*[8] at charges from -4 to $-6\,\mu\text{C cm}^{-2}$ is thus fully consistent with the results obtained by Reeves[12] from those of Hills and Paynes,[16] and with the results obtained by Hills and Hsieh.[19]

The position of minimum net orientation of water molecules at a negative electrode charge has also been evident from the experimental measurement of adsorption of organic compounds as a function of electrode charge. Bockris *et al.*[29] found that maximum adsorption of butanol occurs at $q_m = -2\,\mu\text{C cm}^{-2}$, and hence inferred that at that negative charge, the water molecules do not have any preferred orientation and are easily replaced by the organic molecules. Conway and co-workers[6,52] have recently investigated the molecular orientation in the Hg–water interphase during the adsorption of pyrazine, pyridine, and amylalcohol. They found that the minimum net orientation of water molecules occurs at $q_m = -2\,\mu\text{C cm}^{-2}$. Though the position of minimum solvent orientation has been inferred from organic adsorption to occur at $q_m = -2\,\mu\text{C cm}^{-2}$, while that from entropy measurements occurs at $q_m = -4$ to $-6\,\mu\text{C cm}^{-2}$, the inferences are not inconsistent. The minimum solvent orientation, in any case, occurs at a potential negative to the potential of zero charge (p.z.c.).

The fact, as experimentally observed by a number of workers,[47–52] that the water molecules are preferentially adsorbed with their oxygen end toward the surface at the p.z.c. is thus corroborated by these[8,12,19] experimentally observed entropy maxima at a potential negative to the p.z.c.

*Following an analysis similar to that of Parsons and Zobel,[115] Harrison, *et al.*[24] concluded that the possible variation in interfacial water concentration with electrode charge may account only to about 10%.

III. MODELS OF WATER IN THE INTERFACE

1. The Two-State Water Model

Watts-Tobin and Mott[24,25] were the first to put forward a model of interfacial water in order to analyze quantitatively the role of water in the double layer. Watts-Tobin assumed that the water molecules are adsorbed on the electrode surface either with their oxygen end toward the surface (up dipole) or with their hydrogen end toward the surface (down dipole). This "up–down" model, now commonly known as a *two-state water model*, served as the basis of most of the later development of molecular theories concerning the role of water. Detailed discussion on this model may be obtained elsewhere.[12] However, the essential features of the Mott and Watts-Tobin model[24,25] are that it neglects the lateral interaction between the water dipoles and makes the interpretation that the experimentally observed anodic capacitance hump is due to a high dielectric constant when the field in the inner region is not high enough to give saturation. MacDonald and Barlow[26] considered a multiposition model of water molecules and also attributed the anodic hump to dielectric saturation. Later, Levine *et al.*[27] put forward an alternative treatment based on the two-state water model and attributed the anodic capacitance hump to dielectric saturation. The model of MacDonald and Barlow[26] has recently been discussed in detail by Reeves,[12] and the model of Levine *et al.*[27] has also been discussed by Conway and Dhar[28] and Reeves.[12] However, it is to be pointed out here that if the capacitance hump that occurs on the positive side of the p.z.c. is due to dielectric saturation (i.e., dipole orientation), then the dipole orientation should be minimum at the electrode charge where the hump occurs, i.e., at a potential positive to the p.z.c. But experimental evidence, as discussed in the last section, shows that the minimum orientation of the interfacial dipoles occurs at a potential negative to the potential of zero charge. Therefore, the treatments of Mott and Watts-Tobin[24,25] MacDonald and Barlow,[26] and Levine *et al.*[27] lead to results that are inconsistent with recently established experimental facts.[8,12,19]

Bockris *et al.*[29] developed the two-state model in a way different from that of Mott and Watts-Tobin. They[29] used the Lange and Miscenko[1] equation for the potential difference, i.e.,

$$\Delta_s^m\phi = g_s^m(\text{ion}) + g_s^m(\text{dipole}) \tag{5}$$

where g_s^m(ion) and g_s^m(dipole) are the potential contributions due to charges and dipoles, respectively. Differentiation of Eq. (5) with respect to electrode charge leads to

$$\frac{1}{C_{\text{DL}}} = \frac{1}{C_{\text{charge}}} + \frac{1}{C_{\text{dipole}}} \tag{6}$$

That is, the total double-layer (DL) capacitance is considered to be a combination of two capacitances, one due to charges and the other due to dipoles, connected in series.* According to Watts-Tobin,[24] the potential drop across the interface due to surface charges and water dipoles may be represented by

$$\Delta_s^m \phi = \frac{4\pi q_m d'}{\varepsilon_0} - \frac{4\pi N \mu'}{\sqrt{3}\varepsilon_0} \tanh\left(\frac{\mu'(\Delta_s^m \phi - V)}{\sqrt{3}d'kT}\right) \tag{7}$$

where N, μ', d', and ε_0 are the number of water molecules per unit area, dipole moment of water, thickness of double layer, and dielectric constant, respectively, and V signifies the difference in adsorption energy of the two orientation states of water at the p.z.c.

In the presentation of Eq. (7), the contribution of adsorbed ions has been omitted. Equation (7) may be rearranged to give

$$q_m = \frac{\varepsilon_0 \Delta_s^m \phi}{4\pi d} + \frac{N\mu'}{\sqrt{3}d'} \tanh\left(\frac{\mu'(\Delta_s^m \phi - V)}{\sqrt{3}d'kT}\right) \tag{8}$$

The differential capacity of the double layer is $dq_m/d\Delta_s^m\phi$, i.e.,

$$C_{\text{DL}} = \frac{\varepsilon_0}{4\pi d'} + \frac{N\mu'^2}{3d'^2kT} \operatorname{sech}^2\left(\frac{\mu'(\Delta_s^m \phi - V)}{\sqrt{3}d'kT}\right) \tag{9}$$

or

$$C_{\text{DL}} = C_{\text{charge}} + C_{\text{dipole}} \tag{10}$$

Thus, Watts-Tobin[24] represented the double-layer capacitance as a combination of the capacitances due to charges and those due to dipoles connected in *parallel*. Therefore, the treatment of Bockris *et*

*In fact, the dipole potential is to be subtracted from the potential arising due to charges, and hence Eq. (6) should be written as

$$\frac{1}{C_{\text{DL}}} = \frac{1}{C_{\text{charge}}} - \frac{1}{C_{\text{dipole}}} \tag{6a}$$

al.[29] (BDM) differs from that of Watts-Tobin (WT) in two essential ways: (1) the BDM theory takes into account lateral interaction between the solvent dipoles, which the WT treatment neglects; and (2) BDM considers the capacitances due to charges and dipoles to be connected in series, while WT considers these capacitances to be connected in parallel.

According to Bockris *et al.*, the potential contribution due to water dipole may be expressed as a function of q_m, as follows:

$$g^s(\text{dipole}) = \frac{2\pi N\mu'}{\varepsilon_0} \tanh\left(\frac{\mu' X}{kT} - \frac{Uc'Z}{kT}\right) \tag{11}$$

where

$$Z = \frac{N_\uparrow - N_\downarrow}{N} = \tanh\left(\frac{\mu' X}{kT} - \frac{Uc'Z}{kT}\right) \tag{12}$$

Hence,

$$\frac{1}{C_{\text{dipole}}} = \frac{8\pi^2\mu'^2 N}{\varepsilon_0^2 kT}\left(\frac{1 - Z^2}{1 + (Uc'/kT)(1 - Z^2)}\right) \tag{13}$$

where c' is the number of the nearest neighbor of a water molecule, U is the lateral interaction energy between two adjacent dipoles, and $N_\uparrow$ and $N_\downarrow$ are the number of up and down dipoles, respectively, and $N = (N_\uparrow + N_\downarrow)$. With $c' = 8$ and $U/kT = 0.5$, they found $C_{\text{dipole}} \approx 120\,\mu\text{F cm}^{-2}$ at its minimum; hence, $1/C_{\text{dipole}}$ being small in comparison with the observed $1/C_{\text{DL}}$, they concluded that the water dipoles do not contribute significantly to the double-layer capacitance, and thus the observed anodic capacitance hump is not due to water molecules, a conclusion now known to be consistent with observed entropy results.[8,12,19]

However, it is to be noted that BDM[29] used a factor 2π in Eq. (11) instead of 4π. A corrected BDM formula for C_{dipole} is given in Ref. 30. BDM did not mention what value they used for N, the total number of water molecules present per unit area on the surface. If $N = 10^{15}$ cm^{-2}, $\mu = 1.84$ D, $c' = 8$, $U/kT = 0.5$, and $\varepsilon_0 = 6$, then the corrected BDM formula gives

$$C_{\text{dipole}} = \frac{\varepsilon_0^2 kT}{16\pi^2\mu'^2 N}\,\frac{1 + (Uc'/kT)(1 - Z^2)}{1 - Z^2} = 15.4\,\mu\text{F cm}^{-2} \tag{14}$$

at the minimum where $Z = 0$.

Thus, though the allowance for the lateral interaction reduced the value of g^s(dipole) and thus increased C_{dipole} to some extent, reasonable values of the quantities involved in Eq. (14) give rise to a value of $1/C_{dipole}$ comparable to that of observed $1/C_{DL}$. However, since the Watts-Tobin model leads to configuration in which the dipole capacitance is connected in parallel with other capacitance components, the observed capacitance hump in this model results from a positive dipole capacitance contribution to the charge component of the double-layer capacitance [Eq. (10)]. But the BDM[29] model relates the dipole capacitance to the charge component in a series capacitance configuration [Eq. (6)], and hence the dipole contribution is a negative contribution to the dòuble-layer capacitance. The BDM[29] model should be considered in the light of these assumptions.

2. The Three-State Water Model

The experimental findings of solvent entropy suggest a negligible contribution to C_{DL} by the water molecules; this led Bockris and Habib[31] to introduce the dimer model of solvent dipoles in the interface. According to them, water molecules on the electrode surface are present as dimers and up and down monomers. Support for the presence of associated water molecules on the electrode surface is not new. Parsons[32] and Frumkin and Damaskin[33] suggested that some of the water molecules on the electrode surface may be associated into groups. They did not put forward a specific water configuration. Law[34] determined the entropy of adsorption of water onto mercury from the gas phase and found it to be $150\,J\,K^{-1}\,mol^{-1}$. Since the entropy of water vapor is[35,36] $188.52\,J\,K^{-1}\,mol^{-1}$ at 298°K, the entropy of adsorbed water is $38.52\,J\,K^{-1}\,mol^{-1}$. This low entropy led Law[34] to conclude that the adsorbed water molecules are immobile on the surface and associated into pairs. A similar suggestion had been made by Kemball.[37] Following the findings of Kemball and of Law for the vapor–mercury interface, they[31] postulated that along with up and down dipoles, dimers are also present on the electrode surface. However, for reasons shown below, Bockris and Habib[31] did not assume, as did Kemball and Law, that all the water molecules present on the surface are associated into pairs. The fraction of the surface covered by the dimers and monomers is obtained by the consideration of the following equilibrium cycle[31]:

$$\begin{array}{ccc} 2(H_2O)_{ads} & \underset{k_{-1}}{\overset{k_1}{\rightleftarrows}} & 2(H_2O)_{soln} \\ k_2 \downarrow\uparrow k_{-2} & & k_4 \uparrow\downarrow k_{-4} \\ (H_2O)_{2,ads} & \underset{k_{-3}}{\overset{k_3}{\rightleftarrows}} & (H_2O)_{2,soln} \end{array}$$

The relative amounts of monomers and dimers on the surface follow from

$$\frac{d[H_2O]_{ads}}{dt} = k_1 C_m(1 - \theta_m - \theta_d) - k_{-1}\theta_m^2 + k_{-2}\theta_d - k_2\theta_m^2 \tag{15}$$

and

$$\frac{d[(H_2O)_2]_{ads}}{dt} = k_{-3}C_d(1 - \theta_m - \theta_d) - k_3\theta_d + k_2\theta_m^2 - k_{-2}\theta_d \tag{16}$$

where C_m and C_d are the concentrations of monomers and dimers in solution, respectively, and θ_m and θ_d are the fractions of the surface covered by monomers and dimers, respectively.

For

$$\frac{d[H_2O]_{ads}}{dt} = \frac{d[(H_2O)_2]}{dt} = 0 \qquad \theta_m + \theta_d = 1 \tag{17}$$

$$\frac{\theta_d}{(1 - \theta_d)^2} = \frac{2k_2 + k_{-1}}{2k_{-2} + k_3} \tag{18}$$

where k_{-2} and k_2 are the rate constants for the dissociation of dimers to monomers and association of monomers to dimers, respectively, and k_{-1} and k_3 are those for the desorption of monomers and dimers, respectively. Assuming $k_2 \gg k_{-1}$ and $k_{-2} \gg k_3$,* Eq. (18) may be represented by

$$\frac{\theta_d}{(1 - \theta_d)^2} = e^{-\Delta G_2^0/RT} \tag{19}$$

*These assumptions are equivalent to the neglect of the effect of the solution on the surface equilibrium. This cannot be strictly valid. However, the surface tension of mercury at the mercury–water interface (42.5 μJ cm^{-2})[8] is almost the same as the surface tension of mercury with a monolayer of water adsorbed from the vapor phase (44 μJ cm^{-2})[24,37]. Therefore, it seems likely that the water molecules in a mercury–solution interface are in much the same state as if they were adsorbed in a close-packed monolayer on mercury from the gas phase, and are affected only slightly by the presence of the rest of the water.

where $e^{-\Delta G_2^0/RT} = K_2 = k_2/k_{-2}$ and ΔG_2^0 and K_2 are the standard free energy change and equilibrium constant, respectively, for the formation of dimers on the surface from the adsorbed monomers.

In respect to ΔG_2^0, the ΔH_2^0 is calculated as follows: According to the Bockris and Habib (BH) model,[31] one up and one down water molecules give rise to one dimer with no net permanent dipole moment. The energy of dimers, U_d, comprises the hydrogen-bond energy $U_{\rm HB}$ holding the two monomers together and the dimer metal attraction and repulsion energy. The interaction energy of an adsorbed species with the metal may be calculated from [38–42]

$$U_{\rm dm} = \frac{C}{r^3} = \frac{1}{r^3} \frac{\pi N_{\rm a} mc^2 \alpha_1 \alpha_2}{\alpha_1/\bar{\chi}_1 + \alpha_2/\bar{\chi}_2} \tag{20}$$

where m is the electronic mass, c is the velocity of light, α_1 and α_2 are the polarizabilities, $\bar{\chi}_1$ and $\bar{\chi}_2$ are the diamagnetic susceptibilities of the metal and the adsorbate atom or molecule, respectively, N_a is the number of adsorbent atoms per cm^3, and r is the distance between the centers of the adsorbent atom and the adsorbate molecule. Now[42] $\alpha_{\rm water} = 1.44 \times 10^{-24}$ cm^2, $\alpha_{\rm Hg} = 5.05 \times 10^{-24}$ cm^3, $\bar{\chi}_w = -2.16 \times 10^{-29}$ cm^3, $\bar{\chi}_{\rm Hg} = -5.61 \times 10^{-29}$ cm^3, and $N_{\rm Hg} = 4.26 \times 10^{22}$ cm^{-3} Ref. 44. $\bar{\chi}_{\rm dimer}$ may be obtained from[39]

$$\bar{\chi}_{\rm dimer} = -\frac{e_0^2 a_0^{1/2}}{4mc^2}(n_d \alpha_d)^{1/2} \tag{21}$$

where a_0 is the Bohr radius, $n_d = 36$ is the number of electrons in a dimer, and α_d is the polarizability, given approximately by

$$\alpha_d = \frac{n^2 - 1}{n^2 + 2} \frac{3V_d}{4\pi N_0} = 2.9 \times 10^{-24}\ {\rm cm}^3 \tag{22}$$

where $n = 1.33$ is the refractive index of water, V_d is the molar volume of dimer, and N_0 is Avogadro's number. With these values, and taking $r = r_w + r_{\rm Hg} = (1.38 + 1.5) \times 10^{-8}$ cm, one gets $U_{\rm dm} = -25.1$ kJ mol^{-1}.

Metal–dimer repulsion energy $U_{\rm rep\,dm}$ is represented by A/R^n, where $n \approx 9$. The constant A is evaluated from the equilibrium condition

$$\frac{dU'}{dr} = \frac{d}{dr}\left(-\frac{C}{r^3} + \frac{A}{r^9}\right) = 0 \tag{23a}$$

or

$$A = \frac{Cr^6}{3} \tag{23b}$$

With C from Eq. (20) and r as before, $U_{\text{rep dm}} = 8.36\,\text{kJ mol}^{-1}$. Since[45,46] $U_{\text{HB}} = -18.81\,\text{kJ mol}^{-1}$, one finds $U_d = U_{\text{dm}} + U_{\text{rep dm}} + U_{\text{HB}} = -25.10 + 8.36 - 18.81 = -35.55\,\text{kJ/mol}$ of dimer.

Now, before the dimer formation, the energy of two moles of up and down monomers, U_{m}, is contributed by:

(a) Image interaction U_{image}: This is obtained from

$$U_{\text{image}} = -2\frac{2\bar{\mu}^2}{(2r_w)^3} = -\frac{\bar{\mu}^2}{2r_w^3} \tag{24}$$

Assuming a tetrahedral structure of water,[24] $\bar{\mu} = \mu'/\sqrt{3}$, where $\bar{\mu}$ is the component of dipole moment normal to the surface. With $\mu' = 1.84$ D, and the radius of a water molecule, $r_w = 1.38 \times 10^{-8}$ cm, one gets $U_{\text{image}} = -12.91\,\text{kJ/2 mol}$ adsorbed water.

(b) Metal–water interaction U_{mw}: From Eq. (20), this is found to be $-25.00\,\text{kJ/2 mol}$ of water.

(c) Water–water lateral interaction U_L: Since one water molecule of two forming a dimer is in the up and the other in the down position, U_L is attractive, and is given by $U_L = -\bar{\mu}^2/(2r_w)^3 = -3.22\,\text{kJ/2 mol}$.

(d) Repulsion energy between metal and water and between water and water (lateral), U_{rep}: This is given by $A'/(2r_w)^9$. Putting the equilibrium condition, $dU_A/dr = 0$ (where $U_A = U_{\text{image}} + U_{mw} + U_L$), one gets $A' = 1.274 \times 10^{-64}\,\text{J cm}^9$. Thus, $U_{\text{rep}} = 13.71\,\text{kJ/2}$ mol water.

Therefore, $U_m = -12.91 - 25.00 - 3.22 + 13.71 = -27.42\,\text{kJ/2}$ mol water. Hence, $\Delta H_2^0 = U_{\text{d}} - U_m = -35.55 + 27.42 = -8.13\,\text{kJ/mol}$ dimer.

Change in entropy, ΔS_2^0, on dimerization: ΔS_2^0 may be obtained by considering the formation of 1 mol dimer from 1 mol up monomer and 1 mol down monomer. When these 2 mol of up and down monomers are present on the surface, the configurational entropy is given by

$$S_m = 2k \ln \frac{N_0!}{N_\uparrow!N_\downarrow!} = 2k \ln \frac{N_0!}{\frac{1}{2}N_0!\frac{1}{2}N_0!} = 2kN_0 \ln 2 \tag{25}$$

With N_0 = Avogadro's number, $S_m = 11.5\,\mathrm{J\,K^{-1}}/2\,\mathrm{mol}$ monomer.* All dimers being identical, the configurational entropy of 1 mol dimer is zero. Hence, $\Delta S_2^0 = S_d - S_m = 0 - 11.5 = -11.5\,\mathrm{J\,K^{-1}}$/mol dimer.

Thus, the free energy change for the formation of the dimer is

$$\Delta G_2^0 = \Delta H_2^0 - T\Delta S_2^0 = -4.7\,\mathrm{kJ\,mol^{-1}} \text{ at } 25°\mathrm{C}$$

Substituting this value of ΔG_2^0 in Eq. (19), one obtains $\theta_d = 0.68$ and $\theta_m = 0.32$.

According to the BH[31] model, the dimers do not have any net charge separation within the molecule, and thus do not contribute to g^s (dipole). The monomers are assumed to be in either the up or the down position, their numbers being $N_\uparrow$ and $N_\downarrow$, respectively, so that $N_m = N_\uparrow + N_\downarrow$ = total number of monomers on the surface per unit area.

The potential contributed by the water molecules is

$$g^s(\text{dipole}) = -\frac{4\pi(N_\uparrow - N_\downarrow)\bar{\mu}}{\varepsilon_0} = -\frac{4\pi N_m \bar{\mu} Z}{\varepsilon_0} \tag{26}$$

where

$$Z = \frac{N_\uparrow - N_\downarrow}{N_m} \tag{27}$$

and $\bar{\mu}$ and ε_0 are the effective dipole moment of a monomer and effective dielectric constant of the adsorbed water layer.

With a consideration of the interaction of a dipole with the double-layer field and the surrounding dipoles, Z is given by[30]

$$Z = \frac{\exp[(1/kT)(-\Delta G^c + x)] - \exp[(1/kT)(-\Delta G^c - x)]}{\exp[(1/kT)(-\Delta G^c + x)] + \exp[(1/kT)(-\Delta G^c - x)]} \tag{28}$$

where

$$x = \frac{\bar{\mu}X}{kT} - \frac{Uc'Z}{kT} \tag{29}$$

and X is the double-layer field, c' is the number of nearest neighbors. Water molecules are preferentially adsorbed with their oxygen end

*In Eq. (25), the monomers are taken as being equally in the up and down positions. In reality, they are in this position only at one specific electrode charge near the p.z.c. However, dimer formation is likely to occur only when there is a juxtaposition of up and down dipoles.

toward the metal at the p.z.c.[47–52] This means $|\Delta G^c_{\uparrow}| > |\Delta G^c_{\downarrow}|$, i.e., $-(\Delta G^c - \Delta G^c) = -\Delta\Delta G^c \neq 0$.

Thus, Eq. (28) can be arranged to give

$$Z = \frac{\exp\left(\dfrac{8\pi\bar{\mu}q_m}{\varepsilon_0 kT} - \dfrac{2Uc'Z}{kT} + b\right) - 1}{\exp\left(\dfrac{8\pi\bar{\mu}q_m}{\varepsilon_0 kT} - \dfrac{2Uc'Z}{kT} + b\right) + 1} \tag{30}$$

where the double-layer field X has been replaced by $4\pi q_m/\varepsilon_0$ and

$$b = -\frac{\Delta\Delta G^c}{kT} \tag{31}$$

Substituting Eq. (30) in Eq. (26), one obtains

$$g^s(\text{dipole}) = -\frac{4\pi N_m\bar{\mu}}{\varepsilon_0}\frac{\exp(2x+b)-1}{\exp(2x+b)+1} \tag{32}$$

The capacitance contributed by the water dipoles is then

$$\begin{aligned} C_{\text{dipole}} &= \frac{dq_m}{dg^s(\text{dipole})} \\ &= \frac{\varepsilon_0^2 kT}{64\pi^2 N_m\bar{\mu}^2}\left(\frac{[\exp(2x+b)+1]^2 + (4Uc'/kT)\exp(2x+b)}{\exp(2x+b)}\right) \end{aligned} \tag{33}$$

where x is given by Eq. (29).

The equation for C_{dipole} above is the modified form of the BDM expression for C_{dipole} [Eq. (14)] in that Eq. (33) allows for the preferential adsorption of water at the p.z.c. with their negative ends toward the surface through the parameter b given by Eq. (31), while the BDM formulation does not.

The quantities $g^s(\text{dipole})$ and C_{dipole} may be calculated from Eq. (32) and Eq. (33), respectively, provided $\Delta\Delta G^c$ is known. This quantity $\Delta\Delta G^c$ is made up of the following:

(a) ΔU_{image}: BDM[29] suggested that $\Delta\Delta G^c$ is due to a difference in the image interaction in the two orientations, favoring the case in which the oxygen atom is oriented toward the metal because the dipole is then 0.05–0.1 Å closer to the metal. Thus[29]

$$\Delta U_{\text{image}} \approx -\frac{3}{2}\frac{\bar{\mu}^2 d}{r_w^4} \tag{34a}$$

where r_w is the radius of a water molecule and d is the difference in the distances between the centers of two molecules oriented oppositely. Assuming a tetrahedral structure of surface water,[24] $\bar{\mu} = \mu'/\sqrt{3}$. With $\mu' = 1.84$ D, $d = 0.05$ Å, $r_w = 1.38$ Å, one gets $\Delta U_{image} = -1.42$ kJ mol^{-1}.

(b) ΔU_{disp}: Due to the difference in the distances of their centers from the metal surface, the up and down dipoles will also show a difference in their dispersion interactions with the metal. This is given by[43]

$$\Delta U_{disp} = C\left(\frac{1}{(r-d)^3} - \frac{1}{(r+d)^3}\right) \tag{34b}$$

where C and r are as defined in Eq. (20). C may be obtained from Eqs. (20) and (21), and thus $\Delta U_{disp} = -1.30$ kJ mol^{-1}. Therefore, $\Delta\Delta G^c = \Delta U_{image} + \Delta U_{disp} = -1.42 - 1.30 = -2.72$ kJ mol^{-1} for the Hg–solution interface.

The mutual interaction energy between two water molecules is $U = -\bar{\mu}^2/(2r_w)^3 = -3.22$ kJ mol^{-1}. The average coordination number is 11.[54] Since the dimers are assumed not to react with monomers, the effective coordination number $c' = 11\theta_m = 11 \times 0.32 = 3.52$. The number of monomers present on the surface $N_m = N_s\theta_m = (1/4r_w^2)\theta_m = 0.42 \times 10^{15}$ cm^{-2}, where N_s is the total number of sites on the surface per unit area. With these values of $\Delta\Delta G^c$, U, c' and N_m, the surface potential g^s(dipole) of water is calculated [Eq. (32)] and plotted as a function of q_m (Fig. 3). The plot of C_{dipole} as a function of q_m is demonstrated in Fig. 4. *The C_{dipole} values are too large to affect the observed C_{DL}* [Eq. (6)], in agreement with the implications of the results of Reeves[12] and Harrison *et al.*[8]

The value of $\Delta\Delta G^c$ calculated by Watts-Tobin,[24] who used the quadrupole moments of water[55] is -6.27 kJ mol^{-1}. The dipole capacitance with $\Delta\Delta G^c$ equal to -5.00 kJ mol^{-1} and to -6.27 kJ mol^{-1} are also shown in Fig. 4 for comparison. The minimum in C_{dipole} is shifted toward more cathodic potentials with an increase of $-\Delta\Delta G^c$, but its contribution to C_{DL} remains negligible. *Thus, Bockris and Habib*[31] *concluded that the orientation of water molecules does not cause the anodic hump on the capacitance charge curve for aqueous solutions.*

The more electropositive the metal, the greater is the preferential adsorption of the negative end of the water dipole and the greater is the

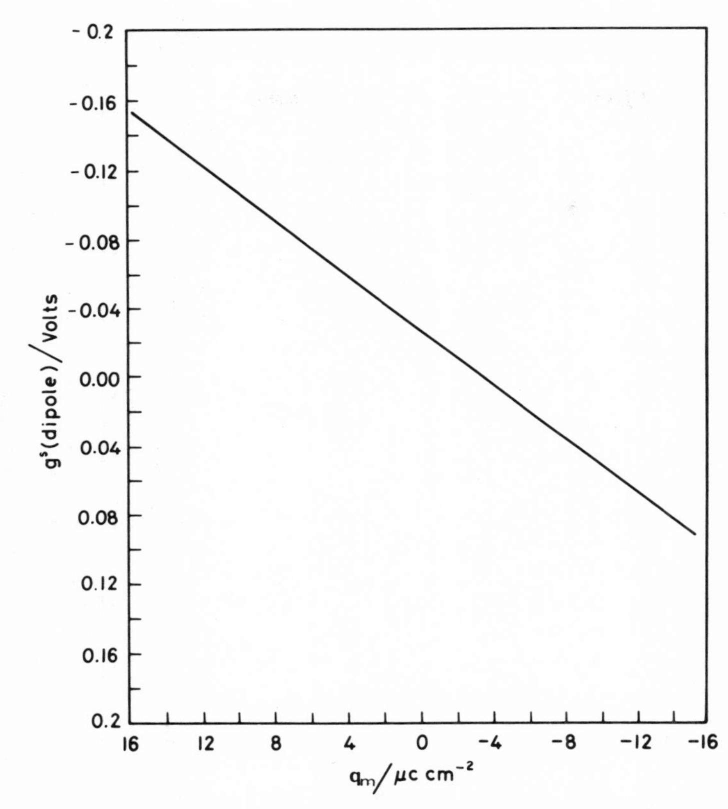

Figure 3. Surface potential of water at the mercury–solution interface as a function of charge density on the metal.[31]

value of $N_{\uparrow} - N_{\downarrow}$. Thus, the surface potential of water should increase at the p.z.c. as the affinity of the metal for oxygen increases[56] [i.e., as the value of $-\Delta\Delta G^c$ increases, cf. Eq. (26)]. This is verified by the plot shown in Fig. 5. With increase of $-\Delta\Delta G^c$, $(N_{\uparrow} - N_{\downarrow})/N_m$ increases, and hence g^s(dipole) also increases [cf. Eq. (24)]. Bockris and Habib[31] also considered Zn and Cd. By using Eq. (22), they obtained $\alpha_{Cd} = 4.42 \times 10^{-24}$ c.c., $\alpha_{Zn} = 3.12 \times 10^{-24}$ c.c., and from (21). $\bar{\chi}_{Cd} = -7.43 \times 10^{-29}$ c.c., $\bar{\chi}_{Zn} = -4.94 \times 10^{-29}$ c.c., $N_{Cd} = 4.64 \times 10^{22}$ cm^{-3} Ref. 44, $N_{Zn} = 6.55 \times 10^{22}$ cm^{-3} Ref. 44, $r_{Cd} = 1.49 \times 10^{-8}$

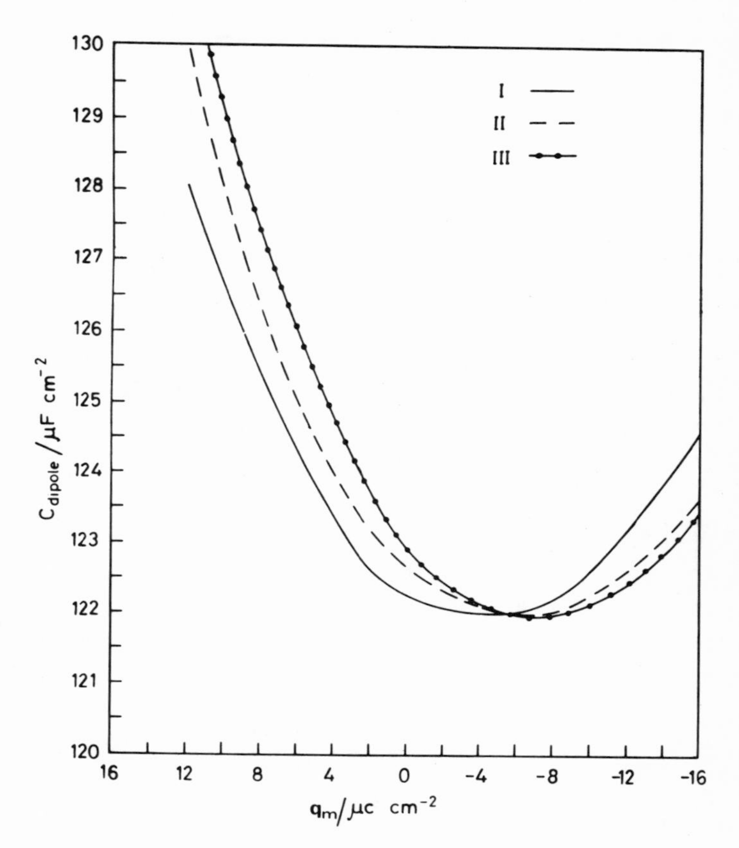

Figure 4. Dependence of dipole capacitance at the mercury–aqueous solution interface on electrode charge: (I) $\Delta\Delta G^c = -2.72$ kJ mol^{-1}; (II) $\Delta\Delta G^c = -5.00$ kJ mol^{-1}; and (III) $\Delta\Delta G^c = -6.27$ kJ mol^{-1}.[31]

cm, and $r_{Zn} = 1.33 \times 10^{-8}$ cm Ref. 44. Using these values, one obtains from Eqs. (20) and (34b). $(\Delta U_{disp})_{Cd-soln} = -1.59$ kJ mol^{-1} and $(\Delta U_{disp})_{Zn-soln} = -1.92$ kJ mol^{-1}. Thus, $\Delta\Delta G^c = \Delta U_{disp} + \Delta U_{image} = -3.01$ kJ mol^{-1} for Cd–solution interface and $\Delta\Delta G^c = -3.34$ kJ mol^{-1} for Zn–solution interface ($\Delta U_{image} = -1.42$ kJ mol^{-1} and same for all metals). Using these values of $\Delta\Delta G^c$ in Eq. (32), Bockris and Habib[31] obtained $g^s_{Cd-soln}(\text{dipole}) = 0.03$ V and $g^s_{Zn-soln}(\text{dipole}) = 0.04$ V at the p.z.c.

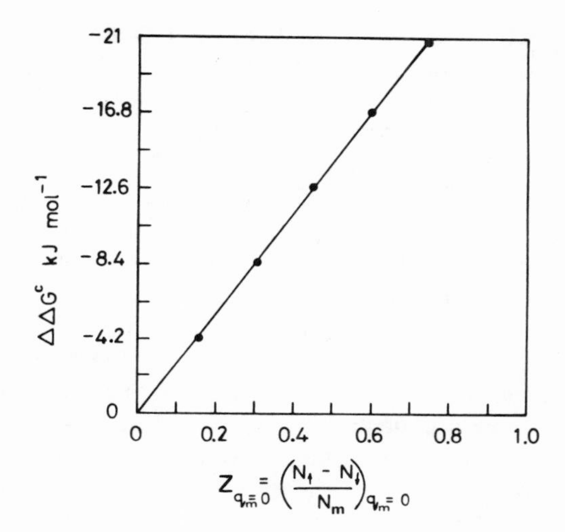

Figure 5. Variation of the degree of orientation $(N_{\uparrow} - N_{\downarrow})/N_m$ of water molecules at the p.z.c. with energy difference $\Delta\Delta G^c$.[31]

According to the molecular approach of Bockris and Habib,[31] the difference in attraction of different metals with the water solvent is attributed only to the difference in dispersion interaction. The metallic properties that cause this difference are the polarizability of the metallic atoms and their diamagnetic susceptibilities. The relationship of these two properties of the metals to the electropositive nature of the metals is not straightforward (see Section IV.2).

The essential feature of the three-state water model (dimer, up and down monomer) is that the model supports a negligible contribution of water dipoles to the double-layer capacitance. Consistency of this model with the experimental solvent excess entropy is discussed in Section III.3.

Another three-state model for interfacial water has been put forward by Damaskin and Frumkin (DF).[57] They assumed that some surface water molecules are associated into small groups and that these small groups have two orientable states, and the rest of the water molecules are chemisorbed. The surface potential contribution due to solvent dipoles is expressed as

$$g^s(\text{dipole}) = -\frac{4\pi N_1 \mu_1}{\varepsilon_0} \tanh\left(\frac{4\pi\mu_1 q_m}{\varepsilon_0 kT}\right) - k' N_2^0 \exp\left(\frac{q_m}{\gamma_2}\right) \quad (35)$$

where N_1 is the number of nonchemisorbed associated groups with dipole moment μ_1 each, N_2 is that of chemisorbed ones at the p.z.c., k' is an adjustable constant, and γ_2 is a parameter, that, according to DF, characterizes the dependence of the adsorption energy on the electrode charge. The first term in Eq. (35) is the potential contribution due to nonchemisorbed orientable associated groups and the second term is that due to chemisorbed ones. The contribution by the chemisorbed groups is always negative for all values of the electrode charge; i.e., the chemisorbed groups do not orient with the variation of electrode charge. The first term, the potential contribution due to nonchemisorbed groups, is obtained by using BDM theory.[29] The molecular structure and nature (or polarity) of the associated groups remain undefined, so it is not clear how an associated group orients under an electric field. Though an associated group is assumed to have a net dipole moment, the lateral interaction energies between the dipole groups are neglected. The orientation of the associated groups leads to the change of sign of the dipole potential with variation of electrode charge. The minimum orientation is given at a charge at which this potential contribution (first term) is zero. The first term is zero only at $q_m = 0$. Thus, according at the DF model,[57] though the total g^s(dipole) may be equal to zero at a charge negative to the p.z.c. [Eq. (35)], it does not signify the minimum orientation at that charge; the minimum orientation is obtained at a charge at which only the first term in Eq. (35) is zero, i.e., at $q_m = 0$ (p.z.c.), and this contradicts the recent experimental results.[8,12,19] Thus, Trasatti's[56] comment that, since according to the DF model the total g^s(dipole) is zero at $q_m = -1.8\ \mathrm{C\ cm^{-2}}$, the minimum net orientation of water occurs at that charge, is not valid.

No reasoning is put forward by Damaskin and Frumkin[57] to support chemisorption of water molecules on the electrode. The total number of water molecules in the monolayer on the surface is not properly accounted for because the concentration of chemisorbed water molecules is allowed to change without requiring any change in the concentration of the water molecules in the clusters.

3. Solvent Excess Entropy: Modelistic Approach

To investigate the effect of the entropy of the first water layer on the total change of entropy during pyridine adsorption, Conway and

Gordon[58] calculated the configurational and librational entropy of the water molecules on the basis of the two-state water model of BDM. The configurational entropy of mixing S_c of the two orientational states was given by

$$S_c = -R(\theta_{\uparrow} \ln \theta_{\uparrow} + \theta_{\downarrow} \ln \theta_{\downarrow}) \tag{36}$$

where $\theta_{\uparrow} = N_{\uparrow}/N_m$ and $\theta_{\downarrow} = N_{\downarrow}/N_T$ are the fractions of the surface covered by the up and down molecules, respectively, N_m is the number of water molecules (monomers) on the surface (all water molecules present on the surface were assumed to be monomers), and R is the universal gas constant. $\theta_{\uparrow}$ and $\theta_{\downarrow}$ were calculated by using Eq. (12) given by the BDM theory. The maximum configurational entropy was found to occur at the p.z.c., which is now disproved by the recent experimental findings. Conway and Gordon[58] brought down the concept of the libration of the solvent dipoles under the influence of the double-layer field, and the entropy associated with the librational energy states in the field was represented by[59]

$$\frac{S_L}{R} = \ln \frac{8\pi^2(8\pi^3 I_1 I_2 I_3 k^3 T^3)^{1/2}}{\sigma h^3} - \ln \frac{U_e}{kT} + \ln \sinh\left(\frac{U_e}{kT}\right) - \frac{U_e}{kT} \coth\left(\frac{U_e}{kT}\right) + \frac{5}{2} \tag{37}$$

where $U_e = -4\pi\mu' q_m/\varepsilon_0$ = field dipole interaction energy, Is are the principal moments of inertia of a water dipole, and σ = symmetry factor. The influence of lateral interaction between the dipoles on libratory motion was not taken into account. Allowance for the preferential adsorption of water molecules with their negative ends toward the surface was not made. Consequently, the librational entropy was found to be a maximum at the p.z.c., in disagreement with experimental findings[8,12,19]

Different components of the solvent excess entropy have recently been evaluated by Bockris and Habib[60] on the basis of a three-state water model[31] proposed by these authors. The configurational entropy of dimers and up and down monomers may be obtained as follows.

Let there be N_d dimers and N_m monomers to occupy $2N_d + N_m = N_s$ sites on the electrode surface (i.e., each dimer occupies two neighboring sites, and each monomer, only one). For $N_m > N_d$, the number of distinct configurations for the dimer molecules is[61]

$$\Omega(N_d, N_m) = \left(\frac{2L}{\sigma_d}\right)^{N_d} \frac{(\frac{1}{2}N_m + N_d)!}{(\frac{1}{2}N_m)!N_d!} \left(1 - \frac{L'(L - L'')}{L} \frac{\frac{1}{2}N_d(N_d - 1)}{2N_d + N_m}\right) \tag{38}$$

where L is the number of nearest-neighbor sites of a surface site, $L' = 2(L - 1)$ are the free nearest-neighbor sites of an adsorbed dimer, $L'' = L - 1$ are the free nearest-neighbor sites of each L', and σ_d is a symmetry factor for a dimer. The number of different configurations of the adsorbed up and down dipoles is represented by

$$\Omega(N_\uparrow, N_\downarrow) = N_m!/(N_\uparrow!N_\downarrow!) \tag{39}$$

Thus, the total number of distinct configurations for the system of adsorbed dimers and up and down monomers is $\Omega_T = \Omega(N_d, N_m)\,\Omega(N_\uparrow, N_\downarrow)$. The entropy associated with these alternative configurations is given by

$$\begin{aligned} S_{\text{conf}} &= k \ln \Omega_T = k \ln \Omega(N_d, N_m)\Omega(N_\uparrow, N_\downarrow) \\ &= S(N_d, N_m) + S(N_\uparrow, N_\downarrow) \end{aligned} \tag{40}$$

where

$$S(N_d, N_m) = k \ln \Omega(N_d, N_m) \tag{41}$$

and

$$S(N_\uparrow, N_\downarrow) = k \ln \Omega(N_\uparrow, N_\downarrow) \tag{42}$$

From Eqs. (38) and (41) and Stirling's approximation, we find

$$\begin{aligned} S(N_d, N_m)/k &= (\tfrac{1}{2}N_m + N_d) \ln(\tfrac{1}{2}N_m + N_d) \\ &\quad - \tfrac{1}{2}N_m \ln(\tfrac{1}{2}N_m) - N_d \ln N_d + N_d \\ &\quad \times \frac{\ln 2L}{\sigma_d} - \frac{L'(L - L'')}{2L} \quad \frac{N_d^2}{2N_d + N_m} \end{aligned} \tag{43}$$

and from Eqs. (39) and (42), we obtain an expression for $S(N_\uparrow, N_\downarrow)$ as represented by Eq. (36).

Since $S(N_\uparrow, N_\downarrow)$ signifies the orientation of the up and down dipoles, it is termed S_{orient}. From Eq. (27), the number of up and down molecules may be expressed in terms of degree of orientation Z, i.e.,

$$N_{\uparrow} = \tfrac{1}{2}N_m(1 + Z) \tag{44}$$

and

$$N_{\downarrow} = \tfrac{1}{2}N_m(1 - Z) \tag{45}$$

where $\theta_{\uparrow}$ and $\theta_{\downarrow}$ have been replaced by $N_{\uparrow}/N_m$ and $N_{\downarrow}/N_m$, respectively.

Substituting these values of $N_{\uparrow}$ and $N_{\downarrow}$ in Eq. (36), we get

$$S_{\text{orient}} = \tfrac{1}{2}kN_m\{2\ln 2 - [(1 + Z)\ln(1 + Z) + (1 - Z)\ln(1 - Z)]\} \tag{46}$$

The degree of orientation, Z, may be expressed as a function of electrode charge as given in Eq. (30). From Eqs. (46) and (30), the orientational entropy has been calculated as shown in Fig. 6. The maximum of the orientational entropy is shown to occur at a potential negative to the p.z.c. ($q_m = -3.5\,\mu\text{C cm}^{-2}$).

The position of the orientational entropy of water molecules may be obtained by differentiating Eq. (46) with respect to q_m and equating $dS_{\text{orient}}/dq_m = 0$, i.e.,

$$\tfrac{1}{2}kN_m\left(\ln\frac{1 - Z}{1 + Z}\right)\frac{dZ}{dq_m} = 0 \tag{47}$$

Since $\frac{1}{2}kN_m \neq 0$, either $\ln[(1 - Z)/(1 + Z)] = 0$ or $dZ/dq_m = 0$. From Eq. (30),

$$\frac{dZ}{dq_m} = \frac{2a_1e^y}{(e^y + 1)^2 + 2a_2e^y} \tag{48}$$

where

$$a_1 = \frac{8\pi\bar{\mu}}{\varepsilon_0 kT}, \quad a_2 = \frac{2Uc'}{kT}, \quad \text{and} \quad y = a_1q_m - a_2Z + b \tag{49}$$

From Eqs. (48) and (49), it is evident that $dZ/dq_m \neq 0$. Hence Eq. (47) is satisfied only when $\ln[(1 - Z)/(1 + Z)] = 0$, i.e., $Z = 0$.

Hence, from Eq. (30), the position of $S_{\max}$ is given by

$$(q_m)_{S=S_{\max}} = -b\varepsilon_0 kT/8\pi\bar{\mu} \tag{50}$$

where the quantity b is given by Eq. (31) and the procedure to obtain

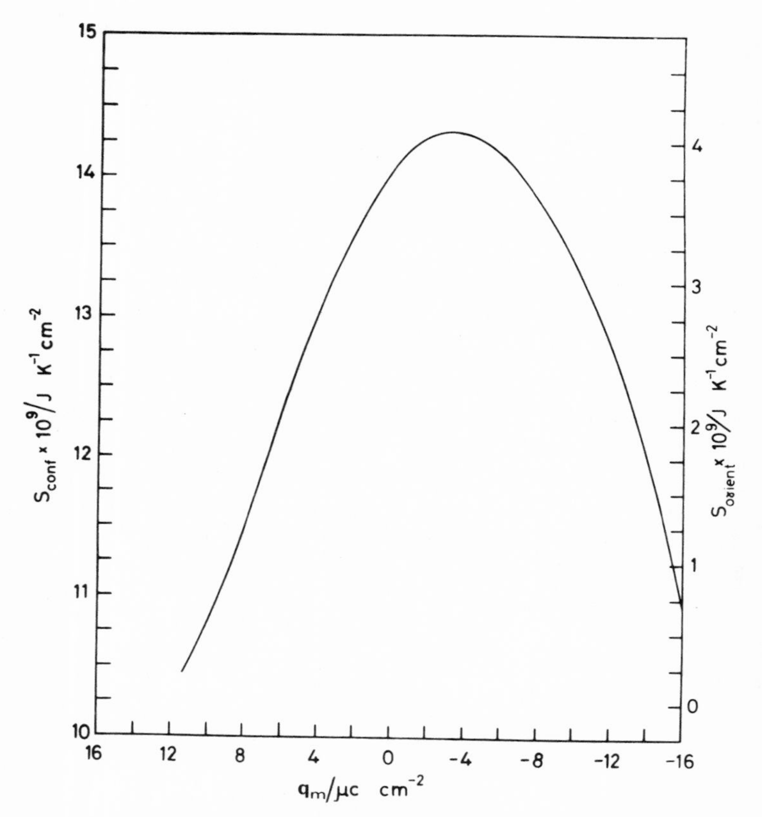

Figure 6. Configurational and orientational entropy of surface water as a function of charge density on metal.[60]

its value is that described in the last section. With $\Delta\Delta G^c = -2.72\ \mathrm{kJ\ mol^{-1}}$, the position of the maximum orientational entropy is found to occur at $q_m = -3.5\ \mu\mathrm{C\ cm^{-2}}$ from Eq. (50) at 25°C. This maximum is independent of U, the lateral interaction energy; N_m, the number of monomers; and c', the number of nearest neighbors [cf. Eq. (50)]. The greater the value of $-\Delta\Delta G^c$, the greater is the attraction of the oxygen end toward the surface, and consequently the more negative should be the position of the entropy maximum. With $\Delta\Delta G^c = -6.27\ \mathrm{kJ\ mol^{-1}}$ (a value used by Watts-Tobin), the $S_{\mathrm{orient,max}}$ is found to occur from Eq. (50) at $q_m = -7.5\ \mu\mathrm{C\ cm^{-2}}$.

Allowance for the preferential adsorption of water dipole with its negative end toward the surface at the p.z.c. makes the entropy maximum occur at a potential negative to the p.z.c. and thus be consistent with experimental findings.[8,12,19]

Now, $N_s = 1/(4r_w^2) = 1.31 \times 10^{15}\,\text{cm}^{-2}$, $N_m = N_s \times \theta_d = 1.31 \times 0.32 \times 10^{15} = 0.42 \times 10^{15}\,\text{cm}^{-2}$, and $N_d = \frac{1}{2} \times 1.31 \times 10^{15}$ $\theta_d = 0.655 \times 0.68 \times 10^{15} = 0.44 \times 10^{15}\,\text{cm}^{-2}$. With these values of N_m and N_d, and with $L = 4$, from Eq. (43), $S(N_d, N_m) = 10.29 \times 10^{-9}\,\text{J}\,\text{K}^{-1}\,\text{cm}^{-2}$. The total configurational entropy, i.e., $S(N_\uparrow, N_\downarrow) + S(N_d, N_m)$, is given in Fig. 6.

The configurational entropy, which is discussed above, is only a part of the total entropy of the interfacial water molecules. In addition to the configurational entropy, other components of the entropy of the interfacial waters discussed below, minus the entropy of water in bulk water, give the solvent excess entropy, which should correspond to the experimental solvent excess entropy published by Reeves[12] and Hills and Hsieh.[19]

The monomers execute libratory motion about the three mutual perpendicular axes with the origin at the center of mass. All the three librational modes are directly or indirectly under the influence of certain fields.

Libration about the two axes parallel to the surface is under the influence of the double-layer field $X = 4\pi q_m/\varepsilon_0$ and the lateral interaction. Therefore, the average energy E_i that influences these two librational modes is given by

$$E_i = \theta_\uparrow E_\uparrow + \theta_\downarrow E_\downarrow \tag{51}$$

where $E_\uparrow = \bar{\mu}X + Uc'(\theta_\uparrow - \theta_\downarrow)$ is the energy of interaction of an up dipole with the double-layer field and the nearest neighbors. With $E_\uparrow = -E_\downarrow$ (where $E_\downarrow$ is the corresponding energy of a down dipole),

$$E_i = (\bar{\mu}X + Uc'Z)Z \tag{52}$$

The librational entropy about the two mutually perpendicular axes but parallel to the surface was then obtained from

$$S_{\text{lib}(y,z)} = \frac{2}{3}kN_m\left[\ln\frac{8\pi^2(8\pi^3 I_x I_y I_z k^3 T^3)^{1/2}}{\sigma h^3} + \ln\left(\sinh\frac{E_i}{kT}\right) - \ln\frac{E_i}{kT} - \frac{E_i}{kT}\coth\frac{E_i}{kT} + \frac{5}{2}\right] \tag{53}$$

Libration about the axis perpendicular to the surface was considered not to be influenced directly by the double-layer field. The later interactions with the nearest neighbors were considered to influence this libration. Therefore, for this case, $E_{\uparrow} = Ec'Z$, and the average energy that influences this librative mode is $E_j = Ec'Z^2$. The librational entropy $S_{\text{lib},x}$ about the axis perpendicular to the surface was obtained from Eq. (53) by replacing E_i by E_j and $\frac{2}{3}$ by $\frac{1}{3}$. The total librational entropy $S_{\text{lib},m} = S_{\text{lib},x} + S_{\text{lib}(y,z)}$ of monomers was thus calculated as shown in Fig. 7.

The only rotations that the absorbed dimers were assumed to execute are those about the axis normal to the surface, and the partition function for this free rotation is

$$f_{\text{rot},d} = 8\pi^2 I_d kT/\sigma_d h^2 \tag{54}$$

where I_d is the moment of inertia of a dimer about the axis

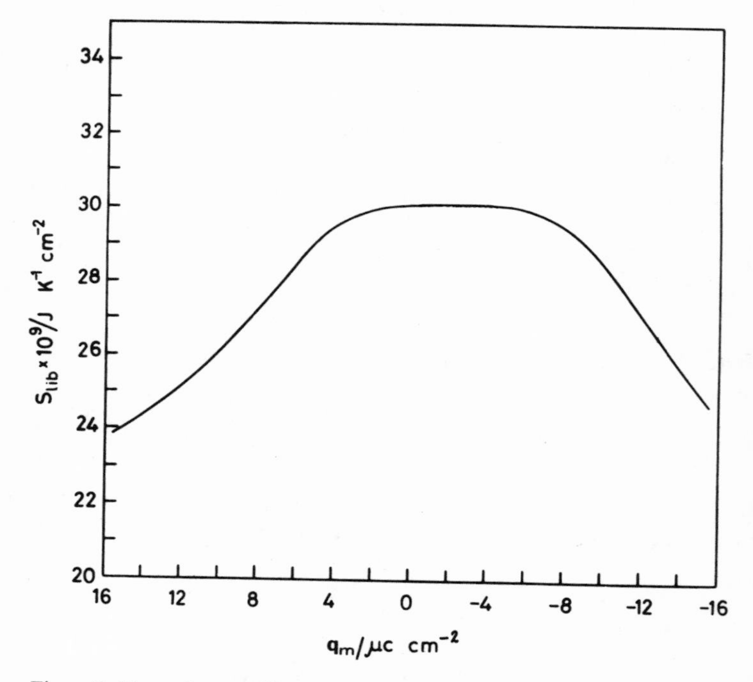

Figure 7. Dependence of libratory entropy of water molecules (monomers) on electrode charge.[60]

perpendicular to the surface and σ_d is a symmetry factor for a dimer. The rotational entropy of the dimers is then given by

$$S_{\mathrm{rot},d} = kN_d[\ln(8\pi^2 I_d kT/\sigma_d h^2) + 1] \tag{55}$$

With $I_d \approx [m_1 m_2/(m_1 + m_2)]r_e^2 \approx 28.5 \times 10^{-40}\,\mathrm{g\,cm^2}$ (where $m_1 = m_2$ = mass of a water molecule $\approx 3 \times 10^{-23}$ g and $r_e = r_w = 1.38 \times 10^{-8}$ cm) and $\sigma_d = 2$, $S_{\mathrm{rot},d}$ was obtained to be $33.2 \times 10^{-9}\,\mathrm{J\,K^{-1}\,cm^{-2}}$.

It is assumed that the internal vibrations of water molecules are little affected by the double-layer field except indirectly by hydrogen-bond bending or breaking effects.[58] The internal vibrational entropy will be small at room temperatures, since the bend and stretch vibrational quanta are between 8 and 16 kT.[58] However, near the electrode surface, water dipoles will execute vibrations normal to the surface, as at ions.[59] The frequency of this vibration, and hence the vibrational entropy as a function of electrode charge, is difficult to evaluate directly. However, indirectly and approximately, it is obtained as follows.

The force constant k for the vibration of a water molecule in the first solvation shell of an ion is obtained from

$$k = 14ze\mu'/r_e^4 \tag{56}$$

where μ' = dipole moment of water, z = valency of ion, and $r_e = r_i + r_w$. The vibrational frequency ν is then calculated from

$$\nu = \frac{1}{2\pi}\left(\frac{k}{m^*}\right)^{1/2} \tag{57}$$

where the reduced mass $m^* = m_i m_w/(m_i + m_w)$, where m_i and m_w are the masses of an ion and a water dipole, respectively. From Eqs. (56) and (57), the frequency ν is calculated for various ions.

The partition function for vibration being given by

$$f_{\mathrm{vib}} = (2\sinh h\nu/2kT)^{-1} \tag{58}$$

the vibrational entropy is then obtained from

$$S_{\mathrm{vib}} = kN_m \ln(2\sinh h\nu/2kT)^{-1} + kN_m(h\nu/2kT)\coth(h\nu/2kT) \tag{59}$$

The dielectric displacement due to an ion at the center of an adjacent water molecule is given by $D_i = e/(r_i + r_w)^2$, and that near an electrode is represented by $D_{\mathrm{DL}} = 4\pi q_m$. D_i and S_{vib} are calculated for various

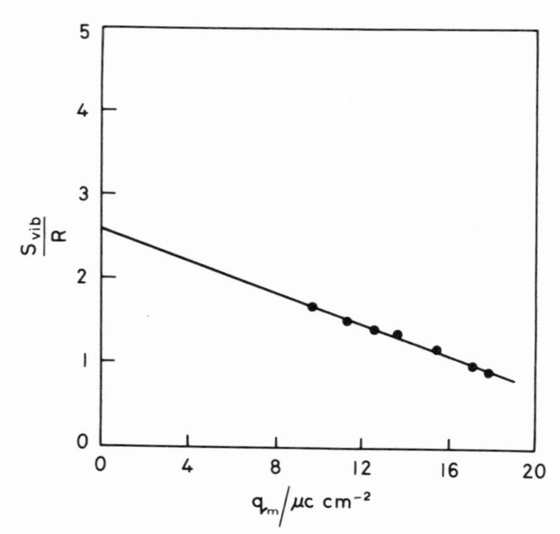

Figure 8. Approximate variation of the vibrational entropy of surface water with charge density on the electrode (k = Boltzmann constant and N_0 = Avogadro's number).[60]

ions. Equating D_i and D_{DL}, corresponding q_m is obtained. Thus, S_{vib} is estimated indirectly as a function of q_m and plotted as in Fig. 8.

The force constant k_d for vibration of dimer with respect to the metal surface was obtained from the double differentiation of the interaction energy of the dimer with the metal with respect to distance, i.e., from Eq. (23a):

$$k_d = \frac{d^2U'}{dr^2} = -\frac{12C}{r^5} + \frac{90A}{r^{11}} \tag{60}$$

From Eq. (60), $k_d = 9 \times 10^3$ dyn cm^{-1}, where A is obtained from Eq. (23b) and C from Eq. (20), with[31] $\alpha_d = 2.9 \times 10^{-24}$ c.c. and $\chi_d = -5.2 \times 10^{-29}$ c.c. The vibrational frequency is then calculated from Eq. (57) with this k_d and $m^* = (m_d m_{Hg})/(m_d + m_{Hg}) = 5.05 \times 10^{-23}$ g, where m_d = mass of a dimer and m_{Hg} = mass of mercury atom. Thus, $\nu = 2.1 \times 10^{12}$ sec^{-1}. Substituting this value of ν in Eq. (59) and replacing N_m by N_d, one obtains $S_{vib,dim} = 11.21 \times 10^{-9}$ J K^{-1} cm^{-2}.

The contribution to the solvent excess entropy due to the breaking of the water structures from the bulk of the solution to the surface of the electrode is estimated indirectly. The dielectric constant is assumed as an indirect measure of the extent of breaking of the water structure. From the experimentally known dielectric constant[61] and the fraction of broken hydrogen bonds[53] in liquid water as a function of temperature, the fraction of broken hydrogen bonds as a function of dielectric constant is estimated. The value of the dielectric constant in different water layers near the electrode surface is taken from Bockris and Habib.[62] Hence, from the estimated broken hydrogen bonds as a function of dielectric constant, the fraction of broken hydrogen bonds in different water layers near the electrode is estimated. In the structure-breaking region, the average fraction of the hydrogen bonds broken is 0.105 mol^{-1}. The change in entropy for the breakage of 1 mol hydrogen bonds is 35.53 $\mathrm{J\,K^{-1}\,mol^{-1}}$ of 0–H.[63] Therefore, in the structure-breaking (SB) region, the entropy change due to breakage of hydrogen bonds $= 0.105 \times 35.53 = 3.73\,\mathrm{J\,K^{-1}\,mol^{-1}}$ of hydrogen bonds. Since 1 mol hydrogen bonds, involves 2 mol water $\Delta S_{\mathrm{SB}} = 3.73 \times 2 = 7.46\,\mathrm{J\,K^{-1}\,mol^{-1}}$ water in the SB region ($7.46\,\mathrm{J\,K^{-1}\,mol^{-1}} = 10.3 \times 10^{-9}\,\mathrm{J\,K^{-1}\,cm^{-2}}$, where $N_w = 0.855 \times 10^{15}\,\mathrm{cm^{-2}}$†).

The entropy of liquid water in bulk water at 25°C is[35,36] $70.0\,\mathrm{J\,K^{-1}\,mol^{-1}} = 99.48 \times 10^{-9}\,\mathrm{J\,K^{-1}\,cm^{-2}}$†.

From modelistic considerations (dimer, monomer models[31]), the solvent excess entropy is therefore given by

$$\begin{aligned} S^* &= (S_{\mathrm{conf}} + S_{\mathrm{Lib},m} + S_{\mathrm{rot,d}} + S_{\mathrm{vib},m} + S_{\mathrm{vib},d} + S_{\mathrm{SB}}) - S_{w-w} \\ &= S_0 + S(q_m) \end{aligned} \tag{61a}$$

where $S_0 = (S_{\mathrm{rot},d} + S_{\mathrm{vib},d} + S_{\mathrm{SB}} - S_{w-w})$ and is assumed to be independent of electrode charge and $S(q_m) = (S_{\mathrm{conf}} + S_{\mathrm{lib},m} + S_{\mathrm{vib},m})$ is the potential-dependent part of the solvent excess entropy. S_0 is shown in Table 1. From Eqs (43), (46), (53), and (59), $S(q_m)$ is calculated, and with S_0 from Table 1, S^* is shown in Fig. 9. Agreement with experiment is good. Allowance has been made for the specific absorption of the ions by writing $N_s = N_T\theta_i$, where θ_i is the ionic coverage, N_T is the total

†The molar volume of water $V_w = 18.054\,\mathrm{cm^3\,mol^{-1}}$ may be represented by $V_w = \frac{4}{3}\pi(r_w + a)^3 N_0$, where r_w is the crystallographic radius of a water molecule and $(r_w + a)$ is the effective radius signifying some dead space surrounding each water molecule. Thus, $a = (3V_w/4\pi N_0)^{1/3} - r_w = 0.55 \times 10^{-8}\,\mathrm{cm}$, where $r_w = 1.38 \times 10^{-8}\,\mathrm{cm}$. Therefore the number of water molecules per square centimeter in bulk solution is $N_w = 1/[\pi(r_w + a)^2] = 0.855 \times 10^{15}$.

Table 1
Different Components of $S_0 \times 10^9$ J K^{-1} cm^{-2}

$S_{rot,d}$	$S_{vib,d}$	S_{SB}	S_{w-w}	S_0
33.2	11.21	10.3	99.48	−44.77

number of sites, and N_s is the number of sites available to water molecules. θ_i is obtained from $\theta_i = (q_{CA}/q_{max})$, where q_{CA} is the specifically adsorbed charge and q_{max} is the maximum possible ionic charge to be specifically adsorbed, given by $ze/(4r_i^2)$, where e is the electronic charge and z is the valency. Values of q_{CA} have been taken from Ref. 66.

The position of the maximum in the solvent excess entropy, calculated from modelistic considerations, occurs at a potential

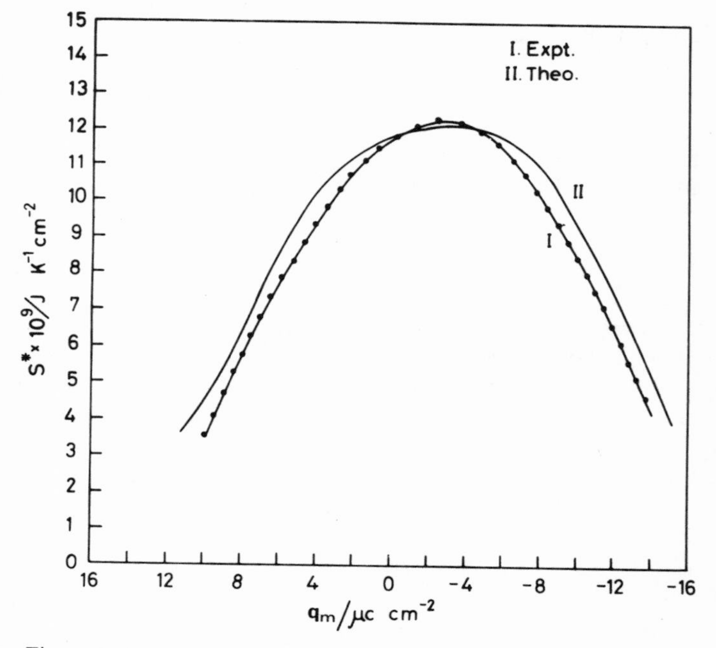

Figure 9. Solvent excess entropy as a function of electrode charge.[60] Experimental curve for 0.1 M NaCl solution.[12]

negative to the potential of zero charge. This maximum is caused by the orientational and libratory contribution, which signifies that the maximum randomness of interfacial water occurs at a negative electrode charge (-4 to $\sim -6\,\mu\mathrm{C\,cm^{-2}}$). An important corollary is that the anodic capacitance hump cannot be attributed to dielectric saturation. These investigations provide evidence in favor of an alternative model of capacitance hump, e.g., the anion repulsion model.[29,62,67–69]

The translational contribution to the entropy has been neglected on the basis of the explanation of the experimental results of Kemball[37] and of Bockris and Law.[34] Since the field–dipole interaction energy ($4\pi\bar{\mu}q_m/\varepsilon_0 \approx 6.46\,\mathrm{kJ\,mol^{-1}}$ at $q_m \approx 16\,\mu\mathrm{C\,cm^{-2}}$) does not exceed the energy of dimerization ($\approx 8.13\,\mathrm{kJ\,mol^{-1}}$), one does not expect potential-dependent dissociation of dimers. However, at higher field strength, the free water dipoles with net dipole moment may tend to displace dimers that have zero dipole moment. The desorption of dimers at higher potentials and consequent adsorption of water dipoles may lower the solvent excess entropy at the potentials far from the p.z.c. to a greater extent than indicated by the present calculations, but this does not affect the position of entropy maximum.

Although interaction effects have been included in the orientation distribution function, the Bragg–William approximation is followed in assuming that the interactions do not appreciably influence the configurational entropy. This is not always a satisfactory assumption, but corrections for nonrandom mixing,[89] when interaction effects are significant, are in fact relatively small and, to a first approximation, are of the order[58]

$$-R\frac{(\theta_{\uparrow}\theta_{\downarrow})^2}{4c'}\left(\frac{Uc'}{kT}\right)^2 \tag{61b}$$

That is, the nonrandomness introduced by interaction contributes, as expected, a negative term in the entropy of mixing. For $\theta_{\uparrow} = \theta_{\downarrow} = 0.16$ (since $\theta_m = 0.32$ according to the BH model), $c' = 3.5$, $Uc'/kT \approx 5$, say, and substituting kN_m for R in Eq. (61b), the maximum value of this entropy correction amounts to only $-0.007 \times 10^{-9}\,\mathrm{J\,K^{-1}\,cm^{-2}}$.

The fitting of experimental data to theoretical curves is a process that requires caution. Some consideration of the errors in the experimental data is essential, but errors in double-layer data are difficult to quantify. However, the fact that the experimental solvent excess entropy has its maximum at a negative electrode charge is now beyond doubt, and the dimer–monomer model is consistent with this essential feature of the solvent excess entropy.

4. The Four-State Water Model

Recently Parsons[70] put forward a four-state model for interfacial water molecules in an attempt to improve the Damaskin–Frumkin model.[57] Parsons assumed that the water molecules remain on the electrode surface as small clusters and also as free molecules. The free molecules (monomers) may adopt one of two orientations with equal and opposite perpendicular dipole components μ'. The clusters may also have one of two orientations, and the components of their dipole moment, perpendicular to the surface, are also equal and opposite in two orientational states. When the positive end of the cluster is toward the solution, the energy of a solvent molecule in a cluster in the presence of a field due to electrode charge is expressed as*

$$U_{c,+} = -\frac{4\pi\rho\mu' q_m}{\varepsilon_0} \tag{62}$$

where

$$\rho = \mu_c/\mu' \tag{63}$$

and μ_c is the dipole component of a cluster expressed per molecule. If the negative end of the dipole is toward the solution, then

$$U_{c,-} = +\frac{4\pi\rho\mu' q_m}{\varepsilon_0} \tag{64}$$

For free molecules with the positive end of the dipole toward the solution, the total energy is

$$U_{+} = -\frac{4\pi\mu' q_m}{\varepsilon_0} + U_{b,+} \tag{65}$$

while that of molecules with the negative end of the dipole toward the solution is

$$U_{-} = \frac{4\pi\mu' q_m}{\varepsilon_0} + U_{b,-} \tag{66}$$

*In Eq. (62) and the following equations, 4π is included to facilitate the comparison with other expressions given earlier. Unit conversion is done in a proper way so that the actual numerical values of the quantities involved remain unaltered.

where $U_{b,+}$ and $U_{b,-}$ are the field-independent part of the energy and signify the energies of bonding with the metal and with bulk solvent or energy due to breaking of bonds in a cluster.

The number of solvent molecules in clusters oriented with the positive end of the dipole toward the solution is expressed as

$$N_{c,+} = N_T \exp(\rho s)/D \tag{67}$$

where N_T is the total number of solvent molecules in the monolayer and

$$s = \frac{4\pi\mu' q_m}{\varepsilon_0 kT} \tag{68}$$

$$D = \exp(\rho s) + \exp(-\rho s) + A_+ \exp(s) + A_- \exp(-s) \tag{69}$$

$$A_+ = \exp(-U_{b,+}/kT) \tag{70}$$

$$A_- = \exp(-U_{b,-}/kT) \tag{71}$$

Similarly,

$$N_{c,-} = N_T \exp(-\rho s)/D \tag{72}$$

and for the free solvents,

$$N_+ = N_T A_+ \exp(s)/D \tag{73}$$

$$N_- = N_T A_- \exp(-s)/D \tag{74}$$

The potential change across the interphase due to the orientation of the clusters and free molecules is then given by

$$g^s(\text{dipole}) = -\frac{4\pi N_{c,+}\rho\mu'}{\varepsilon_0} + \frac{4\pi N_{c,-}\rho\mu'}{\varepsilon_0} - \frac{4\pi N_+\mu'}{\varepsilon_0} + \frac{4\pi N_-\mu'}{\varepsilon_0} \tag{75}$$

If a reduced potential drop is defined by

$$g = \frac{g^s(\text{dipole})\varepsilon_0}{4\pi N_T\mu'} \tag{76}$$

then from Eqs (67) and (72)–(76), we obtain

$$g = \frac{-\rho\exp(\rho s) + \rho\exp(-\rho s) - A_+\exp(s) + A_-\exp(-s)}{D} \tag{77}$$

In the absence of specific ionic adsorption, the potential drop across the inner layer may be represented by

$$\Delta\phi = \frac{q_m}{K_0} + g^s(\text{dipole}) \tag{78}$$

where the first term is the potential due to free charges on the metal and K_0 is the integral capacity of the inner layer at a fixed orientation of solvent dipoles. Differential capacity of the inner region is then given by

$$\frac{1}{C^i} = \frac{d\Delta\phi}{dq_m} = \frac{1}{K_0} - \frac{16\pi^2\mu^2 N_T}{\varepsilon_0^2 kT}$$

$$\times \frac{\rho^2 \exp(\rho s) + \rho^2 \exp(-\rho s) + A_+ \exp(s) + A_- \exp(-s)}{D} - g^2 \tag{79}$$

The inner-layer capacitance as a function of s or q_m is shown in Fig. 10.

The correct shape of the capacitance-charge curve is thus predicted with the help of the parameters ρ, A_+, and A_-, *which are not, however, determined independent of experiment*. The temperature

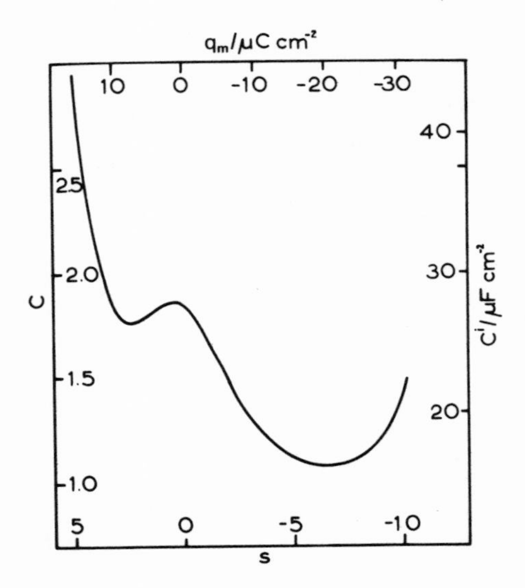

Figure 10. Computed inner-layer capacity charge curve obtained[70] from Eq. (79) with $A_+ = 10^{-2}$, $A_- = 10^{-4}$, $\rho = 0.28$, $(\varepsilon_0^2 kT/16\pi^2\mu N_T K_0) = 0.18$, and $C = C^i/K_0$.

coefficient of the capacitance curve is also predicted quite well. However, though Parsons' model is able to fit the experimental data, it contradicts the diagnostic evidence provided by the position of the maximum entropy; i.e., while Parsons' model predicts minimum dipole orientation at a potential positive to the p.z.c., *the actual position of minimum orientation is at a potential negative to the p.z.c.* Prediction of minimum orientation at a positive charge by Parsons' model may be verified from Eq. (77). With the values of the parameters used in Fig. 10, it is found that $g = 0$ at a positive electrode charge.

IV. INFLUENCE OF THE METAL ON THE SOLVENT PROPERTIES

1. Introduction

The role of the metals in the metal–solution interface has been discussed recently by Reeves,[12] Frumkin and co-workers[57,71,72] and most extensively by Trasatti.[3,10,51,56,74,75] Here, the essential features of the metallic influence on solvent properties will be described briefly, with inclusion of some recent advances.

2. Dependence of Some Electrochemical Properties on the Nature of the Substrate Metal

The properties of a solvent in the interfacial region, when there is no extra charge on the electrode, is different from those in the bulk. This was pointed out by Frumkin[76] long ago from the results of capacity measurements in different solvents. Capacitance measurements for water films adsorbed on solids[77] suggest that the water particles must be ordered. Density measurements of water at interfaces[78] also indicate that the liquid is strongly structured. This difference in the properties of interfacial solvent molecules from those of the bulk must have the effect of the nature of the metals. The surface tension of a metal is derived from the unbalanced forces parallel to the surface plane as a result of lattice interruption at the surface, while the unbalanced perpendicular forces tend to orient particles and therefore to structure the surface region of the phase.

The sudden interruption of the positive field generated by the ion cores results in a trend for the itinerant electrons to spill over the edge of the lattice.[79] Consequently, there is an excess of negative charge outward and a deficiency of negative charges inward. This change in electron density is localized within a region of 1 atomic diameter,[80] and the resulting dipoles give rise to the surface potential χ^m of the metal in contact with a vacuum.

When this metal phase is brought into contact with a solution phase, a certain change $\delta\chi^m$ occurs in its surface potential. Thus, the surface potential of a metal or the electron overlap potential of the metal[81] in contact with a solution phase is represented by

$$g^m(\text{dipole}) = \chi^m + \delta\chi^m \tag{80}$$

For the liquid phase, the unbalanced forces at the surface of the liquid tend to structure the dipole layer on the surface. Consequently, a dipolar double layer is formed[82] at the surface of the liquid phase, which gives rise to a surface potential χ^s of the solution in contact with air. When this solution phase is brought in contact with a metal, then the unbalanced forces on the metallic surface cause some changes $\delta\chi^s$ in the surface potential of solution in contact with air, and thus we obtain

$$g^s(\text{dipole}) = \chi^s + \delta\chi^s \tag{81}$$

where g^s(dipole) is the surface potential of the solvent in contact with a metal.

The total potential drop across a metal–solution interface in the absence of free charges on the phases (i.e., at the p.z.c.) is given by

$$\Delta_s^m\phi = g_s^m(\text{dipole}) = g^m(\text{dipole}) - g^s(\text{dipole}) \tag{82}$$

Since g_s^m(dipole) is not an experimentally measurable quantity, hitherto, in the absence of an acceptable method of measurement or calculation of the absolute potential difference in an interface, it was not possible to obtain a value of g_s^m(dipole).

In 1968, Bockris and Argade[83] proposed a theoretical method by which $\Delta_s^m\phi$ could be approximately estimated, with reservations concerning the calculability of the chemical potential of the electron in the metal. The principle of this approach has been supported by Trasatti,[84] who, however, introduced a modification of the numerical value. Trasatti[84] suggested two changes: one is the revised value of the

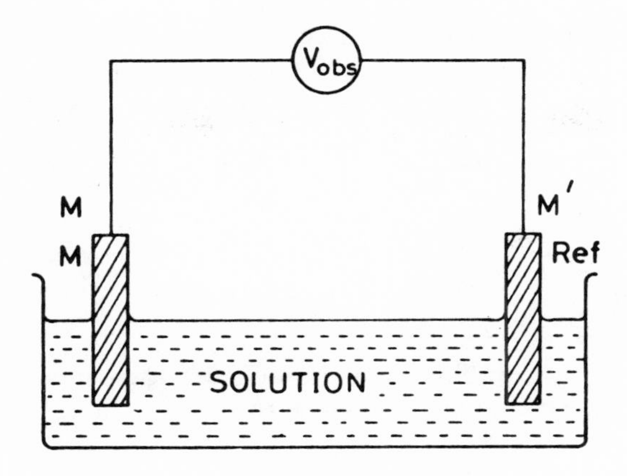

Figure 11. A basic electrochemical cell.

individual heat of hydration of the proton; the other is concerned with the chemical potential of electrons in the metal. Assuming $\Delta_s^m\phi$ is known within certain limits, g^m(dipole) has recently been estimated by Bockris and Habib[81] from the value of $\Delta_s^m\phi$ at the p.z.c. The potential of a basic electrochemical cell (Fig. 11) may be represented by

$$E = \Delta_s^m\phi + \Delta_{\mathrm{ref}}^s\phi + \Delta_m^{\mathrm{ref}}\phi \tag{83}$$

where $\Delta\phi$s are the Galvani or inner potential differences across the various interfaces. The superscripts *m*, *s*, and *ref* represent the metal of the test electrode, the solution, and the metal of the reference electrode, respectively. Expressing $F\Delta_m^{\mathrm{ref}}\phi$ as the difference between the chemical potentials of the metal of the reference electrode and that of the test electrode,[83] one obtains

$$E = \Delta_s^m\phi - \Delta_s^{\mathrm{ref}}\phi + \mu_e^{\mathrm{ref}}/F - \mu_e^m/F \tag{84}$$

where μ_es are the chemical potentials of electrons and F is the Faraday constant.

From Eqs (82) and (84),

$$g^m(\text{dipole}) = E + g^s(\text{dipole}) + \mu_e^m/F + \Delta_s^{\mathrm{ref}}\phi - \mu_e^{\mathrm{ref}}/F \tag{85}$$

The quantity $(\Delta_s^{\mathrm{ref}}\phi - \mu_e^{\mathrm{ref}}/F)$ for the normal hydrogen electrode, calculated by Bockris and Argade[83] by using a thermodynamic cycle, was found to be 4.78 V. From the consideration of a revised value of the

Table 2
Calculated Values of the Electron Overlap Potentials at the Metal–Solution Interfaces at the p.z.c.

Metals	$-\mu_e^m$ / eV	E(p.z.c.) / (N.H.E.)/V	g^s(dipole) / V	g^m(dipole) / V	Φ / eV[a]	χ^m / V	$\delta\chi^m$ / V	$\Delta_s^m\phi$ / V
Cd	2.57	−0.72	0.03	1.05	4.10	1.53	−0.48	1.02
Zn	1.97	−0.63	0.04	1.75	4.12	2.15	−0.40	1.71

[a]The values of the work functions are noted from Ref. 85.

free energy of hydration of the proton and chemical potential of the metal, Trasatti[84] modified this numerical value and suggested it to be 4.31 V.

The solvent dipole contribution to the interfacial potential, g^s(dipole), has been calculated by Bockris and Habib,[31] who found g^s(dipole) = 0.03 V for the Cd–solution interface* and g^s(dipole) = 0.04 V for the Zn–solution interface at the p.z.c.†

Taking the chemical potentials of Cd and Zn from the work of Paasch *et al.*[85] and the potential of zero charge from the recent compilation by Trasatti,[3,73] the method of Bockris and Habib[81] is used to calculate the electron overlap potential g^m(dipole) of Cd and Zn in the interface and shown in Table 2. Equation (82) is then used to get $\Delta_s^m\phi$ (Table 2).

Similarly, g^m(dipole) may be obtained for any metal for which μ_e^m and E at the p.z.c. are known with significant accuracy. The uncertainty in the estimation discussed above lies mainly in its dependence on the validity of the calculation of μ_e^m, and to a lesser extent on the value of the g^s(dipole). The results for μ_e^m reported by Paasch *et al.*[85] are in agreement with those of other authors[86,87] within only about ± 0.3 V. The uncertainty in the estimation of g^s(dipole) is much less in comparison with that in μ_e^m. Consequently, the estimation of g^m(dipole) discussed above involves an uncertainty of up to ± 0.30 V.

*The choice of metals with which these calculations are carried out is made to optimize the accuracy of both p.z.c. and chemical potential values.

†The differences for g^s values calculated for the various metals assume that differences pertain largely to the differences in dispersion interaction of the up (oxygen end toward the metal) and down (hydrogen end toward the metal) water molecules and the relevant metal.

Thus from the work of Bockris and Argade[83] and that of Trasatti,[84] it is possible to obtain g^m(dipole), the surface potential of metal in the interface (Table 2). The surface potential of a metal is more usually defined in terms of its contact with a vacuum. From the theory of work function,[88]

$$F\chi^m = \Phi + \mu_e^m \tag{86}$$

where χ^m is the surface potential of the metal in contact with a vacuum and Φ is the work function.

Knowing Φ and μ_e^m, we can obtain χ^m, but this is different from the interfacial surface potential of metal g^m(dipole) as represented by Eq. (80). With χ^m obtained from Eq. (86), $\delta\chi^m$ has been calculated from Eq. (80) and included in Table 2.

Equations (80), (85), and (86) may be combined to give a relationship between the potential of zero charge and the work function of the metal,

$$E_{q_m=0} = \Phi/F + \delta\chi^m - g^s(\text{dipole}) + K \tag{87}$$

where Φ is the work function and K is a constant.

It was suggested by Ukshe and Levin[90] that the potential of zero charge is made up of two components, one due to the bulk properties of the metal and the other due to its surface properties. The following relationship between the potential of zero charge and the density and atomic weight of a metal has been suggested by these workers[90]:

$$(D'n/A_t)^{1/3} = 0.565 + 0.115 \sin 0.6(E_{q_m=0} + 0.28) \tag{88}$$

where D' is the density, A_t is the atomic weight, and n is the number of free electrons in the metal per nucleus. Equation (88) is purely empirical. It is claimed that it is useful in the range of $E_{q_m=0} = -1.1$ V to 0.6 V (N.H.E.). It has been pointed out by Argade and Gileadi[95] that the free-electron density in the metal should decrease with increase of temperature. The present experimental data are not accurate enough to test whether the variation of $E_{q_m=0}$ with temperature is in agreement with this equation.

Linear relationship between the potential of zero charge and the work function of the metal has been advocated by several authors,[92–95] though on the basis of the experimental data consulted, the slope and intercept of the p.z.c.–work function plot is somewhat different according to various authors, as discussed below.

Frumkin and Gorodetskaya[91] suggested as long ago as 1928 that the difference in the potential of zero charge for two metals is approximately equal to the difference in their work functions. Any deviation from this equality was attributed to different orientation of solvent molecules in the interface. As may be seen from Eq. (87), this may be true if $\delta\chi^m$s for different metals are assumed to be constant, but there is no theoretical reason to justify the validity of this assumption.

Navakovsky *et al.*[92] found from the experimental data at their disposal that the potential of zero charge of metals varies with their work functions according to the relationship

$$E_{q_m=0} = 1.02\Phi - 4.88 \tag{89}$$

A somewhat different relationship was proposed by Vasenin,[93] i.e.,

$$E_{q_m=0} = 0.86\Phi - 4.25 \tag{90}$$

Vasenin attributed the deviation of the slope of a linear $E_{q_m=0} - \Phi$ plot from unity to the specific interactions between the solvent dipoles and the surface.

Frumkin[94] suggested that his experimental data fitted best the relationship

$$E_{q_m=0} = \Phi - 4.72 \tag{91}$$

Another relationship was proposed by Argade and Gileadi,[95] according to whom

$$E_{q_m=0} = \Phi - 4.78 \tag{92}$$

Trasatti[3] has attempted to modify the $\Phi - E_{q_m=0}$ relationship through a critical examination of many experimental data. He suggests[3] that for *sp* metals, except Ga and Zn (Sb, Hg, Sn, Bi, In, Pb, Cd, Tl), the appropriate relationship is

$$E_{q_m=0} = \Phi - 4.69 \tag{93}$$

whereas for the transition metals (Ti, Ta, Nb, Co, Ni, Fe, Pd),

$$E_{q_m=0} = \Phi - 5.01 \tag{94}$$

If $\delta\chi^m$ is assumed to be independent of the metal, then from Eq. (87) it is apparent that any deviation from a linear $E_{q_m=0} - \Phi$ plot of unit slope is to be attributed to g^s(dipole). But from Table 2, $\delta\chi^m$ may vary from metal to metal, and therefore any deviation from linear plot is to be attributable to the variation of $\delta\chi^m - g^s$(dipole) for different metals.

For a group of metals for which either relationship (93) or (94) applies, the quantity $\delta\chi^m - g^s$(dipole) should be constant for all metals of the group. This implies that there is a possible compensation effect between $\delta\chi^m$ and orientation of solvent dipoles. Otherwise, both $\delta\chi^m$ and g^s(dipole) should be constant, though differently, for both *sp* and transition metals, which does not seem probable.

Frumkin *et al.*[72] pointed out, however, that Trasatti, in choosing the values for $E_{q_m=0}$ for metals not adsorbing hydrogen, favors those obtained from the position of the minimum on the differential capacity curve, not always taking proper account whether the conditions ensuring the reliability of these data are satisfied.[71] In the case of the platinum group metals, Trasatti[3] uses the p.z.c. values obtained by the method of Anderson *et al.*[96] in neutral solutions, assuming that unlike the p.z.c. values obtained from the absorption measurements, the former values refer to the metal surfaces free of adsorbed gases. However, the surface renewal without electricity supply from outside in the presence of a solvent can only ensure the vanishing of charges due to applied potential, but not that of the ions present, if any. Frumkin *et al.*[72] also pointed out that Trasatti had not taken into account the dependence of the p.z.c. on the crystal face orientation found for silver[97,98] and bismuth.[99] Thus, the p.z.c. values used by Trasatti[3] still encounter some uncertainty.

Trasatti also suggested that the potential of zero charge for all metals (*sp* and transition) may be related to their respective work functions by a single equation if the dependence of the degree of orientation of solvent dipole at the p.z.c. on the nature of the metal is taken into account. Thus,[3]

$$E_{q_m=0} = \Phi - 4.61 - 0.40\alpha \tag{95}$$

where α is the degree of orientation of water molecules at the p.z.c. and is given by

$$\alpha = (2.10 - x_m)/0.6 \tag{96}$$

and x_m is the electronegativity of the metal surface. Using the electronegativity scale of Pauling,[45] Trasatti thus predicted that the orientation of water increases in the series Au, Cu < Hg, Ag, Sb, Bi < Pb < Cd < Ga. This order is supported by the fact that the hydrophilicity of various metals increases in the sequence[71] Hg < Bi < Sb < Pb < Cd < Ga.

Trasatti's prediction of relative orientation of water on different metals at the p.z.c. is qualitative and based on the empirical relationship (96). The molecular approach of Bockris and Habib[31] correctly predicts that the degree of orientation increases as the attraction of the metal for oxygen increases (see Fig. 5). As mentioned earlier, the calculation of the relative attraction for different metals, $\Delta\Delta G^c$, is not straightforward as far as the properties of the metals are concerned. $\Delta\Delta G^c$ should be directly proportional to the electronegativity of the metals, where electronegativity is defined[45] as the power of an atom in a molecule to attract electrons to itself. Perhaps it would be appropriate to write*

$$\Delta\Delta G^c = A_1 + A_2 x_m \tag{97}$$

where A_1 signifies the relatively greater attraction of the negative end due to image and dispersion interaction [given by Eqs (34a) and (34b)], x_m is the electronegativity, and A_2 is a constant. The greater the attraction of the negative end of the dipole with the metal (as signified by x_m), the greater is the value of $|\Delta\Delta G^c|$, and hence the greater is the degree of orientation [Eq. (30) and Fig. 5].

The surface potential of water at the Hg–solution interface, as given by the BH model,[31] is 0.03 V at the p.z.c. This value is to be compared with that of 0.07 V estimated by Trasatti[51,100] from a consideration of the shift of the potential of zero charge on organic adsorption. This shift is assumed to be brought about by the replacement of water molecules by the adsorbing organic compounds. The adsorption potential shift is given by[74]

$$\Delta_{\mathrm{org}}^{\mathrm{H_2O}} E_{q_m=0}^{M} = g_{(M)}^{\mathrm{org}}(\text{dipole}) - g_{(M)}^{\mathrm{H_2O}}(\text{dipole}) \tag{98}$$

where $g_{(M)}^{\mathrm{org}}$(dipole) and $g_{(M)}^{\mathrm{H_2O}}$(dipole) are the surface potentials of organic compounds and water molecules, respectively, in contact with a metal. Trasatti[74] assumed $g_{(M)}^{\mathrm{org}}(\text{dipole}) = 0$ and attributed the entire shift to the water dipole contribution. The local field and image effects are neglected. If an estimate is obtained for $g_{(M)}^{\mathrm{org}}$(dipole), then $g_{(M)}^{\mathrm{H_2O}}$(dipole), by Trasatti's method, would come out to be less than 0.07 V.

*Values of x_m are given in Pauling's book as positive. To be consistent with the definition, it would be appropriate to use a negative sign in front of x_m while calculating $\Delta\Delta G^c$ from Eq. (97).

V. DISCUSSION AND CONCLUSION

In addition to the experimentally measured solvent excess entropy and the entropy of formation of double layer, there is another experimental quantity, Γ_v, the surface excess volume of the interface, which may bear some relationship to the reorganization of the solvent molecules. The surface excess volume of the electrode solution interface is given by

$$\left(\frac{\partial \gamma}{\partial p}\right)_{T,\mu,E_\pm} = \Gamma_v \tag{99}$$

which may also be expressed as[12]

$$\Gamma_v = V^\sigma - m^\sigma_{\mathrm{Hg}}\overline{V}_{\mathrm{Hg}} - m^\sigma_{\mathrm{H_2O}}\overline{V}_{\mathrm{H_2O}} - \sum m^\sigma_{\mathrm{H_2O}}(m^\sigma_i/m^\alpha_{\mathrm{H_2O}})\overline{V}_i \tag{100}$$

where $\overline{V}$s are the partial molar volumes of the species indicated in subscripts and the other symbols have the same significance as in Eq. (4). If one separates the contribution due to solvent dipoles to Γ_v, then the solvent excess volume may be represented by[12]

$$V^* = \Gamma_v - \sum \Gamma_i \overline{V}_i \tag{101}$$

Reeves[12] derived V^* from the results of Hills and Payne,[16] and found

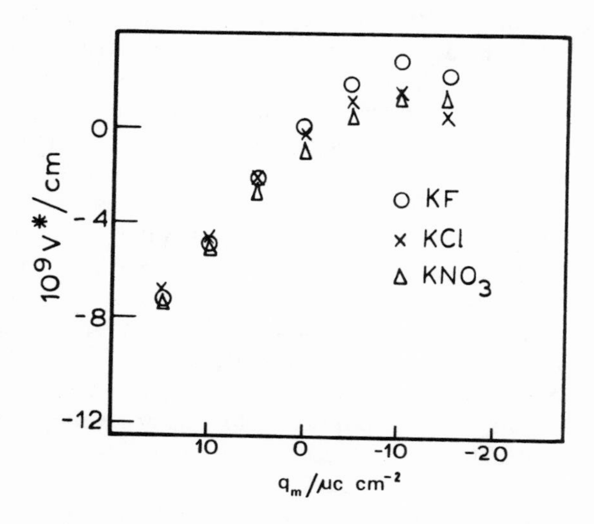

Figure 12. Solvent excess volume for mercury–aqueous solution interfaces as a function of electrode charge.[19]

that the solvent excess volume has a maximum at $q_m \approx -10\ \mu C\ cm^{-2}$. Very recently, Hills and Hsieh[19] have also observed the maximum in V^* at $q_m \approx -10\ \mu C\ cm^{-2}$ (Fig. 12). Once again, it could be argued that the maximum in V^* is principally a function of the mercury, rather than of the aqueous solution. But this argument does not seem to work because the mercury is virtually incompressible, and an independent study[101] has shown that the pressure dependence of the interfacial tension between mercury and a close-packed liquid, such as carbon tetrachloride, is very small.[19] At first, it might be thought that because the highly polar water molecules are strongly influenced by the electric field normal to the surface, as in entropy, the volume will also decrease on either side of zero-preferred orientation, but the observation shows that the maximum in V^* is found to occur at a more negative potential than that of S^*. There is no reason to believe that the variation of water orientation should bear a direct relationship to the variation in its volume. The electrode charge where they occupy maximum volume should not, necessarily, be the same where they have minimum or random orientation. Nevertheless, from both the solvent excess entropy and solvent excess volume measurement, it is certain that a marked reorganization of the interfacial solvent takes place at a negative electrode charge.

So far, most of the theoretical evaluations of the dipole contributions to the double-layer properties such as dipole potential or capacitance involve the first water layer only. An important piece of evidence to support this interpretation, as discussed by Trasatti,[10] is brought in by the measurements of the change of work function of the metal on water adsorption on metals from the gas phase. The work function of Ni and Fe, as measured by Suhrmann and co-workers,[102,103] decreases on adsorption of water until about two molecular layers are formed, as shown in Fig. 13. About 80% of the total change in work function is found to occur during the onset of the first layer, with only 20% during the formation of the second layer. No change in work function is observed on deposition of further layers. A linear change in work function is apparent up to about 70% coverage, whereas a smooth variation is observed for higher coverages up to the saturation value. This may indicate[10] the formation of the second layer before the completion of the first layer because of the strong localization of water molecules on the surface. Therefore, all the work function variation could be attributed to the first layer, whereas the

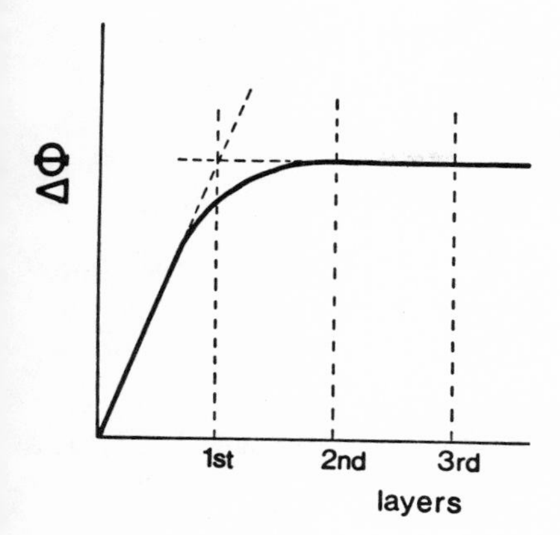

Figure 13. Sketch of the work function drop on water adsorption from the gas phase on transition metals.[102,103]

second layer does not contribute any dipolar potential drop, its role being confined to diluting the total surface potential over about two monomolecular layers. This view may be corroborated by the findings of Suhrmann *et al.*[103] and Huber and Kirk,[104] who observed that in the adsorption of water on Cu and Al, in which water molecules are thought to be more mobile, almost all the work function change occurs during the formation of the first layer.*

Thus, it can reasonably be assumed that the main contribution to the surface dipole potential of the solution side comes from the first water layer, and contributions by the other water layers are negligible.

There may be few exceptions to be included in this generalization. If the interaction with the metallic phase is very strong, as in the case of oxides,[105] at least two more layers of molecules are probably affected. On Ga, the water molecules are found to be adsorbed very strongly,[106] and hence the second layer may be expected to undergo some modifications.

Drost-Hansen[107] has put forward a three-layer model of water in certain aqueous–solid interfaces. In this model, a film of ordered water molecules 5–10 molecular layers thick is assumed to exist adjacent to the solid surface. The ordering is assumed to decrease gradually with distance from the interface. At sufficiently large distances from the

*Prior to Suhrmann and co-workers, Law[34] found while measuring the change in work function of Hg on adsorption of water from gas phase that the major change in work function takes place during the formation of the first layer.

surface, the bulk water structure exists. Thus, a completely disordered region is assumed to exist between the structured and normal bulk water. This view of Drost-Hansen,[107] that there are three different regions of water near the surface, is recurring in studies on water at interface with such particular solid surfaces as mica, cellulose, clays, and others.

In electrochemical systems with electrodes such as Ga with strong attraction toward water dipoles, the Drost-Hansen model[107] may be applicable with restrictions to three monomolecular layers, and not extended to three regions each several layers thick.

Relative values of g^s(dipole) may be derived[74] from inspection of charge-potential curves obtained by the integration of the capacity curves.[64] If $\delta\chi^m$ is assumed to be constant, then one may represent the difference in p.z.c. of two metals from Eq. (87) by

$$\Delta_{\mathrm{Hg}}^{m} E_{q_m=0} = (\Delta_{\mathrm{Hg}}^{m}\Phi)/F - \Delta_{\mathrm{Hg}}^{m} g^s(\mathrm{dipole}) \tag{102}$$

From Eq. (102), one may obtain the values of g^s(dipole) relative to the value for Hg, provided the constancy of $\delta\chi^m$ is valid. At potentials far from the p.z.c., if the orientation of water dipoles is assumed to be saturated, then $\Delta_{\mathrm{Hg}}^{m} g^s$(dipole) is zero; hence, from Eq. (102),

$$\Delta_{\mathrm{Hg}}^{m} E = (\Delta_{\mathrm{Hg}}^{m}\Phi)/F \tag{103}$$

Φ^m values derived[74] from Eq. (103) are in reasonable agreement with directly measured values.[3,65,108] Evaluation of relative g^s(dipole) and $\Delta_{\mathrm{Hg}}^{m}\Phi$ is shown in Fig. 14.

The ability to convert the standard electrode potential to its absolute value has long been a goal in electrochemistry. If one knows the absolute value of the potential of the standard hydrogen electrode, the absolute potential difference (which includes the contribution from solvent dipoles) across the interface may be estimated. It was Bockris and Argade[83] who first estimated numerically the absolute value of the potential of a reference hydrogen electrode and hence calculated the absolute potential across a metal–solution interface. This important breakthrough in electrochemistry has now received its due acceptance.[84] The electron overlap potential of a metal in contact with a solution, not evaluated earlier, has been estimated by using the concept of absolute hydrogen electrode potential by Bockris and Habib.[81] It was Trasatti[3,56,74,84] who brought these important matters into detailed discussion in recent times (see also Refs. 109 and 110).

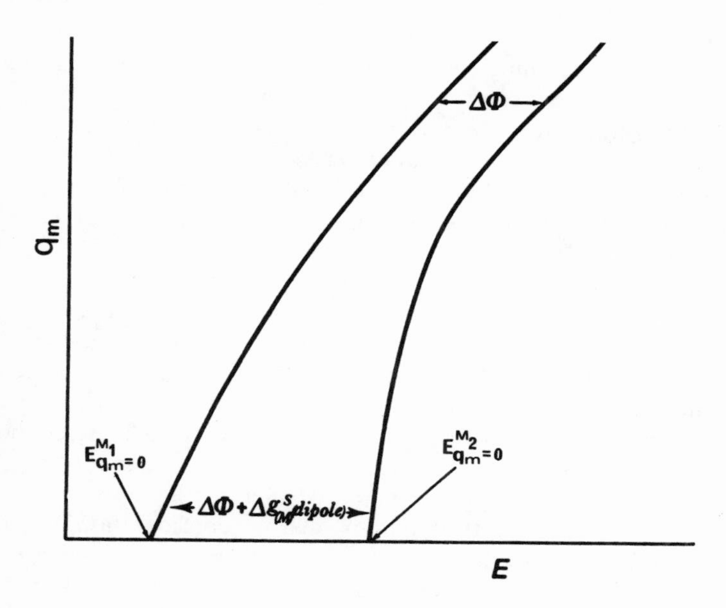

Figure 14. Sketch of charge-potential curves for two metals as obtained by integration of double-layer capacitance data.[10]

The inclusion by Lange and Miscenko[1] of the dipole potential in the total double-layer potential difference marked the existence of the oriented water layer on the electrode. Overbeek and Mackor[111,112] introduced *qualitatively* the consideration of the variation of dipole potential and hence the variation of dipole capacitance with electrode charge, which thus signified the existence of the field-dependent orientation of solvent dipoles. Bockris and Potter[2] applied independently the concept of a variable dipole potential to electrode kinetics. On a molecular level, interpretation of the field-dependent orientation of solvent dipoles was first attempted by Mott and Watts-Tobin.[24,25] This interpretation advanced the introduction of the present concept of interacting oriented dipoles by Bockris *et al.*[29,113] and MacDonald and Barlow,[26] followed by Levine *et al.*[27]

The idea that the water orientation does not cause a major contribution to the capacitance variation was first put forward by

Bockris *et al.*[29] Apart from entropy measurement, important information concerning the orientation of water dipoles has been derived from a series of studies by Conway and co-workers[6,28,52] on organic adsorption at the mercury–water interface. Their interpretation of the Essin–Markov effect led to the conclusion that the water molecules, which are replaced on adsorption of organic molecules, have their zero net orientation at about $q_m \approx -2\,\mu\mathrm{C\,cm}^{-2}$. This agrees closely with the observations of Bockris *et al.*[29,43,113] and of Barradas and Sedlak[114] in their study of organic adsorption on electrodes, and is thus consistent with the idea[29] that water molecules do have net minimum orientation at a negative electrode charge, and is supported by the later theoretical developments on solvent orientation.[31,60] The orientational properties of interfacial water dipoles, now confirmed by the entropy measurements, establishes beyond doubt, one point, that *the double-layer capacitance hump that occurs on the positive side of the p.z.c. is not due to water molecule orientation.* The theoretical and modelistic consideration[31,60] of interfacial dipoles has been shown to support this conclusion precisely. The alternative theory[29,62] of double-layer capacitance based on the anion repulsion concept is thus favored.

Trasatti has also recently published a number of papers that discuss the solvent contribution to double-layer potential difference, but his discussion is mainly confined at the p.z.c. and is based on the analysis of certain experimental data and a modelistic approach, as discussed in Section IV.

As mentioned before, three different molecular models of interfacial water molecules have been advocated by three different schools. The model of Damaskin and Frumkin[57] involves chemisorbed and associated orientable dipoles; the model of Parsons[70] involves nonassociated and associated orientable dipoles. However, it is now known that both these models are inconsistent with entropy results on [8,12,19] Hg (Section III). The dimer–monomer model of Bockris and Habib[31,60] permits consistent interpretation of the observed entropy variation. In the next few years, we must rely on model assumptions that give reasonable interpretation of experimental findings of double-layer properties unless and until some highly developed experimental techniques such as Auger, ESCA, ellipsometry, or other sophisticated devices enable us to see the actual configuration of the species present on the electrode surface.

ACKNOWLEDGMENTS

The author wishes to thank Professor J. O'M. Bockris for his critical reading of the manuscript and for his invaluable comments, which have been helpful in the interpretation of many points. Extension of necessary facilities by the Flinders University is gratefully acknowledged.

REFERENCES

[1]E. Lange and K. P. Miscenko, *Z. Phys. Chem.* **149** (1930) 1.
[2]J. O'M. Bockris and E. C. Potter, *J. Chem. Phys.* **20** (1952) 614.
[3]S. Trasatti, *J. Electroanal. Chem.* **33** (1971) 351.
[4]B. B. Damaskin and A. N. Frumkin, *J. Electroanal. Chem.* **34** (1972) 191.
[5]J. W. Schultze, *Electrochim. Acta* **17** (1972) 451.
[6]B. E. Conway and H. P. Dhar, *Croat. Chem. Acta* **45** (1973) 109.
[7]J. N. De Wit and J. Lyklema, *J. Electroanal. Chem.* **41** (1973) 259.
[8]J. A. Harrison, J. E. B. Randles, and D. J. Schiffrin, *J. Electroanal. Chem.* **48** (1973) 359.
[9]R. Parsons, R. M. Reeves, and P. N. Taylor, *J. Electroanal. Chem.* **50** (1974) 149.
[10]S. Trasatti, *J. Chim. Phys.* **72** (1975) 561.
[11]K. Müller, *J. Res. Inst. Catal. Hokkaido Univ., Jpn.*, **14** (1966) 113.
[12]R. M. Reeves, in *Modern Aspects of Electrochemistry*, Vol. 10, Ed. by J. O'M. Bockris and B. E. Conway, Plenum Press, New York, 1974.
[13]B. B. Damaskin, O. A. Petrii, and V. V. Batrakov, *Adsorption of Organic Compounds on Electrodes*, Plenum Press, New York and London 1971.
[14]R. Payne, *J. Electroanal. Chem.* **41** (1973) 277.
[15]R. Payne, in J. F. Danielli, M. D. Rosenberg, and D. A. Cadenhead, Eds., *Progress in Surface and Membrane Science, Vol. 6*, Academic Press, New York, 1973, p. 51.
[16]G. J. Hills and R. Payne, *Trans. Faraday Soc.* **61** (1965) 326.
[17]J. A. Harrison, Thesis, Birmingham University, 1960.
[18]D. J. Schriffrin, Thesis, Birmingham University, 1965.
[19]G. J. Hills and S. Hsieh, *J. Electroanal. Chem.* **58** (1975) 289.
[20]R. Zana and E. B. Yeager, *J. Phys. Chem.* **71** (1967) 521.
[21]B. E. Conway and L. H. Laliberte, in *Hydrogen Bonded Solvent Systems*, Ed. by A. K. Covington, Taylor and Francis, London, 1968; B. E. Conway, R. E. Verral, and J. E. Desoneyers, *Trans. Faraday Soc.* **62** (1966) 2738.
[22]B. E. Conway, R. E. Verral, and J. E. Desoneyers, *J. Phys. Chem.* **75** (1971) 3031.
[23]J. A. Harrison, J. E. B. Randles, and D. J. Schiffrin, *J. Electroanal. Chem.* **25** (1970) 197.
[24]R. J. Watts-Tobin, *Philos. Mag.* **6** (1961) 133.
[25]N. F. Mott and R. J. Watts-Tobin, *Electrochim. Acta* **4** (1961) 79.
[26]J. R. MacDonald and C. A. Barlow, Jr., *J. Chem. Phys.* **36** (1962) 3062; C. A. Barlow, Jr., in *Physical Chemistry; An Advanced Treatise*, Ed. by H. Eyring, Academic Press, New York, 1970.
[27]S. Levine, G. M. Bell, and A. L. Smith, *J. Phys. Chem.* **73** (1969) 3534.
[28]B. E. Conway and H. P. Dhar, *Electrochim. Acta* **19** (1974) 445.
[29]J. O'M. Bockris, M. A. V. Devanathan, and K. Müller, *Proc. R. Soc. London Ser. A* **274** (1963) 55.

[30]J. O'M. Bockris and A. K. N. Reddy, *Modern Electrochemistry*, Plenum–Rosetta Ed., Plenum Press, New York, 1973, Vol. 2.
[31]J. O'M. Bockris and M. A. Habib, *Electrochim. Acta* **22** (1977) 41.
[32]R. Parsons, *J. Electroanal. Chem.* **8** (1964) 93.
[33]A. N. Frumkin and B. B. Damaskin, *Electrokhimiya* **1** (1965) 63; in *Modern Aspects of Electrochemistry*, Ed. by J. O'M. Bockris and B. E. Conway, Butterworths, Washington, 1964, Vol. 3.
[34]J. T. Law, Ph.D. dissertation, Royal College of Science, London, 1951.
[35]R. W. Gurney, *Ionic Processes in Solution*, Dover Publications, New York, 1953.
[36]W. F. Giaque and J. W. Stout, *J. Am. Chem. Soc.* **58** (1936) 1144.
[37]C. Kemball, *Proc. R. Soc. London Ser. A* **190** (1947) 117.
[38]M. Polanyi, *Trans. Faraday Soc.* **28** (1932) 316.
[39]J. G. Kirkwood, *Z. Phys.* **33** (1932) 57.
[40]A. Muller, *Proc. R. Soc. London Ser. A* **154** (1936) 624.
[41]T. N. Anderson and J. O'M. Bockris, *Electrochim. Acta* **9** (1964) 347.
[42]J. O'M. Bockris and D. A. J. Swinkles, *J. Electrochem. Soc.* **111** (1964) 736.
[43]J. O'M. Bockris, M. Green, and D. A. J. Swinkles, *J. Electrochem. Soc.* **111** (1964) 743.
[44]C. Kittel, *Introduction to Solid State Physics*, 4th Ed., John Wiley and Sons, New York, 1971, p. 39.
[45]L. Pauling, *The Nature of Chemical Bonds*, 2nd Ed., Cornell University Press, Ithaca, N.Y., 1940.
[46]G. H. Haggis, J. B. Hasted, and J. T. J. Buchanan, *J. Chem. Phys.* **20** (1952) 1452.
[47]A. N. Frumkin, Z. A. Iofa, and M. A. Gerovich, *J. Phys. Chem. USSR* **30** (1956) 1445.
[48]J. E. B. Randles and K. S. Whitely, *Trans. Faraday Soc.* **52** (1956) 1509.
[49]J. E. B. Randles and D. J. Schiffrin, *J. Electroanal. Chem.* **10** (1965) 480.
[50]R. G. Barradas, P. G. Hamilton, and B. E. Conway, *J. Phys. Chem.* **69** (1965) 3411.
[51]S. Trasatti, *J. Electroanal. Chem.* **64** (1975) 128.
[52]B. E. Conway, H. P. Dhar, and S. Gottesfield, J. Colloid Interface Sci., **43** (1973) 303.
[53]G. Nemethy and G. A. Scheraga, *J. Chem. Phys.* **36** (1962) 3382.
[54]J. Topping, *Proc. R. Soc. London Ser. A* **114** (1927) 67.
[55]A. D. Buckingham, *Discuss. Faraday Soc.* **24** (1957) 151.
[56]S. Trasatti, *J. Chem. Soc. Faraday Trans. 1* **70** (1974) 1752.
[57]B. B. Danaskin and A. N. Frumkin, *Electrochim. Acta* **19** (1974) 173.
[58]B. E. Conway and L. G. M. Gordon, *J. Phys. Chem.* **73** (1969) 3609.
[59]D. D. Eley and M. G. Evans, *Trans. Faraday Soc.* **34** (1938) 1093.
[60]J. O'M. Bockris and M. A. Habib, *J. Electroanal. Chem.* **65** (1975) 473.
[61]C. G. Malmberg and A. A. Maryott, *J. Res. Natl. Bur. Stand.* **56** (1956) 1.
[62]J. O'M. Bockris and M. A. Habib, *J. Res. Inst. Catal. Hokkaido Univ., Jpn.*, **23** (1975) 47; *Z. Physik. Chem. Neue Folge* **98** (1975) 43.
[63]G. E. Walrafen, in *Proceedings of a Symposium on Equilibrium and Reaction Kinetics in Hydrogen-Bonded Solvent System*, University of Newcastle upon Tyne, 10–12 January 1968, Ed. by A. K. Covington and P. Jones, Taylor and Francis, London, 1968.
[64]A. Frumkin, N. Polianovoskaya, N. Grigoryev, and I. Bagotskaya, *Electrochim. Acta* **10** (1965) 793.
[65]S. Trasatti, *Chim. Ind.* (*Milan*) **53** (1971) 559.
[66]H. Wroblowa, Z. Kovac, and J. O'M. Bockris, *Trans. Faraday Soc.* **61** (1965) 1523.
[67]M. A. V. Devanathan and B. V. K. R. S. A. Tilak, *Chem. Rev.* **65** (1965) 635.
[68]M. A. V. Devanathan, *Trans. Faraday Soc.* **50** (1954) 373.

[69]R. Parsons, in *Symposium on Some Aspects of Electrochemistry*, National Institute of Science, India, Oct. 6th, 1961.
[70]R. Parsons, *J. Electroanal. Chem.* **59** (1975) 229.
[71]A. N. Frumkin, B. B. Damaskin, N. Grigoryev, and I. Bagotskaya, *Electrochim. Acta* **19** (1974) 69.
[72]A. N. Frumkin, B. B. Damaskin, N. Grigoryev, and I. Bagotskaya, *Electrochim. Acta* **19** (1974) 75.
[73]S. Trasatti, *Chim. Ind.* **53** (1971) 559.
[74]S. Trasatti, *J. Electroanal. Chem.* **54** (1974) 437.
[75]S. Trasatti, *Z. Physik. Chem. Neue Folge* **98** (1975) 75.
[76]A. N. Frumkin, *Z. Phys. Chem.* **103** (1923) 43.
[77]M. S. Metsik, V. D. Perevertaev, V. A. Liopo, G. T. Timoshtchenko, and A. B. Kiselev, *J. Colloid Interface Sci.* **43** (1973) 662.
[78]G. E. Van Gils, *J. Colloid Interface Sci.* **30** (1969) 272.
[79]C. Herring and M. H. Nichols, *Rev. Mod. Phys.* **21** (1949) 185.
[80]J. Horiuti and T. Toya, in *Solid State Surface Science*, Ed. by M. Green, Marcel Dekker, New York, 1961, Vol. 1, p. 1.
[81]J. O'M. Bockris and M. A. Habib, *J. Electroanal. Chem.* **68** (1976) 367.
[82]J. E. B. Randles, *Adv. Electrochem. Electrochem. Eng.* **3** (1963) 1.
[83]J. O'M. Bockris and S. D. Argade, *J. Chem. Phys.* **49** (1968) 5133.
[84]S. Trasatti, *J. Electroanal. Chem.* **52** (1974) 313.
[85]G. Paasch, H. Eschrig, and W. John, *Phys. Status Solid B* **51** (1972) 283.
[86]A. O. E. Animalu and V. Heine, *Philos. Mag.* **12** (1965) 1249.
[87]T. Schneider, *Phys. Status Solidi* **32** (1969) 323.
[88]R. Parsons, in *Modern Aspects of Electrochemistry*, Ed. by J. O'M. Bockris and B. E. Conway, Butterworths, London, 1954, Vol. 1, republished 1968.
[89]R. Parsons, *Proc. R. Soc. London Ser. A* **261** (1961) 79.
[90]E. Ukshe and A. Levin, *Zh. Fiz. Khim.* **29** (1955) 219.
[91]A. Frumkin and A. Gorodetskaya, *Zh. Phys. Chem.* **136** (1928) 215.
[92]V. Navakovsky, E. Ukshe, and A. Levin, *Zh. Fiz. Khim.* **29** (1945) 1847.
[93]R. Vasenin, *Zh. Fiz. Khim.* **27** (1953) 878; **28** (1954) 1672.
[94]A. N. Frumkin, *Sven. Kem. Tidsk.* **77** (1965) 300.
[95]S. D. Argade and E. Gileadi in *Electrosorption*, Ed. by E. Gileadi, Plenum Press, New York, 1967.
[96]T. Anderson, J. Anderson, and H. Eyring, *J. Phys. Chem.* **73** (1969) 3562.
[97]E. Sevastyanov, T. Vitanov, and A. Popov, *Elektrokhimiya* **8** (1972) 412.
[98]S. Valette and A. Hamelin, *C. R. Acad. Sci. C* **272** (1971) 602; **273** (1971) 320.
[99]A. Frumkin, N. Grigoryev, M. Pyarnoya, and U. Palm, *Elektrokhimiya* **10,** No. 5 (1974).
[100]A. De Battisti and S. Trasatti, 4th *International Symposium on Electrochemistry of Interfacial Phenomena*, June 24–July 3, 1975, Cavtat. Yugoslavia.
[101]G. J. Hills and D. Posadas, to be published.
[102]R. Suhrmann, J. M. Heras, L. Viscido de Heras, and G. Wedler, *Ber. Bunsenges. Phys. Chem.* **68** (1964) 511.
[103]R. Suhrmann, J. M. Heras, L. Viscido de Heras, and G. Wedler, *Ber. Bunsenges. Phys. Chem.* **72** (1968) 854.
[104]E. E. Huber and C. T. Kirk, *Surface Sci.* **5** (1966) 447.
[105]E. McCafferty, V. Pravdic, and A. C. A. C. Zettlemoyer, *Trans. Faraday Soc.* **66** (1970) 1720.
[106]E. V. Osipova, N. A. Shurmovskaya, and R. K. Burshtein, *Elektrokhimiya* **5** (1969) 1139.

[107]W. Drost-Hansen, *Ind. Eng. Chem.* **61** (1969) 10.
[108]J. C. Riviere, in *Solid State Surface Science*, Ed. by M. Green, Marcel Dekker, New York, 1969, Vol. 1.
[109]A. Frumkin and B. Damaskin, *J. Electroanal. Chem.* **66** (1975) 150.
[110]S. Trasatti, *J. Electroanal. Chem.* **66** (1975) 155.
[111]E. L. Mackor, *Rec. Trav. Chim.* **70** (1951) 747.
[112]E. L. Mackor, *Rec. Trav. Chim.* **70** (1951) 763.
[113]J. O'M. Bockris, E. Gileadi, and K. Müller, *Electrochim. Acta* **12** (1967) 1301.
[114]R. G. Barradas and J. M. Sedlak, *Electrochim. Acta* **16** (1971) 2091.
[115]R. Parsons and F. G. R. Zobel, *J. Electroanal. Chem.* **9** (1965) 333.

4

Preparation and Characterization of Highly Dispersed Electrocatalytic Materials

K. Kinoshita* and P. Stonehart†

Advanced Fuel Cells Research Laboratory
Power Systems Division
United Technologies Corporation
Middletown, Connecticut

I. INTRODUCTION

Catalysis can be defined as the science of maximizing chemical reaction rates for the formation of preferred products. Catalysis is achieved by using high-surface-area materials that may have specific surface features (i.e., preferred reaction sites). A secondary area is the study of adsorbed species on the catalytic surface, catalyst deactivation by poisoning, and loss of surface area due to sintering.

One branch of heterogeneous catalysis that has received revived attention in the last decade is electrocatalysis,[1] the study of heterogeneous catalytic reactions involving reactant species that transfer electrons through an electrolyte–catalyst interface. The increased interest in this area of catalysis stems from the development of more sophisticated energy-conversion devices based on electrochemical reactions, namely, fuel cells and batteries. Although many electrochemists are thoroughly conversant with the study of electrochemical reaction mechanisms occurring on electrode–electrocatalytic surfaces, these investigations have been generally restricted to measurements on low-surface-area catalysts

*Current address: Argonne National Laboratory, Argonne, Illinois 60439.
†Current address: Stonehart Associates, Madison, Connecticut 06443.

(e.g., flat sheets, wires, evaporated films). There is some prevailing controversy whether high-surface-area catalysts possess the same electrocatalytic properties as low-surface-area electrocatalysts. This question can be answered only by a thorough understanding of the physical characterizations and preparations of high-surface-area electrocatalysts. To promote further work on electrocatalysis with highly dispersed materials, we feel that it is important that these techniques be reviewed and widely disseminated. The preparations and characterizations are an integral part of the overall picture of electrocatalysis and, as in gas-phase and liquid-phase heterogeneous catalysis, are necessary considerations for understanding the performance of energy-conversion devices. Theoretical or kinetic aspects of electrocatalysis will not be emphasized, since there are available reviews on these subjects,[2–4] although they do not deal with high-surface-area electrocatalysts or the incorporation of such electrocatalysts into porous electrode structures.

To limit the scope of this review, the electrocatalytic materials discussed are inevitably subject to our personal biases, and so are restricted primarily to noble metals and supports for energy-conversion devices operating with aqueous electrolytes. Much of the expertise for preparing and analyzing highly dispersed catalysts exists in the gas-phase catalysis literature. We have therefore drawn on and discussed this literature where it is pertinent to our understanding of electrocatalytic materials.

The best-studied noble metal for technological applications is platinum, which will therefore be the focal point of this review. Both supported and unsupported noble metal blacks with surface areas greater than 20 $m^2\,g^{-1}$ noble metal and average noble metal crystallite sizes less than 140 Å have technological significance. The ultimate dispersion of Pt might be considered to be crystallites in which every Pt atom is a surface atom and available for reaction, which would obtain at approximately 10 Å diameter. The Pt surface area would correspond to about 280 $m^2\,g^{-1}$ Pt.

II. ELECTROCATALYST SUPPORTS

High surface-area electrocatalysts are generally supported on porous substrate, which must be chemically stable to the electrolyte environment. Furthermore, the substrate must be electrically conducting since electron transfer occurs at the electrolyte–electrocatalyst interface. This limitation, which is not present for gas-phase heterogeneous catalysts, restricts the number and variety of materials available as electrocatalyst supports. In some instances, unsupported electrocatalysts of lower surface areas have been used in electrochemical systems where economic considerations or electrochemical reaction rates are not the limiting factors.

1. Carbon

Carbon and graphite have been used extensively as electrode materials in the chlor–alkali industry. Besides being relatively inert chemically in most electrolytes, carbons can be obtained as high-surface-area powders as well as in solid electrode form. For the preparation of high-surface-area electrocatalysts, carbon has many attractive properties, i.e., chemical stability, high surface area, electrical conductivity, and both macroscopic and microscopic porosity.

There are three different forms of carbon, namely, diamond, amorphous carbon, and graphite, but there are an infinite variety of carbons available between an ungraphitized material and a fully graphitized carbon. The degree of graphitization affects the carbon surface groups as well as the surface microstructure and carbon particle porosity. An ideal graphitic structure consists of layers of carbon atoms arranged in hexagonal rings stacked in the sequence ABAB.... The carbon–carbon bond length is 1.42 Å in the hexagonal rings and 3.35 Å in the direction of the *c*-axis, which is perpendicular to the rings. Electrical resistivity along the *c*-axis is approximately 10^{-2} Ω-cm and approximately 10^{-4} Ω-cm parallel to the *c*-axis.[5] For powder samples, such as ungraphitized carbon blacks and charcoals, the contact resistance between particles may be as great as the carbon intrinsic electrical resistivity.[6] The resistivity of carbon blacks varies from a few tenths to 1 Ω-cm when measured under a pressure of 150 psi.[7] Besides the compaction pressure, the amount of oxygen complex present on the carbon black surface greatly influences the measurement of the electrical resistivity. Carbons prepared from heavy oils (furnace blacks) and acetylene carbon blacks are low in surface oxygen and are good electronic conductors, whereas carbons prepared from oil well flare gases and natural gas (channel blacks) with relatively high oxygen content are poorer conductors.

A wide variety of carbon–oxygen surface complexes have been identified on porous carbons[8–11]; these complexes are represented schematically in Fig. 1. The most common carbon–oxygen surface complex groups have been identified by chemical reactions with specific reagents, as well as by spectroscopic and electrochemical techniques. These so-called surface oxides are classified as basic surface oxides (formed when the carbon surface is freed from all surface compounds by heating in an inert atmosphere or vacuum and then exposed to oxygen only after cooling to room temperature) or acidic surface oxides (formed when carbon is exposed to oxygen at temperatures near its ignition point). The presence of carboxyl, phenolic, and lactone gives rise to the acid groups on the carbon surface, and the acidity[12] measured by neutralization with barium hydroxide or sodium hydroxide decreased in the order: color black > channel black > furnace black. In general, the concentrations of organic functional groups comprising the surface oxide may be different according to the nature of the carbon and the oxidation process to which the carbon is subjected. The surface oxides exercise a considerable influence on the behavior and surface reactions of carbons used as electrocatalysts or electrocatalyst supports, and these observations will be discussed briefly.

At high potentials (> 1 V) in aqueous electrolytes, carbon blacks are oxidized to CO_2 and surface oxides,[13] and hence their application as electrocatalyst supports is restricted. Heat treatment and the resulting graphitization, however, render the carbon more resistant to oxidation at high potentials.[14] The influences of carbon surface area and heat treatment on the corrosion rate of various carbon blacks in

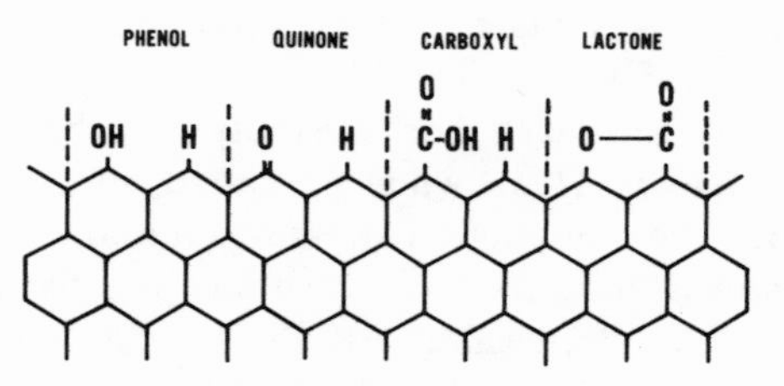

Figure 1. Schematic representation of oxygen surface functional groups on carbon.

hot concentrated H_3PO_4 were examined by Kinoshita and Bett.[14] The specific corrosion rate of a carbon black heat-treated at 2700°C was lower than that observed on the starting material. These results suggested that graphitization decreased the electrochemical oxidation of the carbon by an amount greater than the decrease in surface area, which indicates a decrease in the "activity" of the graphitized relative to the nongraphitized carbon surface.

Two anodic processes were observed during the electrochemical oxidation of carbon blacks in 96 wt% H_3PO_4 at 135°C—formation of carbon surface oxides and CO_2 evolution.[13] Both reactions decreased at different rates under potentiostatic control, and eventually CO_2 evolution became the major process. Binder *et al.*[15] also observed these two anodic reactions during oxidation in H_3PO_4 and H_2SO_4, and in addition, they reported that an intercalation compound, graphitic oxide, was formed during oxidation of electrode carbon in 40 wt% H_3PO_4 at 100°C. Graphitic oxide prepared by Kinoshita and Bett[14] was found to be unstable in 96 wt% H_3PO_4 at 130°C. Humic acid has also been reported as a reaction product formed during electrolysis at carbon electrodes, and is recognized by the appearance of a brown discoloration in clear electrolytes.[16] The isolation of artificial humic acid has not been successful, and thus controversy prevails as to the existence of such a compound.[17]

Carbon surface oxides formed during electrochemical oxidation increase the electrical resistance of the carbon support, and could be a limiting factor for high-current-density operation of a supported electrocatalyst. Binder *et al.*[15] measured an increase in resistivity from 0.053 to 0.247 Ω-cm when surface oxides were formed on electrode carbon in 10 wt% H_3PO_4 at 100°C. Potentiodynamic analysis of oxidized carbon blacks indicated the presence of a surface quinone–hydroquinone-type species.[18] Current–potential curves for a graphitized carbon black showed[18] no indication of a surface electroactive species, but oxidation of the carbon with hot chromic acid yielded an electroactive species identified as a quinone–hydroquinone couple.

Carbon has been examined both as an electrocatalyst support and as an electrocatalyst in its own right for a gas diffusion oxygen electrode. The open circuit oxygen potential on graphite and activated carbon electrodes in various concentrations of KOH was measured by Berl.[19] These carbon electrodes obeyed the Nernst equation for the reaction

$$OH^- + HO_2^- = O_2 + H_2O + 2e^- \quad (1)$$

which had a measured equilibrium potential of 0.0416 V at 27°C. Berl concluded that the formation of the hydroperoxide ion did not require the dissociation of the oxygen–oxygen bond. Further reduction to water, however, required the breaking of the oxygen–oxygen bond. Many years later, the conclusion reached by Berl concerning the mechanism for the reduction of O_2 on carbon was confirmed by Davies *et al.*[20] using the O^{18} isotope. Carbon surfaces do not decompose peroxide easily, which allows the reversible couple in Eq. (1) to be maintained. Mercury surfaces also do not decompose peroxide. Stonehart[21] used Faradaic impedance techniques to measure the rate constant for the reversible two-electron reaction of oxygen on mercury in base and found $k_s \sim 10^{-2}$ cm sec^{-1}. Provided that the hydroperoxide concentration was maintained at a low value, so that catalytic decomposition was small, by increasing the oxygen concentration by pressurization, Bowen and Urbach[22,23] were able to show that the reversible oxygen–hydrogen peroxide reaction was also maintained on platinum surfaces.

Yeager and Kozawa[24] studied the kinetics of the oxygen reaction on carbon and graphite electrodes in KOH. They suggested that the reducible organic groups that may be present on carbons and graphites could participate as intermediates, but the absence of large variations in the exchange current densities on various carbons and graphites led to the conclusion that the rate-determining step does not involve any of these organic functional groups.

Various carbons have been subjected to heat treatments in different gaseous environments with the aim of altering the carbon surface. Activated carbon heat-treated in NH_3 produced an electrode yielding higher activity for O_2 reduction in 12 *M* KOH than activated carbon heat-treated in H_2.[25,26] This difference in behavior was explained as arising from the formation of nitrogen groups (bound to the carbon surface during heat treatment in NH_3), which are good peroxide-decomposing catalysts, and are considered as one requirement for obtaining a better-performing oxygen electrode.[27] In another paper, Mrha *et al.*[28] used a triangular potential sweep technique to study the electrocatalytic activity of carbon electrodes.

Porous carbon oxygen electrodes for sulphuric acid fuel cells have been described by Luft *et al.*[29] The carbon was prepared by pyrolyzing polyacrylonitrile and had a Brunauer, Emmet, and Teller (BET) surface area between 1000 and 1400 $m^2 g^{-1}$. It was found that with an increase in the pyrolysis temperature, the nitrogen content of carbon decreased together with a concomitant increase in conductivity. Optimum performance was obtained for carbons progressively pyrolyzed up to 1000° C.

2. Other Supports

Considerable effort has been directed toward the discovery of nonnoble-metal electrocatalysts for fuel cell applications and also to develop electrocatalyst supports other than carbon. Meilbuhr[30] reported that boron carbide may be a useful support for the cathode in the base (KOH) fuel cell. Boron carbide, tantalum boride, and silicides of tungsten and titanium have been used as support for electrocatalysts in an acid environment.[31,32] Niedrach and McKee[31] observed that these materials were corrosion-resistant in 85% H_3PO_4 (150°C) and 6 *N* H_2SO_4 (85°C), and also possessed high electrical conductivity. Other carbides such as TaC and NbC were not corroded at potentials lower than 2 V in 6 *N* H_2SO_4 at room temperature,[33] and may prove useful as electrocatalyst supports.

Electrochemical measurements by Laitinen *et al.*[34] on vapor-deposited SnO_2 on glass clearly showed that the incorporation of Sb into the lattice structure of SnO_2 enhanced the electrical conductivity of the oxide. Tseung and Dhara[35] have demonstrated the application of Sb-doped SnO_2 as an electrocatalyst support. Using a procedure similar to the one outlined by Vincent and Weston,[36] they were able to prepare Sb-doped SnO_2 with a surface area of 55 $m^2 g^{-1}$. The Sb-doped SnO_2 showed satisfactory corrosion resistance over the range 0–1.2 V in 150° C, 85% H_3PO_4 and should therefore deserve consideration as an alternative electrocatalyst support for C in acid media.

Metal oxides having a perovskite structure have been investigated for their electrocatalytic properties as oxygen reduction or evolution electrodes in aqueous electrolytes. The general formulas for perovskites are (1) ABO_3 for the ternary type, in which A is a large metal cation and B is a smaller metal cation; and (2) $A(B'_xB''_y)O_3$ for the complex type, in which B′ and B″ are two different elements in different oxidation states and $x + y = 1$. The tungsten bronzes are examples of metal oxides with perovskite structures that have received considerable attention as fuel-cell electrocatalysts.[37–40] It has been concluded that the bronzes have little intrinsic activity as electrocatalysts,[38] but the addition of Pt appears to produce a synergistic influence for the electrocatalysis of fuel-cell reactions. (This aspect will be discussed in Section II.3.) In the sense that the tungsten bronzes can act as supports for Pt electrocatalyst crystallites, these materials deserve consideration in this section. In general, the nonstoichiometric oxide bronzes possess metallic conduction properties that are necessary for electrocatalyst supports. The chemical stability of the bronzes, on the other hand, varies widely, depending on the particular bronze. These observations have been reviewed by McHardy and Stonehart.[38]

The Sr-doped lanthanum cobaltites ($La_xSr_{1-x}CoO_3$) are perovskites that display near-metallic conductivity for $x > 0.15$ and good corrosion resistance in 75 wt% KOH at 230°C.[41] Meadowcroft[42] reported that sintered disks of $La_{0.8}Sr_{0.2}CoO_3$ showed greater electrocatalytic activity than platinum disks for oxygen reduction in 5 *M* KOH at 25°C. Pinnington,[43] on the other hand, concluded that $La_{0.8}Sr_{0.2}CoO_3$ did not possess any steady electrocatalytic activity for oxygen reduction. It is evident that the experimental conditions and the nature of the perovskite are responsible for the conflicting conclusions obtained with $La_{0.8}Sr_{0.2}CoO_3$; nevertheless, the good electronic conductivity and corrosion resistance of the $La_xSr_{1-x}CoO_3$ perovskites make these materials attractive as supports for more active electrocatalysts.

The electrocatalytic properties of complex perovskites of the general formula $A(B'_{0.5}B''_{0.5})O_3$ were examined by ESSO[44] for potential application in fuel cells. Perovskites with A = Li, Cs, Ca, Sr, Ba, La, Tl, Pb, and Bi, B′ = V, Cr, Mn, Fe, Co, Ni, Ru, Rh, Ir, and Pt, and B″ = Ti, Zr, Sn, Hf, V, Nb, Ta, Mo, and W were synthesized with the B′ ions chosen for their known catalytic properties and B″ selected for the corrosion resistance that they impart to the compound. In general, most of the perovskites examined were poor conductors, but were quite acid-resistant. The conductivity of the perovskites could be increased by incorporation of oxygen vacancies, but then the acid resistance was not so good. If the electronic conductivity and the corrosion resistance of the perovskites could be improved, these materials would offer an alternative to carbon as an electrocatalyst support.

3. Electrocatalyst–Support Interactions

The type of carbon used to support Pt electrocatalysts has been observed to produce large differences in the Pt catalyst apparent activity. Hillenbrand and Lacksonen[45] concluded from their studies using electron spin resonance (ESR) spectroscopy that a chemical interaction existed between the Pt and the C substrate. The Pt–C interaction was believed to be an electronic interaction that altered the Pt catalytic activity for hydrogen oxidation. ESR measurements indicated a large decrease in the number of unpaired electrons after application of Pt to the C support. By subjecting various carbons to different heat treatments, it was concluded that the Pt–C chemical interactions occurring on oxidized C surfaces were more desirable than those with clean C surfaces for catalysis of the hydrogen oxidation reaction.

Nicolau and co-workers[46,47] concluded from ESR studies of Pt supported on C (charcoal) catalysts that a "pseudosandwich" complex in which the Pt atom donates electrons to the C is formed. It was further concluded that a relationship existed between the ESR absorption intensity and the catalytic activity of Pt supported on C for certain reactions. This observation certainly supports the findings reported by Hillenbrand and Lacksonen.[45]

More recently, Ross *et al.*[48] used electron spectroscopy for chemical analysis (ESCA) to study Pt supported on carbon black. The photoelectron spectrum indicated a net electron transfer from Pt to the C support, in agreement with the ESR results.

Garten and Weiss[49] suggested that two types of activated carbons and carbon blacks can be formed. These are the oxidized (*H* carbon) and reduced (*L* carbon) forms; the former carbon acquires a quinonoid structure that creates a number of fixed olefinic bonds, and the latter carbon acquires the aromatic structure of hydroquinones. The electrokinetic behavior of *H* and *L* carbons was investigated for platinized and unplatinized carbons.[49] Platinized C heated at 980°C in hydrogen was electrokinetically negative in contact with hydrogen at room temperature, whereas it was positive when exposed to oxygen. This behavior was explained by assuming that the presence of the Pt enabled the C to be oxidized or reduced with oxygen or hydrogen at room temperature. Thus, a platinized *H* carbon can be reduced to an *L* carbon, and the latter can be reoxidized with oxygen to an *H* carbon at room temperature. Both the results summarized by Hillenbrand and Lacksonen[45] and those summarized by Garten and Weiss[49] suggest that the presence of Pt on C influences the chemical properties of C surfaces.

A general survey on the role of the electronic interaction between metal catalyst particles and supports has been discussed by Solymosi.[50] This review discusses primarily metal catalysts and their interaction with semiconducting oxide supports, but the interpretations have some significance for electrocatalyst–support interactions. At thermodynamic equilibrium, the Fermi level of the electrons must be the same at the interface, so if the work function of the electrons in the metal is larger than in the support, electrons will migrate from the support into the metal until equilibrium is attained. This flow of electrons sets up a positive space charge in the support and an induced negative charge on the surface of the metal. If the relative work function of electrons in the metal and support is reversed, electrons will migrate from the metal to the support. Electronic interactions arising in this manner offer an interpretation for electron transfer observed with Pt supported on carbons.

Sagert and Pouteau[51] observed a metal–support interaction for hydrogen–water–deuterium exchange over Pt metals on C supports. The interactions between various heat-treated carbon blacks and the noble metals lower the chemical activation energies and increase the rates of reaction under conditions in which the support appears to donate electrons to the noble metal. It was suggested that the C support with a lower work function donates electrons to the Pt catalyst so that the platinum–hydrogen bond $Pt^{\delta+}$–$H^{\delta-}$ is weakened. This in turn leads to a lower activation energy. Arora[52] has reviewed the role of catalyst supports in the activity and stability of reforming catalysts, but these aspects will not be discussed here.

Platinum supported on WO_3 is a prime example of the synergistic effect between an electrocatalyst and its support when used as a hydrogen electrode in acid solution.[53–56] WO_3 does not possess any activity toward H_2 oxidation, but in contact with Pt, the activity of Pt–WO_3 is greater than Pt on a graphite or TaC substrate.[54,55] This synergism is believed to arise from the migration of dissociated H_2

from the Pt surface to the oxide by a "spillover"* mechanism, according to Boudart and co-workers.[57,58] Hobbs and Tseung[53] suggest that the effective reaction area is increased by the presence of WO_3, which forms a blue H_2–W bronze by hydrogen spillover, acting as an oxidation site. WO_3 does not act as a passive support, but instead may be considered as an active support enhancing the catalytic activity of the system.

On adding trace amounts of Pt to sodium tungsten bronzes, Bockris and McHardy[60] observed that the rate of oxygen reduction on the Pt–bronzes in H_2SO_4 and $HClO_4$ increased by the third power of the Pt concentration. This behavior was consistent with a spillover model in which surface diffusion of reaction intermediates from the Pt to the bronze took place. The lower coverage of reaction intermediates on the Pt surface decreased the activation energy for oxygen adsorption, and thereby accelerated the rate-determining step for oxygen reduction. This synergistic effect for Pt on sodium tungsten bronze does not mean that the support is active in the sense discussed by Hobbs and Tseung.[53]

Platinum supported on carbon exhibits hydrogen spillover in the gas phase.[61,62] This is the term given to the process whereby the platinum provides dissociation sites for hydrogen molecules and the hydrogen atoms formed migrate to the carbon support surface by surface diffusion. This leads to a ratio higher than unity for hydrogen atom adsorption per surface Pt atom on the Pt crystallites.

There is some controversy as to whether hydrogen spillover is seen with Pt on C in electrolytic systems, since the Russian investigators claim to have observed it,[63–66] whereas the authors of this review have not.[67–69] The latter workers obtained supported Pt crystallite sizes from X-ray line-broadening, CO chemisorption (from the gas phase and in electrolytes), hydrogen atom chemisorption (from the gas phase and in electrolytes), and dark-field electron microscopy.[70] Carbon-supported platinum crystallite sizes determined by hydrogen adsorption were not different from the crystallite sizes obtained by the other techniques, thus indicating no hydrogen spillover. It may be that the support used by the Russian workers is unique, so that hydrogen spillover cannot be unequivocally

*See Sermon and Bond[59] for a review of hydrogen "spillover."

denied. On the other hand, hydrogen spillover in the gas phase[61] for Pt on C is observed only at elevated temperatures (300–400°C) and not at room temperature, so it would not be expected for the room-temperature electrochemical cases.[63–66]

III. PREPARATION OF ELECTROCATALYSTS

The preparation of catalysts has in some instances been considered an art rather than a science. There are, however, some technical aspects that place catalyst preparation on a more scientific level. These findings will be discussed in this section. The wealth of information published on the various catalyst-preparation techniques cannot be adequately covered in this review, but the salient features of electrocatalyst preparation will be considered, with the emphasis being on noble metal electrocatalysts. Extensive discussions on catalyst preparations that are important to the petrochemical industry have been presented by Sinfelt,[71] Morikawa *et al.*,[72] and Ciappetta and Plank.[73]

1. Impregnation

There have been numerous studies concerned with the preparation of highly dispersed noble metal catalysts supported on inert substrates. The physical nature of Pt supported on silica[74–79] and alumina[75–83] has been investigated extensively due to its wide application as an industrial petrochemical catalyst. The more recent interest in fuel-cell development has led to the investigation of Pt supported on carbon[84,85] as an electrocatalyst.

Of the many procedures to prepare supported metal catalysts, the most commonly used has been an impregnation of the support material with a solution of the catalyst metal salt, followed by heating to dryness and subsequent gas-phase reduction to the metal. In this method, the interaction of the metal salt in solution with the support surface is of vital importance in determining the final crystallite size of metal obtained.

If the metal salt does not adsorb on the support, the salt will concentrate in the pores of the support during the drying step, and the resulting catalyst will yield metal crystallites, the number and hence the metal surface area, of which are determined by the pore structure of the support. Due to the capillary forces, the smallest pores and those at the center of the support agglomerate will be enriched in metal salt. Clearly, this result is contrary to the ideal, since it is advantageous to concentrate the metal catalyst particles at the periphery of the support agglomerate to minimize mass transfer of the reactants and products. Dorling and Moss[78] described a system in which silica gel was impregnated with chloroplatinic acid of various concentrations. It yielded catalysts with a constant number of Pt crystallites that could be related to the pore structure of the gel. The surface area of Pt per gram of catalyst followed the relationship

$$A = N^{1/3}\rho^{-2/3}W^{2/3} \tag{2}$$

where A is the surface area of Pt per gram of catalyst, N is the number of crystallites, ρ is the density of Pt, and W is the Pt content of the catalyst. This relationship is to be expected for the growth of a fixed number of crystallites of regular shape.

A graphitized carbon black (graphitized Vulcan XC-72, $70\,m^2\,g^{-1}$ C) was impregnated with increasing concentrations of $Pt(NH_3)_2(NO_2)_2$ solution, evaporated to dryness, and then thermally decomposed by heating in air for 3 hr at 300°C.[87] The Pt surface areas of these electrocatalysts were determined by an electrochemical technique described later in this review. Figure 2 shows the relationship between Pt surface area and Pt content for this series of impregnation electrocatalysts. The nonlinear dependence is readily explained by the pore concentration mechanism, and indeed the data gave an approximate fit to the equation derived by Dorling and Moss[78] for this model.

Grubb and McKee[32] used boron carbide as an alternative support for C to support Pt electrocatalysts. It was prepared by impregnation using $Pt(NH_3)_2(NO_2)_2$ solution, drying, and reducing to the metal in hydrogen at an elevated temperature. The Pt surface areas obtained for a series of supported electrocatalysts containing 0.5–9.1 wt% Pt are also plotted in Fig. 2. The relationship between Pt surface area and Pt content using boron carbide as an electrocatalyst support is similar to the results obtained with Pt supported on a graphitized carbon black.

A recent patent[88] describes an impregnation technique for maintaining relatively high Pt surface areas, even with high Pt

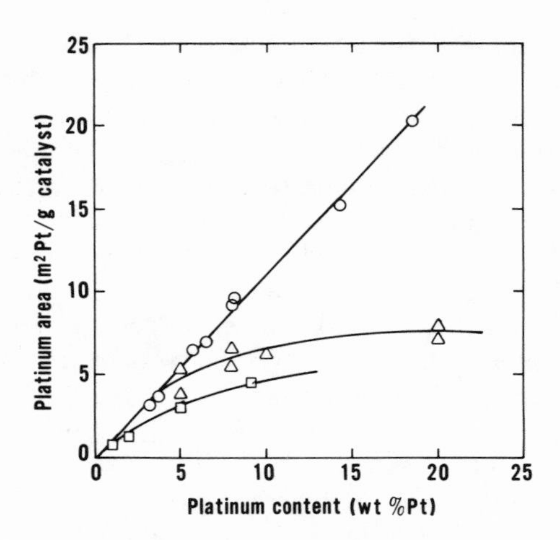

Figure 2. Variation of Pt surface area of impregnated and ion-exchanged electrocatalysts with Pt content: (△) impregnated graphitized carbon support; (○) ion-exchanged graphitized carbon support; (□) impregnated boron carbide support.[32]

contents. The patent claims that high-surface-area Pt electrocatalysts can be prepared by sequential catalyzation, which involves repetitive impregnation steps using low concentrations of Pt salt solutions until the final Pt content is reached. For example, to prepare a supported 20 wt% Pt electrocatalyst, the C support was impregnated with a $Pt(NH_3)_2(NO_2)_2$ solution such that initially 5 wt% Pt was deposited on the C. The Pt salt was decomposed by heating in air at 500°F for 3 hr. The electrocatalyst was impregnated again with a Pt salt solution to bring the Pt content up to 10 wt%, and after drying, the heat treatment step was repeated. After four repetitive catalyzations involving the additions of 5 wt% Pt, the final Pt content of 20 wt% was reached. If the impregnation with Pt salt solution was done in a single step, the 20 wt% Pt electrocatalyst had Pt crystallites of the order 30–40 Å diameter, whereas sequential catalyzation using four additions of 5 wt% Pt yielded a 20 wt% Pt electrocatalyst containing Pt crystallites on the order of 20 Å diameter.

Grubb *et al.*[89] also demonstrated that silver supported on boron carbide was a good electrocatalyst for oxygen reduction in alkaline electrolytes. The electrocatalyst was prepared by impregnating the boron carbide with a silver acetate solution. It was subsequently dried, and the silver salt was reduced in hydrogen.

The impregnation technique has been applied to prepare supported catalysts containing metal salts that can be thermally decomposed to the metal oxide in air or reduced to the metal in a reducing environment. For example, the following metal salts, which are reduced in hydrogen to the metal, have been used to prepare supported catalysts: $HAuCl_4$,[90] $AgNO_3$,[91] $NiNO_3$,[92] $RhCl_3$,[93] $PdCl_3$,[93] H_2IrCl_6,[93] and H_2PtCl_6.[74,77] The advantages of using nitrate or chloride salts are that they are very soluble in water and the decomposition products on reduction in hydrogen are volatile, which minimizes contaminants on the electrocatalyst surface.

The impregnation technique lends itself to supported catalyst production on a commercial scale.[94] Generally, two approaches are taken. In the "dipping and soaking" technique, the support is immersed in the impregnating solution, and the variables—time, temperature, and solution pH—are controlled. The solution concentration, to a large extent, controls the metal loading on the support. The other main approach is a "spraying" technique, in which a fixed amount of support material is sprayed with a fixed amount of solution of known concentration. This latter technique yields better control over metal loading and also produces a catalyst that is fairly uniformly impregnated from particle to particle.

2. Ion-Exchange and Adsorption

On substrates that adsorb metal ions, there may be specific interactions between the ion and the surface. For instance, chloroplatinic acid adsorbs on acid silica–alumina, as well as on heavily oxidized carbons,[86] but not on silica gel surfaces. The most successful technique for adsorbing metal ions on catalyst supports has been to ion-exchange the metal amine complex $[M(NH_3)_n]^{z+}$, containing the noble metal M, for a proton on the support material (R). Exchange occurs by the following reaction:

$$zRH + [M(NH_3)_n]^{z+} \rightleftharpoons R_zM(NH_3)_n + zH^+ \qquad (3)$$

Quantitative exchange can be controlled by altering the pH of the solution; increasing the pH drives the reaction to the right. This technique[72,75,76,79,95,96] has been used to achieve high dispersions of precious metals on silica and alumina, and also for exchanging a wide variety of metal cations (Pd,[96] Co,[97] alkali and alkaline earth metals[98]) on carbon surfaces.

The Pt complex cation $Pt(NH_3)_4^{++}$ was used by Kinoshita[87] for adsorbing Pt ions on C, since this cation was successfully used on silica and alumina.[75,76] Since the amount of exchange could be controlled by adjusting the solution pH, the highly alkaline base $Pt(NH_3)_4(OH)_2$ was used. This base was prepared from a tetramine platinum chloride solution, which was slowly passed through the hydroxyl form of an anion-exchange resin (Rohm and Haas Company, Amberlite IRA-400) to convert the chloride to the hydroxide.[75]

In the presence of a strong base, the cation acted as the acid form of an ion-exchange resin, and the proton sites on the C surface could be titrated with the strongly basic platinum tetramine hydroxide (Fig. 3).

To increase the amount of cation exchange on the carbon

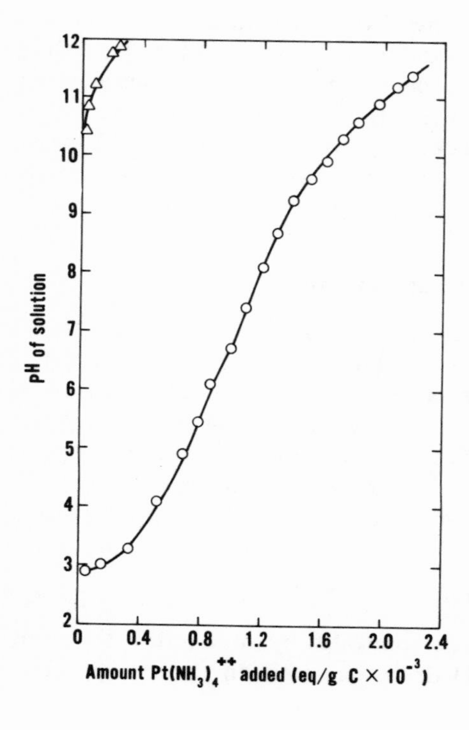

Figure 3. Ion-exchange adsorption of $Pt(NH_3)_4^{++}$ on graphitized Vulcan XC-72: (△) as received; (○) chromic acid oxidized (120° C, 2 hr).

support, it was necessary to increase the number of available exchange sites. The low concentration of proton sites on the C surface generally requires an oxidation step before exchange at higher concentrations can occur. Nitric acid,[96,97] ammonium persulfate,[97] and air[98] oxidation have been used to increase the concentration of exchangeable proton sites, most likely in the form of hydroxyl or carboxyl groups attached to the C surface.[8,10,99,100]

The pH titration curves in Fig. 3 show that the unoxidized graphitized C does not exchange an appreciable amount of $Pt(NH_3)_4^{++}$. After oxidation in chromic acid (100°C for 2 hr), the oxidized sample shows a substantial amount of cation exchange. A potentiodynamic sweep curve of chromic acid-oxidized graphitized Vulcan XC-72 clearly showed an increased concentration of quinone–hydroquinone species, where the hydroquinone-type species are potential exchange sites.

The ion-exchange electrocatalyst yielded a linear relationship between Pt surface area and Pt content (see Fig. 2), in agreement with the results obtained by Benesi *et al.*[75] for ion-exchanged Pt on silica gel. This linear relationship indicates that the average Pt crystallite size remains constant (corresponding to $100\,m^2\,g^{-1}$ Pt), independent of Pt content in the ion-exchange electrocatalyst. On the other hand, the average Pt crystallite size increases with Pt content for impregnated electrocatalysts.

Blurton[101] described the preparation of highly dispersed Pt on C by pyrolyzing the C form of a cation-exchange resin (Amberlite IRC 50) in which Pt cations were exchanged. It was claimed that Pt crystallite sizes of less than 15 Å were prepared, but it was concluded from comparison with the Pt crystallite size calculated from the Pt surface area that some Pt particles were not accessible to surface-area measurements, being "buried" in the pyrolyzed resin. Considering the economic necessity for increasing the available Pt electrocatalyst surface accessible to reactants, highly dispersed Pt prepared by the ion-exchange technique on C will probably result in a better utilization of noble metal than the ion-exchange procedure starting with the resin precursor. Moreover, there is relatively little control over the C morphology in the latter case.

A related form of noble metal catalyst support that has received much attention lately is the so-called "carbon molecular sieve."[102–108] A noble metal–carbon catalyst possessing molecular sieve properties is prepared by carbonizing an organic material, which is impregnated with the catalytically active metal. Cooper[102] patented the preparation of Group VIII noble metal–carbon molecular sieve catalysts. Briefly, the procedure involved adding an aqueous solution of the metal chloride to furfuryl alcohol, and then adding formaldehyde to reduce the metal cation. After metal catalyst particles were formed, the mixture was heated to polymerize and cure the furfuryl alcohol. The polymer was then crushed and heated at high temperatures ($>500°C$) in an oxygen-free atmosphere to carbonize the catalyst. Carbon molecular sieve catalysts have been investigated as hydrogenation catalysts,[103–105] but their application as electrocatalysts has not been pursued.

3. Colloidal Electrocatalyst Particles

Supported catalysts can be prepared by the adsorption of colloidal metal particles on the support, rather than by the deposition of an ionic species directly. Bond[109] described a procedure that yielded colloidal Pt particles of uniform size (about 15 Å diameter). These small Pt particles were obtained by the slow reduction of H_2PtCl_6 solution (100°C) using a dilute 1% sodium citrate solution. To this stable colloidal suspension was added a silica or alumina support, which adsorbed the Pt particles. Similarly, colloidal Pt on graphitized C was prepared in this laboratory,[87] but the average Pt crystallite size was about 40 Å, considerably larger than the crystallite size observed in solution as described by Bond. Colloidal Pt has also been prepared by passing an electric current between two Pt wires immersed in distilled H_2O. Modifications of this procedure are well documented by Ciapetta and Plank.[73]

A common technique for preparing metal electrocatalyst particles is to dissolve the salt containing the catalyst in an appropriate solvent, and then chemically reduce to the metal. There are numerous recipes in the literature for obtaining noble metal blacks of various surface areas. Wilenzick *et al.*[110] described recipes involving the reduction of a H_2PtCl_6 solution by a variety of reducing agents: sodium citrate, formaldehyde, acrolein, hydrogen peroxide, phosphorus in ether, tannin, and hydrazine hydrate. Sodium citrate reduction produced the smallest uniform Pt crystallite (40 ± 20 Å), and tannin reduction produced uniform crystallites of large dimensions, 500 Å. The preparation of high-surface-area Pt blacks or other colloidal dispersions requires extreme precaution as to water purity, glassware cleanliness, and other factors that can significantly affect the size and morphology of the crystallites.

Reduction of an aqueous solution of H_2PtCl_6 by KBH_4 produced smaller Pt crystallites than the catalyst prepared by hydrogen reduction.[111] The use of dilute KBH_4 as a reducing agent produced small average Pt crystallites, approximately 30–37 Å. Carefully controlled experiments were performed by Giner *et al.*[112] to prepare unsupported Pt blacks by the formaldehyde reduction of H_2PtCl_6. Pt blacks produced in this manner were tested for their activity for the anodic oxidation of propane in hot concentrated H_3PO_4. The most important factor controlling the preparation of high-activity Pt blacks appeared to be the order in which the H_2PtCl_6, HCHO, and NaOH were added. If a solution of H_2PtCl_6–HCHO was added to NaOH, an active Pt black was produced. Furthermore, a surface area of 22–32 $m^2\,g^{-1}$ Pt was obtained by this procedure, in contrast to the 2.5–7 $m^2\,g^{-1}$ Pt produced when HCHO was added to a H_2PtCl_6–NaOH mixture.

There appears to be a rather wide range of Pt surface area that can be produced by the reduction of H_2PtCl_6 in solution. Surface areas of 8–43 $m^2\,g^{-1}$ Pt have been reported for Pt blacks prepared by the reduction of H_2PtCl_6 with formaldehyde, borohydride, and hydrogen.[113] Another important factor in determining the Pt surface area is the concentration of H_2PtCl_6, which is reduced in solution. A decrease in H_2PtCl_6 concentration from 20 to 1 g/liter resulted in an increase in Pt surface area from 5 to 44 $m^2\,g^{-1}$ when $NaBH_4$ was the reducing agent.[113]

Recently, a technique to prepare high-surface-area colloidal Pt was described.[114] Chloroplatinic acid was neutralized in solution with Na_2CO_3, then $NaHSO_3$ was added to give a pH of approximately 4. The pH was subsequently readjusted to 7 with the addition of Na_2CO_3, which resulted in the formation of a white precipitate claimed to be $Na_6Pt(SO_3)_4OH$. The Na^+ cation was exchanged with Dowex 50, and after concentration by evaporation, the product was reported to be $H_3Pt(SO_3)_2(OH)_2$. This compound was then heated in air at 135°C for 1 hr to give a black glassy substance, which when dispersed in water yielded a stable colloidal Pt sol of particle size 15–25 Å diameter.

Turkevich and Kim[115] have investigated extensively the preparation and catalytic properties of colloidal Pd. Various reducing agents were used to reduce $PdCl_2$ in 3 *N* HCl, and the colloidal Pd was examined by electron microscopy. The Pd particle size was strongly dependent on the pH of the Pd salt solution when sodium formate was the reducing agent. The average Pd particle size increased from 200 to 325 Å in diameter with a decrease in pH from 11.2 to 2.1. Similarly, the average particle size of Pd formed with sodium citrate as the reducing agent was dependent on pH, with a pH of 6.1 producing the smallest average particle size (75 Å diameter).

Turkevich and co-workers[116,117] have also studied the colloidal chemistry of gold sols formed by the reduction of chloroauric acid ($HAuCl_4 \cdot 3H_2O$) by sodium citrate. More recently, Frens[118] was able to control the size of the gold particles in the range 150–1500 Å diameter. It was observed that the number of gold nuclei initially formed depended on the amount of sodium citrate added to the gold salt solution. Small amounts of citrate reducing agent produced relatively few nuclei, which then grew to form large particles as reduction proceeded. Large amounts of added citrate, on the other hand, produced more nuclei, which tended to grow less and produced smaller gold particles. The important factor governing the particle size was demonstrated to be the control over the relationship between nucleation and growth, not the amount of gold reduced from solution.

A novel procedure was described by Booth *et al.*[119] to prepare unsupported Pt and other metals. One equivalent of the metal halide was added to 15 equivalents of monosodium naphthalide dissolved in diglyme at ambient temperature. After 16 hr, the excess reagent was destroyed by ethanol, and the metal particles were separated from the solution by centrifuging. Gold, silver, copper, platinum, palladium, rhodium, ruthenium, iron, nickel and cobalt prepared in this manner were reported to have unusual properties. No crystallite size obtained with Pt was reported, but Pd catalysts with crystallite sizes of the order 60 Å diameter were produced. If ethanol was excluded from the preparation, smaller metal crystallites were produced; e.g., small Pd crystallites of 30 Å diameter were prepared. The authors also reported that other organic reagents were also successful in reducing the metal halides.

Noble metals other than Pt and Pd have also been prepared by the aqueous reduction* of suitable salt solutions (usually halides). Typical ranges of values reported in the literature are listed in Table 1 for Pt, Rh, Ru, Au, Ir, Ag, and Pd. A wide range of surface areas was obtained

Table 1
Surface Area of Noble Metal Blacks Prepared by Reduction of Salts in Solution

Noble metal	Reducing agent	Surface area ($m^2 g^{-1}$ BET)	Reference
Ru	$NaBH_4$	74.4	187
	$NaBH_4$	27	310
Au	$NaBH_4$	5.0	187
	HCHO	3.55	150
Rh	Diphenyl silane	40	311
	$NaBH_4$	28.1	187
	$NaBH_4$	35	146
	$NaBH_4$	26	312
	$NaBH_4$	15.4	310
	HCHO	100	146
Pd	$NaBH_4$	14.3	187
	$NaBH_4$	17.8	310
	HCHO	31.8	150
Ir	$NaBH_4$	8.3	310
Pt	KBH_4	13.4	313
	$NaBH_4$	4.9	187
	$NaBH_4$	44	146
	$NaBH_4$	5.0–43.8	113
	$NaBH_4$	16	312
	$NaBH_4$	2.7	310
	HCHO	20.0	146
	HCHO	8.0–30	113
	HCHO	32.9	150
	HCHO	8.9	313
	H_2	10	113
	H_2	22.5	313
	Li	26.9	313
Ag	HCHO	0.19	150

*The reduction of Ru salts by borohydride solutions results in an explosive Ru black when dried,[120] and this reducing agent should be avoided. Another dangerous combination is the preparation of Au blacks in the presence of NH_4OH, since,[120] apparently, shock-sensitive Au–N compounds are formed. See also Cusamano.[121]

for Pt; independent of the reducing agent, which again suggests that the experimental procedure and conditions involved in the preparation are important factors.

Supported noble metal blacks have been prepared by reduction of the noble metal salt in a solution that also contains the support material in suspension. This technique, with H_2 gas[122] as the reducing agent, was used to prepare Pt supported on C catalysts. Electron micrographs confirmed that the Pt was dispersed on the support. The use of this technique to prepare supported electrocatalysts is not widespread due to the difficulties in controlling the metal particle size, but it is an easy method for reclamation and extraction of noble metals from spent solutions.

4. Adams' Method

The so-called "Adams' method"[123] has also been widely used to prepare unsupported noble metal blacks.[124–130] In this preparative procedure, the noble metal oxide is formed by the fusion of the noble metal salt and an alkali metal nitrate. For example, $NaNO_3$ is mixed intimately with H_2PtCl_6 and then fused at 300–700°C.[123] The probable chemical reactions occurring under these conditions are

$$6NaNO_3 + H_2PtCl_6 \rightarrow 6NaCl + Pt(NO_3)_4 + 2HNO_3 \quad (4)$$

$$Pt(NO_3)_4 \rightarrow PtO_2 + 4NO_2 + O_2 \quad (5)$$

After being heated, the fused mass is cooled and washed thoroughly with water to remove nitrates and chlorides. Noble metal surface areas obtained after reduction of the oxides are summarized in Table 2.

Table 2
Surface Area of Noble Metal Blacks Prepared by Adams' Method[123]

Noble metal	Surface area ($m^2\,g^{-1}$ BET)	Reference
Pt	118	126
Pt	10–51	129
Pt	72.9	130
Rh	223	126
Ir	132	126
Ru	114	127

Adams' method appears to yield noble metal surface areas considerably higher than those obtained by the reduction of noble metal salt solutions (see Table 1). The lower surface areas obtained by reduction of noble metal salt solutions may arise from the inability to control the particle growth step that follows the initial nucleation process. Giner *et al.*[112] visualized the reduction of H_2PtCl_6 solution by formaldehyde as occurring by a two-step mechanism. First, nucleation of Pt occurs, followed by an electrochemical deposition of Pt, with the electrochemical anodic oxidation of HCHO occurring on the same Pt particles. This second growth process is considered to occur as a mixed electrode process. In order to form high-surface-area Pt (small Pt crystallites), it is necessary to inhibit the electrogrowth step in the reduction process. Adams' method, on the other hand, involves the intimate mixing and solidification of a metal oxide trapped in a matrix (nitrate salt). Physical separation of the metallic component is achieved by immobilizing the metal oxide in a solid matrix, and thereby minimizing coalescence among particles. Removal of the nitrate matrix by leaching is carried out at low temperatures, so sintering of the metal oxides is not encountered. One point of precaution is worth noting; the Pt catalyst usually contains residual amounts of Na,[131] even after extensive washings. This impurity may affect electrocatalytic reactions on the metal catalyst, so this possibility should be considered when using an Adams' preparation. Cahan and Ibers[132] also investigated the composition of platinum oxide prepared by the Adams' method and based on X-ray diffraction analysis; they concluded that the platinum oxide is a mixture of Pt, α-PtO_2 (possibly hydrated), and $Na_xPt_3O_4$, a sodium platinum bronze.

5. Raney's Method

Another technique commonly used to prepare unsupported catalysts is "Raney's method," named after Murray Raney,[133] who patented this method of catalyst preparation. This procedure, which was used extensively to prepare unsupported Ni catalysts (original preparations have been reviewed by Lieber and Morritz[134] and Ciapetta and Plank[73]) and electrocatalysts,[135,136] has been used more recently to make noble metal and noble metal alloy electrocatalysts.[137–142] The catalyst material is alloyed with aluminum or another base metal, and subsequently the base metal is leached from the alloy structure.

Smith *et al.*[143] studied the preparation and properties of Raney nickel catalysts obtained by varying the temperature at which the aluminum was leached out of the Ni–Al alloy structure by NaOH and the temperature of the subsequent digesting and washing. The Ni surface area varied between 18.2 and 37.5 $m^2 g^{-1}$ Ni, depending on the reaction temperature of the alloy and NaOH (−5 to 60°C) and the temperature of the subsequent treatment (35–70°C).

Burshtein *et al.*[144] prepared high-surface-area electrocatalysts starting with an alloy of 50% Ni, 48% Al, and 2% Ti. The specific BET surface areas of the electrocatalysts after leaching in 5 *M* KOH were 80–90 $m^2 g^{-1}$ and remained practically unchanged on heating in H_2 up to 800°C. Owing to the high surface areas, these catalysts are highly pyrophoric and must be handled with caution.

Krupp *et al.*[142] prepared Raney alloys by alloying Cu, Co, Ni, Pd, Pt, and Rh with Al at temperatures in the neighborhood of 700°C. If the Al content was present in a proportion greater than 1:1 atomic ratio, then the Al was difficult to leach out, but the Al content needed to be greater than about 20 wt%, otherwise the alloy became too ductile for preparing powdered specimens. The Al in the Raney alloy was removed by leaching in concentrated KOH at 80°C. The Al residue that remains from this process was finally removed by anodic dissolution in a KOH solution. Binder *et al.*[145] showed that the Pt surface area increased as the Al content in the Pt–Al alloy increased, with BET surface areas of 23.0, 33.0, and 43.9 $m^2 g^{-1}$ Pt obtained with $PtAl_2$, $PtAl_3$, and $PtAl_4$ alloys, respectively. Pt–Li alloys have been used[146] as the starting material to obtain unsupported Pt electrocatalysts with surface areas of 18–20 $m^2 g^{-1}$ Pt. Other electrocatalysts (Ag, W, Mo) have also been prepared by Raney's method for consideration in fuel-cell electrodes.[147]

The composition of the Raney Ni catalyst varies with the method of preparation. The presence of residual amounts of Na in Adams' catalyst has its parallel in Raney catalysts, in which the residual Al content is a function of the duration of the treatment with alkali.[134]

6. Thermal Decomposition

In general, the preparation of metal oxide and supported noble metal electrocatalysts involves a heat-treatment step in an inert or a reactive gas environment. Unsupported metal catalysts are also prepared by the direct thermal decomposition of metal salts. Supported noble metal catalysts are generally heat-treated in a flowing hydrogen stream at temperatures sufficiently elevated to reduce the metal salt to the metallic state. Even though this is the accepted procedure, the temperatures at which the decomposition step is conducted and the gaseous environments employed to optimize the noble metal dispersion have not been systematically investigated. Newkirk and McKee[147] determined the decomposition temperatures of rhodium, ruthenium, and iridium chlorides in air and hydrogen by thermal gravimetric analysis. They suggested that the catalyst support had an influence on the reduction temperatures of these noble metal salts, with the supported salt crystals decomposing at temperatures different from bulk crystals. More recently, Kinoshita *et al.*[148] used thermal gravimetric analysis and differential thermal analysis to study the decomposition of Pt salts (platinum diamino dinitrite, tetramine platinous hydroxide, tetramine platinous chloride, platinum acetylacetonate, and chloroplatinic acid) in hydrogen, air, and argon. The thermal stability of the Pt salts was sensitive to the gas environment, and the Pt complexes were more stable in air than in hydrogen, in which decomposition was observed for all Pt salt samples at temperatures below 200°C. By determining the lower temperature limits to reduce supported noble metal salts, one can avoid high temperatures, at which loss of surface area due to sintering is more likely to occur.

Temperature-programmed reduction (TPR) was applied by Robertson *et al.*[149] to determine the reducibility and to identify the alloying of Cu–Ni on silica catalysts. A H_2–N_2 gas mixture was passed over a copper- and nickel-nitrate-impregnated silica support, which was gradually heated at a programmed rate. The consumption of H_2 due to the reduction of species on the support was monitored by a thermal conductivity cell. This technique permitted a "fingerprint" of the catalyst reduction, and these characteristics were sensitive enough to study highly dispersed, low-loaded supported catalysts, which cannot be detected by X-ray diffraction. A further characteristic was that alloying could be detected by monitoring the H_2 uptake during the TPR. Although this technique has not been applied to catalyst systems of interest in electrocatalysis, it is apparent that electrocatalysts can also be similarly investigated. In effect, this technique is a gas-phase analogue of potentiodynamic scanning of electrocatalysts, and can be expected to see more use in the gas-phase arena as the facility of the technique is appreciated.

The thermal decomposition of unsupported noble metal salts (H_2PtCl_6, $PdCl_2$, and $HAuCl_4$) at 500°C in a nitrogen environment for 20 min produced very low BET surface area electrocatalysts[150]: 0.18, 0.20, and 0.02 $m^2 g^{-1}$, respectively. The thermal decomposition of silver oxalate at 375–500°C produced silver powder with surface areas in the range 2–6 $m^2 g^{-1}$.[151] The low surface areas are not too surprising in view of the rapid sintering rate of unsupported Pt black at low temperatures in hydrogen.[152]

Metal oxide electrocatalysts have also been prepared by the thermal decomposition of metal salts. Tseung *et al.*[153] prepared Li-doped nickel oxide by the high-temperature decomposition (950°C, 3 hr) of a physical mixture of lithium and nickel nitrate salts. The BET surface area was 0.4–0.5 $m^2 g^{-1}$, considerably lower than the surface area obtained on freeze-dried Li-doped metal oxide.[154] Cobalt molybdate with a BET surface area of 2.2 $m^2 g^{-1}$ was prepared by heating a physical mixture of ammonium molybdate [$(NH_4)_6Mo_7O_{24} \cdot 4H_2O$] and cobalt carbonate ($CoCO_3$) at increasing temperatures–500, 750, and 1000°C–for various times.[155] The thermal decomposition of tungstic acid in air at 600°C in a muffle furnace produced WO_3 with a BET surface area of 11.5 $m^2 g^{-1}$, but the original surface area of the tungstic acid precursor was not known. This is a critical factor in preparing high-surface-area materials.

Metal oxide bronzes have received considerable attention as electrocatalysts for fuel cells. Since a very comprehensive review dealing with the electrochemistry of bronzes[38] is available, the electrocatalytic property of this class of materials will not be discussed. The preparation of bronzes as *powders* deserves mention in the context of this review. A versatile method is the solid-state reaction of a metal tungstate, tungstic acid, and tungsten powder in an alumina crucible at 850°C under vacuum (10^{-5} Torr) for 2 hr.[156] The bronze powder was prepared according to the reaction

$$\frac{x}{2}M_2WO_4 + \frac{(3-2x)}{3}WO_3 + \frac{x}{6}W \rightarrow M_xWO_3 \qquad (6)$$

Excess reagents were subsequently leached out after heat treatment. Dickens and Whittingham[156] obtained BET surface areas in the range 0.2–2.3 $m^2 g^{-1}$ for alkali metal tungsten bronzes prepared by the solid-state reaction above. Niedrach and Zeliger[39] reported surface areas of 4–6 $m^2\,g^{-1}$ for Na_xWO_3 prepared by the procedure described above, followed by ball-milling. Finally, Broyde[40] prepared U, Na, K, Pb, Ba, Tl and Cd tungsten bronzes by reacting solid materials at 800–1000°C. The BET surface areas were all within 15% of 5 $m^2\,g^{-1}$.

Tungsten carbide has been studied extensively as an anode catalyst in acid fuel-cell systems.[157–160] There are numerous techniques available to prepare tungsten carbide that involve heat treatment at elevated temperatures. Mund *et al.*[157] prepared WC by reducing tungstic acid in hydrogen at 500°C for 5 hr and 700°C for 5 hr, followed by carburizing in CO for 6 hr at 860°C. Using a similar technique, Schulz–Ekloff *et al.*[158] obtained WC with a surface area of 17 $m^2\,g^{-1}$. Von Benda *et al.*[159] prepared WC by thermally decomposing an intimate mixture of tungsten powder, carbon black, and ammonium carbonate at 1300°C in hydrogen. The $(NH_4)_2CO_3$ was initially decomposed under reduced pressure at 200°C before the high-temperature treatment, and was added to produce a porous WC structure. Svata and Zabransky[161] obtained WC by carburizing tungsten, WO_3, or tungstic acid with CO at 900°C. By decreasing the carburizing times from 5.5 to 1.5 hr, they were able to increase the surface area of WC from 5.9 to 8.2 $m^2\,g^{-1}$. Alternatively, WC can be formed from colloidal suspensions of graphite in ammoniacal tungstic acid solutions that have been freeze-dried, and the resulting WO_3 can be reduced with hydrogen to produce tungsten powder mixed with carbon.[162] The mixture was heated to 1500°C *in vacuo* for 15 min to produce WC with a mean particle size of approximately 350 Å. Very fine WC particles have been prepared by reacting gaseous compounds in a hydrogen plasma.[163] CH_4 and WCl_6 in a mole ratio of 10:1 were introduced into a hydrogen plasma to produce WC particles with an average diameter of approximately 50 Å by electron microscopy and 50–80 Å by X-ray diffraction line-broadening.

It is clear that the formation of WC *per se* is not enough to form an active electrocatalyst for hydrogen oxidation. Stoichiometric WC is inactive, but when the surface of the WC is rendered carbon-deficient (by gentle oxidation in the presence of N_2H_4), tungsten sites become available for hydrogen molecule dissociative chemisorption[164] and promote the electrode reaction.

A binary compound containing Pt that has been reported to be an electrocatalyst for hydrogen oxidation in acid fuel cells is platinum phosphide, PtP_2, which Rhodes[165] prepared by a technique somewhat similar to Adams' method. In this technique, 4 mol Al were combined with 6 mol red phosphorus and 1 mol platinum black. The mixture was heated for 48 hr at 850°C in an evacuated sealed quartz ampul. After being cooled, the reacted sample was stirred by refluxing for 72 hr in 8 N H_2SO_4 in an oxygen-free atmosphere to extract all the aluminum and excess phosphorus. The BET surface area obtained was 0.24 $m^2 g^{-1} \pm 20\%$.

7. Freeze-Drying

Freeze-drying has been used in ceramics,[166–169] metallurgy,[170–173] and catalysis[174–177] to prepare a wide variety of high-surface-area materials. This technique generally involves four steps to reach the final product: (1) The salts containing the elements present in the final product are dissolved in a suitable solvent, commonly water. This step produces a homogeneous solution. (2) The solution is sprayed into a cold environment, which produces flash freezing of the solvent. The solutions are generally sprayed as fine droplets into cold hexane[178] or liquid nitrogen;[179] rapid freezing occurs, and homogeneity of the salt mixture is attained. (3) Sublimation of the solvent is then carried out. If the solvent is water, the structure obtained is very open due to the removal of H_2O in the ice structure. (4) The freeze-dried material is converted to the desired product by reduction or heat treatment.

Tseung and Bevan[41,180] prepared, by freeze-drying, mixed metal oxides that were active as oxygen reduction electrocatalysts in KOH. Semiconducting oxides are generally prepared by the high-temperature heat treatment of metal salts, and sintering usually results in metal oxides having very low surface areas ($<1\ m^2 g^{-1}$). Freeze-drying, however, yields finer particles, which require lower temperatures to form the mixed metal oxide. Tseung and Bevan[180] prepared freeze-dried Li-doped nickel oxide by starting with the acetate salts dissolved in 500 ml H_2O. The solution was sprayed as a fine jet into a dish containing liquid nitrogen, which froze the liquid droplets instantaneously. The solid sample was transferred to a vacuum bell jar and subjected to a vacuum of 10^{-1} Torr to remove all traces of water. The freeze-dried salts were decomposed *in vacuo* at 250°C, and then subsequently heat-treated in air at 300–1000°C for 3 hr to form the metal oxide. BET surface areas of Li-doped nickel oxide prepared by freeze-drying were on the order of 60 $m^2 g^{-1}$, considerably higher than those obtained by the conventional sintering of metal salts. King and Tseung[181] also prepared high-surface-area nickel–cobalt oxides (67 $m^2 g^{-1}$) by freeze-drying. The spinel Co_2NiO_4 was prepared starting from the nitrate salts, which were freeze-dried and then heat-treated in air at 300°C for 24 hr. Catalytically active materials with a perovskite structure have also been prepared by freeze drying.[176,177] Tseung[176] prepared Sr-doped $LaCoO_3$ by dissolving the nitrate salts [8.5 g $Sr(NO_3)_2$, 23.30 g $Co(NO_3)_2 \cdot 6H_2O$, and 17.81 g $La(NO_3)_3 \cdot 6H_2O$] in 500 ml distilled H_2O and then spraying the solution into liquid nitrogen as a fine jet. The solid was evacuated at 10^{-2} Torr for 48 hr and then decomposed at 250°C, followed by further heat treatment in air at 500°C for 3 hr. X-ray diffraction analysis indicated that a perovskite phase was formed, and the surface area measured by BET was 38 $m^2 g^{-1}$.

Freeze-drying techniques are not restricted to preparing catalyst supports and unsupported catalysts. Rowbottom[174] prepared supported catalysts by using freeze-drying to dry the porous solid impregnated with the catalyst salt. The patent claims that freeze-drying immobilizes the dissolved metal salt, and thereby enhances the catalyst distribution. Tseung and Wong[182] prepared freeze-dried platinum and silver supported on carbon electrocatalysts starting with an impregnated carbon substrate. Electron-microscopy examinations indicated that the freeze-dried Pt electrocatalyst had a smaller particle size and a more uniform distribution of catalyst particles than the conventionally prepared impregnated catalysts. On the other hand, supported Ag catalysts showed no difference in dispersion between the two techniques.

8. Binary Alloy Catalysts

Techniques described in the previous sections have also been applied to prepare alloy electrocatalysts. Unsupported binary alloys are commonly prepared by Adams' method,[31,130,183,184] Raney's method,[138,139,141,185] or the simultaneous reduction of a catalyst salt solution[186–194] containing the two metal components. The surface areas of the binary alloys appear to depend on the technique, with reduction of the salt solutions yielding the lowest values (Table 3). The presence of an alloy phase was determined by X-ray diffraction analysis, which technique was used generally to characterize the binary systems summarized in Table 3. Supported binary alloy catalysts are generally prepared by the coimpregnation of the support with metal salts and then reduction in hydrogen gas.[195–200]

A study of Pt–Rh alloys supported on graphitized carbon black (graphitized Vulcan XC-72) as an anode electrocatalyst was conducted by Ross *et al.*[198] The electrocatalysts were prepared by the coimpregnation of a graphitized carbon black ($70\,m^2\,g^{-1}$) using appropriate amounts of H_2PtCl_6 and $RhCl_3$ solutions. The impregnated C was evaporated to dryness and then reduced at 200°C for 2 hr in flowing H_2. The presence of single-phase alloys was determined by X-ray diffraction analysis, and only those catalysts with diffraction patterns having symmetrical peak profiles were considered as consisting of a single-phase alloy. Furthermore, the lattice parameters for the Pt–Rh alloys yielded a bulk alloy composition that was in good agreement with the chemical composition analysis. X-ray diffraction line-broadening yielded average alloy crystallite sizes in the range 50–100 Å in diameter. McKee *et al.*[126] have also prepared supported Pt–Rh alloy electrocatalysts by the coimpregnation

Table 3
Preparation of Unsupported Binary Alloys

Preparative technique	Binary alloy	Surface area ($m^2\,g^{-1}$ alloy)	Reference
Adams' method	Pt–Ni	35–48	130
	Pt–Ru	50–70	31
	Pt–Ru	50–120	184
	Pt–Rh	46–245	31
	Pt–Rh	90–120	184
	Pt–Ir	71–111	31
	Pt–Ir	90–160	184
	Pt–Pd	10–90	184
Reduction of salt solution	Au–Pd	40	189
	Au–Pd	8–18	187
	Pt–Ru	9–13	187
	Pt–Rh	16–35	187
	Pt–Rh	29–43	192
	Pt–Pd	3–6	187
	Pt–Pd	3–10	191
	Pd–Rh	3–10	191

technique, but the characterization of the alloy phases was not reported.

Supported binary alloys of Pt–Fe on graphitized carbon black (Graphon, $87\,m^2\,g^{-1}$) were prepared by coimpregnation with H_2PtCl_6 and $Fe(NO_3)_3$ dissolved in a 4:1 mixture of benzene and absolute ethanol.[196] This particular alloy system was one that could not be dried in air at high temperature (500°C), since phase separation forming α-Fe_2O_3 and Pt was observed. The presence of Pt–Fe alloys was determined by X-ray diffraction, and also by magnetic susceptibility measurements. This alloy system allowed the use of another analytical tool, Mossbauer spectroscopy, to further indicate that no free metallic Fe was present, but was alloyed with the Pt. Chemisorption measurements and electron-microscopy observations showed that alloy particles of 30–40 Å diameter were formed.

When the binary alloy content of a supported catalyst is sufficiently low or the metal crystallite size is sufficiently small, then it is difficult to determine by physical methods whether an alloy has been formed. This was clearly a problem in a recent investigation by Sinfelt,[199] who prepared silica-supported Ru–Cu and Os–Cu catalysts. The miscibility of the Ru–Cu and Os–Cu systems in the bulk state is extremely low, and the term *bimetallic cluster* was therefore used to refer to these binary metallic systems. The catalyst particle sizes were too small for resolution by X-ray diffraction, so the presence or absence of alloying could not be determined. In this situation, a catalytic reaction itself was used to serve as the probe to obtain evidence of synergistic interaction between the atoms of the two components. This is discussed further in Section IV.

An interesting approach to the formation of binary alloy electrocatalysts has been described by Pott[201] for the direct oxidation of methanol in sulfuric acid. It was claimed that group VIII metals alloyed with cocatalysts such as Pb, Tl, Sn, As, Sb, Bi, and Re are beneficial methanol electrooxidation catalysts. In particular, Pt–Sn catalysts were specified. The catalyst preparation involved depositing Pt in a highly dispersed form on a substrate, adsorbing H_2 on the Pt surface, and exchanging the adsorbed H_2 for cocatalyst metal ions in solution in a quantitative manner, e.g.,

$$4PtH + Sn^{4+} \rightarrow Pt_4Sn + 4H^+ \tag{7}$$

This reaction is, of course, akin to the ion-exchange technique [Eq. (3)]

discussed earlier. Other than Sn as a cocatalyst, claims were made for Ti, Ta, Re, Ru, and Os, although it is unlikely that *metal* alloys of Ti, Ta, and Re could in fact be produced from an aqueous solution by this method. Whereas exchange reactions as proposed in Eq. (7) look good on paper, it has been the experience of the authors that when conductive substrates such as C are used, the coupled oxidation of hydrogen atoms and reduction of metal salt can occur at distinctly different sites. Nonconductive substrates do not have this problem. The quantitative aspects of the deposition process given in Eq. (7) are no doubt maintained, but the exact distribution of cocatalysts on the noble metal crystallites is open to question.

IV. CHARACTERIZATION OF ELECTROCATALYSTS

There are several experimental tools available to characterize electrocatalysts. The three most common techniques are (1) electrocatalyst and support surface-area measurements by physical adsorption and chemisorption, (2) X-ray analysis, and (3) electron-microscopy examinations. Adsorption measurements on unsupported catalysts yield information on structural aspects (i.e., porosity) as well as on the surface area. For supported catalysts, on the other hand, only surface-area values have been measured, and no information on the microstructure of the electrocatalytic particles has been obtained by adsorption techniques. X-ray techniques are of two types, depending on the angle at which the X-ray striking the specimen is monitored. At small angles ($<2°$), continuous X-ray scattering occurs, and the information obtained is different from that determined at higher angles. The so-called "small-angle X-ray scattering" (SAXS) and the more common X-ray diffraction line-broadening technique are both applicable to the characterization of electrocatalysts. Use of the electron microscope complements and provides information on electrocatalyst particles that cannot be obtained by the other two techniques; i.e., the electron microscope allows the observer to "see" the electrocatalyst particles and extract information on size, shape, and distribution of the electrocatalyst.

1. Adsorption Methods

The most common techniques for measuring surface areas of catalysts are adsorption methods. These methods can be classified into two general categories: (1) physical adsorption, involving van der Waals forces; and (2) chemisorption, involving adsorbate–adsorbent interactions similar to those associated with chemical bond formation. Both of these adsorption processes have been extensively studied with regard to surface-area measurements.[202–205] A comprehensive review by Farranto[206] on the use of gas-phase chemisorption techniques for measuring surface areas of catalysts complements this.

(i) Physical Adsorption

(*a*) *Surface Area.* The amount of gas molecules adsorbed on the catalyst surface is related to the catalyst surface area. Since the interaction forces between the adsorbate and adsorbent are van der Waals forces, the extent of physical adsorption is related to the adsorbate boiling point, and the energy of adsorption is comparable to that of liquefaction. The temperatures must therefore be low relative to those used in chemisorption measurements.

Brunauer, Emmett, and Teller (BET) in 1938 formulated a theory[207] for the adsorption of gases on solids based on the Langmuir adsorption isotherm.[208] From this analysis was derived the well-known BET equation:

$$\frac{P}{V_{\text{ads}}(P_0 - P)} = \frac{C-1}{V_m C}\left(\frac{P}{P_0}\right) + \frac{1}{V_m C} \tag{8}$$

where V_{ads} is the total volume of adsorbed gas on the surface of the absorbent, V_m is the volume of adsorbed gas required for monomolecular coverage, C is the constant related to the net adsorption energy, P is the partial pressure of the adsorbing gas, P_0 is the saturation pressure of the adsorbing gas, and P/P_0 is the relative pressure.

For a given adsorbate gas–adsorbent system, C and V_m are constants, so a plot of the left-hand side of Eq. (8) vs. P/P_0 should yield a straight line with a slope of $(C - 1)/V_m C$ and a y intercept of $1/V_m C$. From the slope and the y intercept, the two unknowns V_m and C are calculated. The specific surface area S is then obtained from the equation

$$S = \frac{V_m N \sigma}{V_0} \tag{9}$$

where V_0 is the molar volume of gas at STP, N is Avogadro's number, and σ is the cross-sectional area of the adsorbing gas molecule in Å^2. The cross-sectional area σ of the adsorbing gas was determined by the following equation suggested by Emmett and Brunauer[209]:

$$\sigma = (4)(0.866)\left(\frac{M}{4/2N\rho}\right)^{2/3} \tag{10}$$

where M is the molecular weight of the adsorbate gas and ρ is the density of the liquefied adsorbate. The value of σ for N_2 at $-195°C$ calculated by Eq. (10) is 16.2 Å^2. Cross-sectional areas of molecules adsorbed on solid surfaces have been presented recently by McClellan and Harnsberger,[210] and these values of σ can be used in the BET equation.

Instead of using Eq. (8) to calculate V_m, Genot[211] suggested a new, simple way of determining the BET surface area by use of log–log plots of the variables V_{ads} and $(1 - P/P_0)^{-1}$. The coordinates of the tangent to the log–log plot having a slope $[d \ln V_{ads}/d \ln (1 - P/P_0)^{-1}] = 2$ represents the point corresponding to monolayer coverage by the adsorbed gas, V_m. Then the surface area can be computed in a fashion similar to that previously described.

A thorough examination of the nitrogen BET specific surface areas of various porous powders was carried out by 13 laboratories in England[212] with the objective of setting a standard for surface-area determinations. From the nitrogen adsorption measurements, various factors affecting the accuracy of nitrogen BET surface area measurements were noted. The more important factors can be briefly summarized:

1. The pretreatment and outgassing procedures must be controlled and reproducible.
2. The level of liquid nitrogen surrounding the sample must be kept constant to within a few millimeters, and the sample must be immersed in liquid nitrogen to a depth of at least 5 cm.
3. The temperature of the liquid nitrogen must be monitored.
4. Nitrogen used for adsorption must be of a purity greater than 99.9%.
5. Gas burettes and exposed leads containing appreciable volumes of gas must be thermostatted (preferably to $\pm 0.1°C$ or better).

The BET equation has been applied extensively to measure

surface areas of porous materials with a wide variety of adsorbate gases, but only a few examples will be considered here to illustrate this technique for characterizing electrocatalysts. Emmett and Cines[213] measured the surface areas of numerous carbon blacks using nitrogen, carbon disulfide, and butane as adsorbates. The latter two gases yielded surface areas approximately 20–50% lower than those obtained by the use of nitrogen. Kr and N_2 have been used to measure the surface area of unsupported Pd,[214] and Ar, N_2, Kr, and Xe have been used to measure the surface area of unsupported Pt.[215] Surface areas of Pt electrocatalysts have also been measured by the low-temperature nitrogen–BET technique. The "continuous flow method"[216,217] and conventional static adsorption measurements were made on a series of unsupported electrocatalyst Pt blacks before and after sintering in 96% H_3PO_4 at 150°C.[218] Discrepancies were observed in Pt black surface areas determined by the two methods, which were ascribed to the variabilities in the states of aggregation of the samples due to differences in sample preparation (see Table 4). The Pt black surface areas[218] were calculated from the low adsorption region ($<0.4\ P/P_0$) of the N_2 adsorption–desorption isotherm (see Fig. 4).

Adsorption from solution has been applied to obtain surface areas for powdered samples. These methods are generally simpler and more rapid than the conventional gas adsorption methods. Adsorption isotherms of phenol in aqueous solution on carbon blacks have been measured, and with the use of the BET equation, the phenol

Table 4
Surface Area, Multimolecular Layer Adsorption, and Capillary Condensation for Nitrogen at 78°K on Pt Black Electrocatalysts[a]

Pt black sample	Surface area ($m^2\ g^{-1}$ Pt) Continuous flow BET	Static BET	*t* curve
"As received"	30.5	37	32
Sintered in 96% H_3PO_4, 150°C, 128 hr at 800 mV	12.0	20	17

[a]From Kinoshita *et al.*[218]

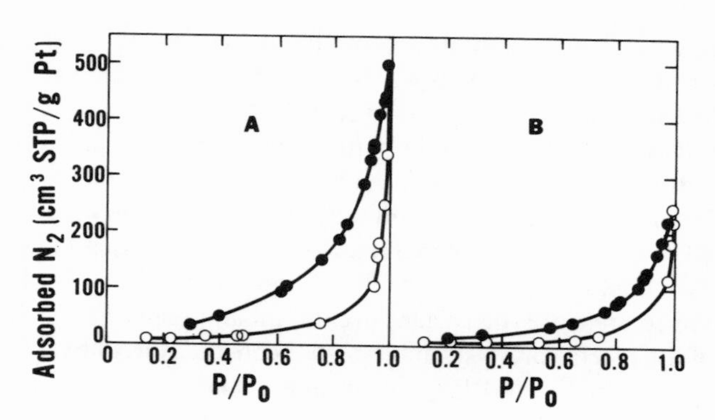

Figure 4. Nitrogen adsorption–desorption isotherms for unsupported Pt blacks at 78° K[218]: (A) as received, (B) sintered at 0.8 V for 128 hr in 96% H_3PO_4, 150°C, (○) adsorption, (●) desorption.

monolayer coverage was calculated.[219] Assuming that one molecule of phenol covers 41.2 $Å^2$ (corresponding to 2.64 $m^2\ mg^{-1}$ phenol), good agreement was reached between surface areas determined by nitrogen and phenol adsorption. Similarly, several organic dyes (e.g., methylene blue, safranine T, orange II, naphthalene red J) have been used to determine the surface areas of carbon powders.[220] Earlier work by Smith and Hurley[221] demonstrated that fatty acids (capric, lauric, palmitic, and stearic acids) were also useful adsorbates for determining surface areas of carbon blacks. Surface areas calculated assuming that the cross-sectional area of the acid molecule in the adsorbed layer was 20.5 $Å^2$ yielded reasonable agreement with areas calculated by other methods. Surface areas determined by fatty acid adsorption have also been demonstrated[222] for other porous samples (silica–alumina, cadmium, glass beads, iron, nickel, titanium dioxide, and zinc powders).

(*b*) *Catalyst Microstructure* (*Porosity*). At relative pressures at which nitrogen monolayer adsorption occurs, it is possible for nitrogen to condense in the small pores of porous powders. The empirical *t*-curve method of deBoer *et al.*[223] gives direct information on the surface area, capillary condensation, and distribution of pore sizes for microporous adsorbents. It is therefore possible to obtain information on the microstructure of microporous catalysts such as noble metal blacks from physical adsorption measurements. Analysis by the t-curve method requires that the multimolecular adsorption curve be measured over its whole range of pressures. The data are then expressed in terms of an average thickness of the adsorbed layer in Angstrom units. Considering the case of nitrogen adsorption, deBoer[223] used the thickness of one statistical layer as 3.54 Å and the thickness $t = 3.54n$, where n is the number of layers calculated from the ratio of the adsorbed volume v and the volume of a monolayer v_m. When the volume v is plotted vs. the thickness t, the slope of the line gives the specific surface area, S_t ($m^2\,g^{-1}$), by the equation

$$S_t = 15.47\frac{v}{t} \tag{11}$$

If the t plot is linear and passes through the origin, the adsorbent may be assumed to be nonporous. The Pt surface areas calculated in this fashion from the isotherms in Fig. 4 yielded values somewhat larger than the surface areas calculated from the continuous flow measurements (see Table 4). The t-curve method has been applied by Smith and Kasten[224] to examine the porosity of carbon blacks. The t-plots of various carbon blacks are plotted in Fig. 5, where the original carbon black (curve △) and the graphitized version (curve □) yield t-plots that pass through the origin and are therefore nonporous. The oxidized thermal black (curve ○), on the other hand, shows a positive intercept for the t-plot indicative of porosity arising from the oxidation treatment.

A simple pore size distribution was presented by Kinoshita *et al.*[218] for the electrocatalyst Pt black samples the adsorption–desorption isotherms of which are presented in Fig. 4. The relative pressure of the adsorbing gas (N_2 at -195°C) can be related to the pore diameter r through the Kelvin equation[225]

$$r = \frac{-4.14}{\log P/P_0} \tag{12}$$

which is uncorrected for the thickness of the multimolecular adsorption for a plane surface. Capillary condensation—represented by $(V_{des} - V_{ads})\,0.001558$, where V_{des} and V_{ads} are the amounts of N_2 adsorbed on the desorption and adsorption branch of Fig. 4—is plotted as a function of the uncorrected Kelvin radius in Fig. 6. These results show that capillary condensation increases with pore diameter,

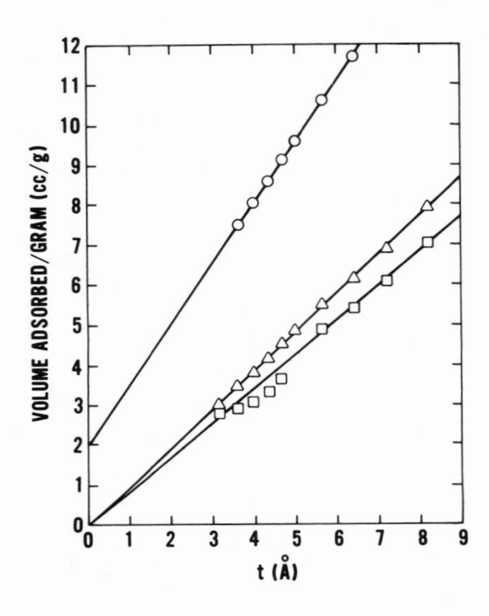

Figure 5. t-Plots for fine thermal (FT) C blacks using N_2 as absorbate at 77°K[224]: (△) standard FT black; (□) graphitized FT black; (○) air-oxidized FT black (600°C for 10 min).

and the total interparticle porosity of the "as received" Pt black is about 3–4 times larger than the electrochemically sintered Pt black. Furthermore, not only is the pore volume decreased on sintering, but also the pore-size distribution has shifted to larger diameters, which indicates an elimination of the small pores. The semiqualitative results obtained on the porosity of these Pt black samples was confirmed by electron-microscopy observations, which indicated that the "as received" Pt black underwent densification and the elimination of surface irregularities on sintering.[218]

Pore size distributions have been determined for porous carbon blacks[226] using a rigorous method developed by Barrett *et al.*[225] This technique involves the computation of pore volumes and area distributions directly from the desorption isotherms for porous materials, with correction of the pore radius for the thickness of the adsorbed multilayer. Recently, Viswanathan and Sastri[227] suggested a method for computing pore size distributions in terms of surface areas, rather than on pore volume, that was simpler than the method proposed by Barrett and co-workers. Other methods have also been described that allow calculation of the pore size distribution[228] and surface area distribution[229] of porous materials starting with the N_2 adsorption. These methods are outside the scope of this review, and the reader is directed to the references cited for more details.

Pore-size measurements using N_2 adsorption are limited to pore diameters of approximately < 300 Å, since P/P_0 becomes very close to unity for larger pores, and it may take days for equilibrium to be attained.[228] For larger pore diameters, the Hg porosimeter can be used if Hg does not wet or react with the porous solid. Pore diameters up to 10^4 Å can be readily measured.[230] The lower limit in pore diameter measurable by the Hg porosimeter (approximately 10–50 Å) is set by the highest pressure at which the Hg can be forced into the pores of the powder sample. This technique presents a difficulty, in that the high pressures employed can disrupt the pore system to be measured. Nevertheless, Paxton *et al.*[231] were able to measure the pore-size distributions of porous carbons of widely varying physical properties.

Another technique developed by Benesi *et al.*[232] allows the measurement of micropore volume at room temperature using conventional laboratory equipment. Benesi suggested the use of CCl_4 as the adsorbate, with various amounts of cetane added to control the

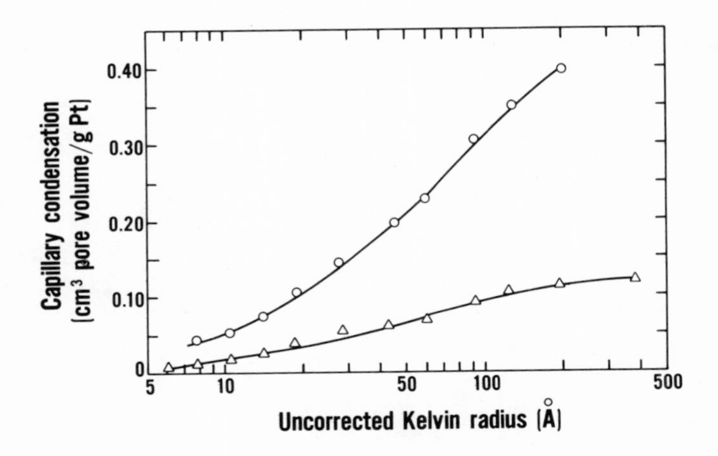

Figure 6. Capillary condensation as a function of Kelvin radius for unsupported Pt black[218]: (○) as received; (△) sintered at 0.8 V for 128 hr in 96% H_3PO_4, 150°C.

partial pressure of adsorbing gas. The porous sample is evacuated in a desiccator containing the CCl_4–cetane mixtures. After equilibration, the weight gain by the sample represents the volume of condensed CCl_4. This procedure has the advantage of simplicity. Furthermore, pore-volume measurements to higher Kelvin radii (800 Å at $P/P_0 = 0.95$) than obtained by nitrogen BET adsorption can be routinely attained. There are numerous other techniques available to determine pore structures and surface areas of porous solids.[202,203]

(ii) Chemisorption

Physical adsorption cannot be used to measure the surface areas of supported electrocatalysts, since the adsorbate gas forms a layer irrespective of the adsorbent. Chemisorption, on the other hand, can yield selective adsorption of gases and hence a separation of the surface area of an electrocatalyst from its support. Table 5 lists a variety of metals and common gases that form chemisorbed layers. Noble metal electrocatalysts, except for Au, all form chemisorbed layers with H_2, O_2, and CO. These gases have been used to determine the surface areas of supported noble metal electrocatalysts. Surface areas of unsupported noble metal electrocatalysts have also been determined by chemisorption.

Surface areas of electrocatalysts have been measured by both gas-phase and liquid-phase chemisorption techniques.[67] Gas-phase chemisorption involves measuring the pressure and volume and

Table 5
Metals and Gases That Form Chemisorbed Layers[a]

Metal	Gases[b]					
	N_2	H_2	O_2	CO	C_2H_4	C_2H_2
W, Mo, Zr, Fe	+	+	+	+	+	+
Ni, Pt, Rh, Pd	−	+	+	+	+	+
Au	−	−	−	+	+	+
Cu, Al	−	−	+	+	+	+
Zn, Cd, Sn, Ag	−	−	+	−	−	−

[a]Adapted from Whyte.[205]
[b](+) gas chemisorbed; (−) gas not chemisorbed.

relating these parameters to the amount of adsorbed gas, whereas in the liquid-phase, electrochemical techniques involving the measurement of the coulombic charge associated with chemisorbed gases are applicable.

Gas-phase chemisorption has been thoroughly investigated as a tool for surface-area determination.[67,74,233–239] Both the continuous-flow[233–237] and the static volumetric[67,237–239] methods have been applied to measure the surface areas of supported catalysts. In general, the continuous-flow method involved the chemisorption of H_2[233–235] or CO[236,237] and relating the amount of chemisorbed gas to the metal surface area. The volume of adsorbed gas was determined by passing an inert gas (He or N_2) over the catalyst sample after chemisorption of H_2 or CO and measuring downstream the amount of desorbed gas. H_2 chemisorption overestimates the surface area of a supported metal in catalysts exhibiting H_2 spillover.[59]

A typical procedure[67] to obtain surface areas of Pt supported on C electrocatalysts is as follows: The sample was first pretreated at 300°C for 3 hr under H_2 to ensure reduction of surface Pt oxides, followed by evacuation at 10^{-5} mm Hg at 350°C for 3 hr to desorb H_2 from the Pt surface. Then chemisorption measurements were made at 24 ± 1°C using H_2 or CO at gas pressures between 30 and 400 mm Hg. Some reversible adsorption of CO and H_2 associated with the C support was observed, which was compensated for by extrapolation to zero pressure. H_2 chemisorption can be similarly used to determine the surface areas of Rh,[93,198,240] Ir,[93,241] and Ru.[93,242,243] Pd dissolves H_2, and therefore complicates the surface-area measurement by chemisorbed H_2,[93] but as Aben[244] has demonstrated, H_2 chemisorption may be used if the experimental conditions are chosen to minimize errors due to palladium hydride formation (1 mm Hg H_2 pressure and 70°C). CO chemisorption has also been used to measure the surface areas of Pd,[93] Rh,[93,240,245,] and Ir[93,241] catalysts. Sinfelt and Yates[93] suggested that CO chemisorption was applicable for surface-area measurements of supported Ru, contrary to more recent studies reported by Dalla Betta.[242] Taylor[246] expanded on the chemisorption techniques to determine Ru surface areas, and showed the feasibility of using O_2 adsorption at ambient temperatures and H_2–O_2 titrations at 100°C.

Before the amount of CO or H_2 gas chemisorbed on the catalyst surface can be related to a surface area, the stoichiometry of the chemisorption bond between adsorbate and adsorbent must be known. Freel[233] and Wilson and Hall[239] concluded that the saturation coverage for H_2 on small Pt crystallites supported on silica or alumina corresponds to one H atom per surface Pt atom at room temperature. For atomically dispersed Pt, the H: Pt (total) ratio should be 1, and values of H: Pt = 2 should be treated with suspicion.[247] The H chemisorption stoichiometry H: M (surface), where M is Pd, Rh, Ru, or Ir, is also 1. CO, on the other hand, yielded less than one chemisorbed CO molecule per surface Pt atom (0.83–0.96) for Pt crystallites less than 60 Å.[247] Dorling and Moss[78] concluded that the ratio CO: Pt (surface) depends on Pt crystallite size, decreasing from 2 (bridge-bonded CO) on large crystallites to 1 (linear-bonded CO) on the highly dispersed Pt. CO and H_2 chemisorption measurements on Pt black and Pt supported on graphitized C electrocatalysts by Bett *et al.*[67] yielded CO :H stoichiometry close to unity. To avoid the controversy between linear- and bridge-bonded CO, it is sufficient to assume an adsorbed CO molecule surface area of 11 $Å^2$ (25°C). Multiplying by the measured saturation CO molecule coverage yields a surface area for the Pt catalyst.

Another technique for determining surface areas of supported and unsupported Pt involves the titration of preadsorbed O_2 on Pt with H_2 at room temperature.[215,239,248] The titration reaction[215] on supported Pt was represented as

$$\text{PtO(surface)} + \frac{3}{2}\text{H}_2\text{(g)} \rightarrow \text{PtH(surface)} + \text{H}_2\text{O(support)} \quad (13)$$

It is evident that three H atoms are required for each Pt surface atom. This results in a threefold increase in accuracy over H_2 chemisorption. For unsupported Pt black, titration with H_2 can be similarly described by Eq. (13), except that the water is not chemisorbed. The stoichiometry represented by Eq. (13) has been the subject of much controversy. A close examination of the titration technique by Wilson and Hall[247] led to the conclusion that the stoichiometry was dependent on Pt crystallite size. For impregnated Pt–silica catalysts, the ratio H_2 chemisorption: O_2 chemisorption: H_2 titration changed from about 1.6:1:3.8 to about 1:1:3 as the particle size increased, reflecting a change in the stoichiometries of the chemisorptions. Thus, the titration reaction represented by Eq. (13) would apply to supported Pt that was not atomically dispersed [H: Pt (total) = 0.5]. Recently,

Kikuchi *et al.*[248] verified Wilson and Hall's observations on the variable stoichiometry for the H_2 titration reaction on supported Pt. Mears and Hansford[249] reported chemisorption ratios of 2:14, which they explained by the variable stoichiometry of the O_2 chemisorption. Despite the obvious advantages of the enhanced volume uptake of H_2 by titration of oxidized Pt, the dubious O_2-equilibrium-coverage-to-surface-Pt ratio (which may be crystallite-size-dependent) is a drawback to the use of this technique for Pt surface-area measurements. An analogous surface titration technique has been described by Wentrcek *et al.*[250] using CO to determine the surface area of Pt catalysts. The chemisorbed O_2 reacted with CO at room temperature according to the stoichiometric equation

$$\text{PtO (surface)} + 2\,\text{CO} \rightarrow \text{PtCO (surface)} + \text{CO}_2 \qquad (14)$$

By determining the CO consumed or the CO_2 formed or both, a quantitative measure of the number of chemisorbed O atoms was obtained. It was found on bulk as well as on supported Pt catalysts that the maximum value of O: Pt (surface) $= \frac{1}{2}$, so the Pt surface area could be obtained from CO titration.

The degree of dispersion of other noble metals, Ir,[241] Rh,[245] Pd,[251,252] and Ru[246] have been measured by H_2 titration of the noble metal surface oxide. The stoichiometry for the titration of chemisorbed oxygen on Ir was found to be the same as observed on Pt, [Eq. (13)]. Rh, on the other hand, has a significant uncertainty with regard to the stoichiometry of chemisorbed O_2, except at high Rh dispersions, where the ratio was O:Rh = 1.5. Thus, H_2 titration is not recommended for surface-area analysis of Rh. The H_2 titration of oxidized unsupported Pd can be represented by the reaction[252]

$$\text{Pd–O (surface)} + \tfrac{3}{2}\text{H}_2 \rightarrow \text{Pd–H} + \text{H}_2\text{O (gas)} \qquad (15)$$

Since this technique relies on the water produced from the H_2–O_2 titration desorbing, it will not be applicable to supported Pd systems in which water is adsorbed on the support. Boudart and co-workers[251] pointed out that H_2 titration of supported oxidized Pd is feasible if precautions are taken to ensure complete removal of the water formed in the reaction. In principle, if Pd is supported on a hydrophobic material such as highly graphitized C, the addition of small amounts of previously dried alumina, silica, or molecular sieve mixed in with the catalyst could scavenge the water generated during H_2 titration.

H_2–O_2 titration on Ru[246] was assumed to have the same stoichiometry as expressed for Pt in Eq. (13), but in all instances, the *caveat* regarding equilibrium oxide stoichiometry on very small crystallites of noble metals must be considered.

By taking advantage of the ion-exchange properties of porous materials, one can determine surface areas by a simple method that is somewhat analogous to the gas-phase titration technique described previously. In general, high-surface-area metal oxides contain surface complexes that can react with cations or anions via an ion-exchange mechanism. The ion-exchange properties of silica were used to prepare highly dispersed supported Pt catalysts as discussed in a previous section. Sears[253] was also able to determine the specific area of colloidal silica by an ion-exchange titration with sodium hydroxide in the relatively short time of 10–15 min. Similarly, the ion-exchange adsorption of zinc and copper ions on silica was shown by Kozawa[254] to offer a method of surface-area determination. The surface areas of RuO_x,[255] an electrocatalyst used in the chloralkali industry, and MnO_2,[256] used in batteries, has also been measured by the ion-exchange adsorption of zinc. The Zn^{2+} exchanges with protons bound to surface oxides, and the amount of exchange is determined by titration of the Zn^{2+} concentration in solution using EDTA with Eriochrome Black T as the indicator. At saturation coverage, the occupied area per adsorbed Zn^{2+} is taken to be 17 Å^2.

The chemisorptive properties of carbon were used by Kronenberg[257] to develop a method for determining the relative surface areas of carbon powders. The powder samples were suspended in an acidic or alkaline electrolyte saturated with oxygen. After saturation, the electrolyte was purged with an inert gas to minimize the dissolved oxygen content. The oxygen chemisorbed on the carbon surface was reduced electrochemically, and the quantity of oxygen reduced at the electrode was determined by the coulombic charge from the current–time curves and related to the concentration of active adsorption sites. Whereas this technique is qualitative, the measurements do give an indication of the relative surface areas of carbon powders.

The heterogeneous nature of graphite surfaces has been well established, and the existence of two distinct surface sites, basal plane and crystallite edges, has been well documented. The relative concentrations of these two forms of carbon atom sites are important in determining the corrosion behavior of carbons used in electrochemical systems. The preferential adsorption properties of organic molecules on graphite "edge" and basal plane sites allow a determination of the surface area associated with the two crystallographic sites. Groszek and Andrews[258] demonstrated this measurement for various graphitized C samples. The surface areas associated with basal plane sites were determined from adsorption isotherms of *n*-dotriacontane from solution in *n*-heptane. It was assumed that one molecule of *n*-dotriacontane occupies 178 $Å^2$ when adsorbed at monolayer coverage. One surface property of graphite is the preferential adsorption of polar solutes onto the polar "edge" sites. This behavior allowed the selective adsorption technique to be used without interference in the adsorption measurement at basal plane sites. The surface area associated with the graphite edge sites was determined by the use of *n*-butanol adsorption from *n*-heptane, under the assumption of 27 $Å^2$ per molecule of *n*-butanol.

Electrochemical techniques have also been used to measure the surface areas of supported and unsupported electrocatalysts. Bett *et al.*[67] made a comparison of gas-phase and electrochemical measurements in 1 *M* H_2SO_4 for chemisorbed CO and H_2 on Pt crystallites supported on graphitized C. They found that the stoichiometries for the ratios of H:Pt and CO:Pt were the same in both the liquid- and gas-phase environments. Application of a potentiodynamic sweep[259,260] to a Pt electrode in H_2SO_4 yields a current–potential curve from which the quantity of chemisorbed H_2 can be determined. A potentiodynamic current–potential curve for a Pt black electrocatalyst is shown in Fig. 7. It should be noted that potentiodynamic curves obtained on high-surface-area unsupported Pt black electrocatalysts fabricated into porous flooded electrode structures show extremely good resolution of the current peaks and good interelectrocatalyst reproducibilities. This is due to the extremely high electrocatalyst-surface-to-electrolyte-volume ratio within the porous electrode structure. (An unsupported Pt black, 24 $m^2\ g^{-1}$ Pt, 65% porosity, or 30 $m^2\ g^{-1}$ Pt, 70% porosity, has $3.7 \times 10^{-7}\ cm^3$ electrolyte/cm^2 Pt.) Effectively, the electrocatalyst surface is maintained in an extremely clean electrolyte environment. The adsorbed H_2 is formed during the cathodic potential sweep (0.4–0.05 V), and can be represented in an acid medium by the charge-transfer reaction (Volmer reaction):

$$Pt + H^+ + e \rightleftharpoons Pt\text{–}H_{ads} \quad (16)$$

Monolayer coverage of H_2 has been shown to be constant for sweep rates as high as 800 V/sec.[261] At potentials less than about 0.05 V in dilute H_2SO_4 at room temperature, H_2 evolution resulting from the Tafel reaction

$$2\,Pt\text{–}H_{ads} \rightleftharpoons 2\,Pt + H_2 \quad (17)$$

becomes appreciable. This causes H_2 molecules to dissolve in the electrolyte within the porous electrode structure, and produces an anomalously high value for the electrocatalyst surface area. Saturation

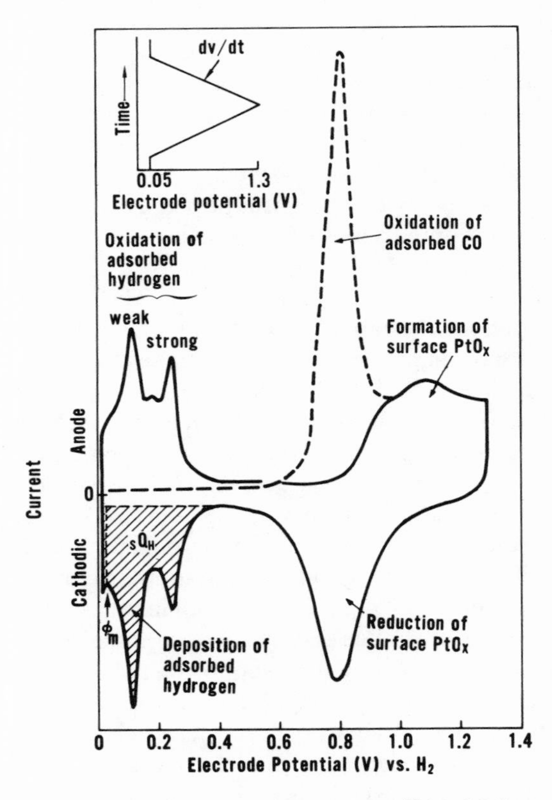

Figure 7. Potentiodynamic sweep of Pt[67]: Solid line: 20% H_2SO_4, N_2; dashed line: 20% H_2SO_4; CO, 30 min; N_2, 10 min; 24°C.

or monolayer coverage by adsorbed H_2 requires a distinct separation of charge associated with reactions 16 and 17. The monolayer coverage ${}_sQ_H$ is usually taken as the minimum ϕ_m in the current–potential curve, following cathodic deposition of the weakly bonded hydrogen species[261,262] or at $\eta = 0.005$ V.[67,68,263] Woods[264] evaluated the error associated with the assumption of defining the monolayer coverage of H_2 on Pt by the potential ϕ_m. This results in an underestimation of ${}_sQ_H$ for Pt. To obtain greater precision in the determination of ${}_sQ_H$, Woods measured H_2 adsorption isotherms on Pt in 5 *M* H_2SO_4 at $-72°C$. At this low temperature, the rate of evolution of molecular H_2 was lowered so that the current minimum ϕ_m was shifted to 0.01 V, which rendered a more accurate estimate of ${}_sQ_H$. The integrated charge ${}_sQ_H$ (corrected for double-layer charging) can be related to the Pt surface area by assuming 210 μC/real cm^2 Pt.[67,68,265] Based on the crystallographic geometry of Pt surface atoms on the three common crystal faces, the numbers of surface atoms[266] are 1.5×10^{15}, 1.3×10^{15}, and 0.92×10^{15} atoms/cm^2 for the (111), (100), and (110) crystal faces, respectively. If one H_2 atom adsorbs per Pt surface atom, the corresponding charges are 242, 210, and 150 $\mu C/cm^2$, respectively. Platinum surface areas of supported and unsupported Pt electrocatalysts determined by gas-phase techniques were in good agreement with results obtained by the potentiodynamic sweep technique in H_2SO_4[67] using 210 μC/real cm^2 Pt.

Potentiodynamic stripping of chemisorbed CO on Pt has also been used to determine Pt electrocatalyst surface areas.[67,267] Typically, CO molecules are adsorbed on Pt by holding the electrode potential constant at 0.05 V (vs. N.H.E.) while flushing the H_2SO_4 electrolyte with CO.[67] When saturation coverage by CO is achieved, the electrolyte is purged with N_2 to remove any dissolved CO. The potential is increased linearly, and the resultant current–potential curve for the oxidation of CO on Pt shown in Fig. 7 is obtained. Whereas one electron was transferred during the oxidation of adsorbed H_2, two electrons were transferred[67] when oxidizing CO to CO_2 on Pt. There have been considerable discussions in the literature on whether CO can be bridge-bonded on Pt. This point is trivial at saturation coverages at which CO may be used to calibrate noble metal electrocatalyst surface areas. Due to steric effects, not every surface atom can accommodate one CO molecule. Bett *et al.*[67] showed that

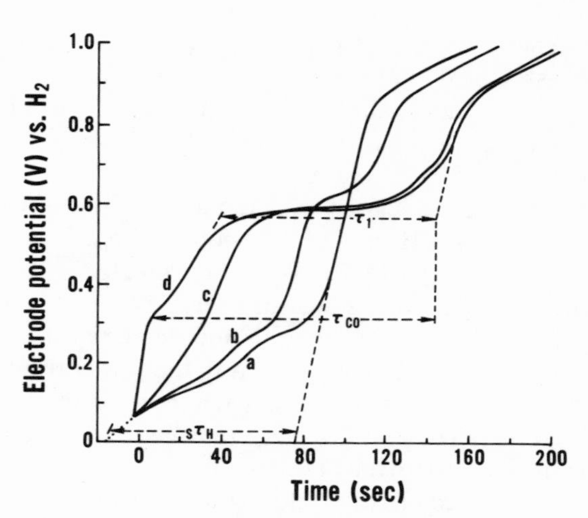

Figure 8. Charging curves on platinized Pt ($50 \, cm^2$ apparent surface) at 50 mA in 0.5 M H_2SO_4 after different times t_{ad} of CO adsorption at open circuit[270]: (a) $t_{ad} = 0$; (b) $t_{ad} = 100$ sec; (c) $t_{ad} = 500$ sec; (d) $t_{ad} = 2000$ sec.

electrochemical determinations of CO–H atom ratios were 0.87–0.89 for unsupported Pt blacks and 0.89–0.90 for Pt supported on graphitized Spheron 6 (a channel black).

Galvanostatic stripping of adsorbed O_2 or H_2 on Pt is another usual electrochemical technique for measuring Pt surface areas.[268,270] Figure 8 shows a galvanostatic charging curve for the oxidation of adsorbed H_2 on platinized Pt in 0.5 M H_2SO [270] The adsorbed H_2 layer is formed at low potentials ($\sim$0.05 V) and then oxidized by applying a constant current of 50 mA. Potential transients show three distinct regions: (1) H_2 oxidation up to about 0.4 V, (2) double-layer charging from 0.4 to 0.9 V, and (3) oxide film formation at $\gtrsim$0.9 V. The transition time $_s\tau_H$ and the current I yield a measure of the surface coverage by adsorbed H_2, $_sQ_H = I_s\tau_H$, and hence the Pt surface area. Connolly *et al.*[268] showed that the surface area of Pt-supported C electrocatalysts could be determined by galvanostatic stripping if the charge consumed by C was compensated. Thus, the Pt surface area S_{Pt} is

$$S_{Pt} = \frac{Q_{Pt,H}}{k_H} = \frac{Q_{T,H} - Q_{C,H}}{k_H} \tag{18}$$

where $Q_{Pt,H}$, $Q_{T,H}$, and $Q_{C,H}$ are the charges consumed on the Pt, total electrode, and C support, respectively, during the H_2 oxidation transition time. k_H is the proportionality constant relating charge to real Pt surface area, and was determined to be 190 μC/real cm^2 Pt,[268] less than the generally accepted value of 210 μC/real cm^2 Pt discussed above.

Similarly, Connolly and co-workers[268,269] showed that the transition time for the galvanostatic stripping of O_2 on Pt could be related to the Pt surface area of supported electrocatalysts, again compensating for the charge associated with the C support. The constant analogous to k_H above, in this case, was equal to 450 μC/cm^2 Pt. Galvanostatic stripping of adsorbed CO on Pt, shown by the potential transient in Fig. 8 (curve d), also allows a calculation of the CO surface coverage. Assuming 210 μC/real cm^2 Pt for adsorbed H_2 oxidation charge and a CO/H of 0.9,[67] the charge equivalent for CO oxidation was 378 μC/real cm^2 Pt.

There are many comparisons that can be made concerning the gas-phase and the electrochemical H_2 chemisorption methods for Pt surface-area determinations.[67] In both techniques, the Pt surface must be carefully pretreated to remove any impurities that may inhibit the chemisorption of hydrogen atoms. This is generally achieved in the gas phase by heating the catalyst sample to 200–400°C in H_2 for a few hours, followed by evacuation. For electrochemical removal of impurities, the electrode potential is increased anodically to oxidize any undesirable adsorbed species. The electrochemical measurements require only a small sample ($\sim$10 mg) compared to gas-phase measurements ($\sim$500 mg), due to the high accuracy with which current parameters are measurable. The electrochemical surface-area measurements, however, cannot be made on Pt supported on C electrocatalysts containing only a small weight fraction of Pt, due to interference from the double-layer charging current associated with the C support.[67,268,271] Connolly *et al.*[268] and Urisson *et al.*[271] investigated the effect of the surface concentration of Pt deposited on C supports ($S_C = 0.4$–$1200\ m^2\ g^{-1}$) on the specific surface area of Pt, S_{Pt}. They reported that the Pt surface area could be obtained by an electrochemical method (galvanostatic stripping of adsorbed H_2) if $S_{Pt}/S_C \geq 0.006$–0.01, which corresponds to greater than approximately 1 wt % Pt supported on C. Bett *et al.*[67] reported that the potentiodynamic sweep technique did not yield meaningful values of

S_{Pt} for their Pt–C electrocatalysts containing 1% Pt. On the other hand, the gas-phase chemisorption measurements for low-loaded Pt-supported electrocatalysts (1% Pt) have been verified.[67]

Besides Pt, the surface areas of Rh, Ru, and Ir electrocatalysts have been obtained by galvanostatic and potentiostatic techniques. If an accurate measure of the surface coverage by chemisorbed H_2 is known at a given potential, then dividing the H_2 charge determined to that potential by the fractional coverage allows a method for determining the monolayer coverage, and hence the real surface area. For these noble metals, however, there is some controversy over the potential variation of the coverage by chemisorbed H_2. On Rh, the surface coverage by chemisorbed H_2 at 0.05 has been reported to be[272] $\theta_H = 0.5$ and[146] 0.80 in 1 N H_2SO_4, based on galvanostatic and potentiostatic methods, respectively. For Rh in more concentrated H_2SO_4, Breiter[273] reported $\theta_H = 0.75$ at 50 mV in 2.3 M H_2SO_4 (30°C). Ross *et al.*[198] determined the surface area of a 17.5% Rh supported on graphitized C black electrocatalyst using gas-phase chemisorption of H_2. A surface area of 60 $m^2\,g^{-1}$ Rh was obtained assuming 7.6 $Å^2$/Rh surface atom.[240] This value compared favorably with the 54 $m^2\,g^{-1}$ Rh obtained from the anodic portion of the H_2 oxidation current recorded during a potentiodynamic sweep in 1 M H_2SO_4. The lower limit for the potential sweep was 0.04 V, and a value of 210 μC/real cm^2 Rh was assumed. The lower limit of the potential sweep was not extended to avoid the evolution of molecular H_2 interfering with the measurement of the adsorbed H_2 oxidation current. This study[198] suggested that the surface area of Rh would be determined from the H_2 charge integrated to 0.04 V in 1 M H_2SO_4, although the surface coverage by H_2 is less than unity at this potential.

The measurement of the H_2 monolayer coverage on Ir in H_2SO_4 at room temperature cannot be performed easily due to interference from molecular H_2 evolution at potentials near the reversible H_2 potential. By decreasing the temperature to −72°C, Woods[264] was able to separate the potentiodynamic sweep current due to H_2 chemisorption and molecular H_2 evolution on Ir, and thus obtain the saturation H_2 coverage ${}_sQ_H$, in 5 M H_2SO_4. Using this value and measuring the fractional hydrogen coverage on Ir as a function of potential in 5 M H_2SO_4 at 20°C, he was able to calculate Ir surface areas. At 20°C, the H_2 coverage on Ir measured at 0.06 V corresponded to $\theta_H = 0.65$, assuming 1 cm^2 Ir was equivalent to 218 μC. The error in this method may be very significant, as Woods points out, since the slope of the hydrogen adsorption isotherm on Ir is greatest at this potential, and differences in the isotherms on different Ir surface structures will be important.

Electrochemical surface-area measurements by the potentiodynamic sweep technique on Ru electrocatalysts are also very difficult to obtain due to the lack of separation between the currents associated with the deposition of adsorbed H_2 and the evolution of molecular H_2. Binder *et al.*[139] proposed that the oxidation current measured in the potential range 0–0.3 V (vs. N.H.E.) during an anodic potential sweep in 3 N H_2SO_4 can be attributed to the oxidation of adsorbed hydrogen on Ru. Since the formation of an oxide on Ru occurs at low potentials (relative to oxide formation on Pt), an overlap in the oxidation currents for H_2 and oxide formation exists, and the adsorbed H_2 charge cannot be easily determined. It is questionable whether the H_2 coverage measurement described by Binder and co-workers is a meaningful method for determining surface area of Ru electrocatalysts.

Electrochemical surface-area measurements on Pd using H_2 deposition require the separation of charges associated with H_2 adsorption on the surface Pd sites from those due to absorption and molecular H_2 evolution. Burshtein *et al.*[274] used galvanostatic stripping to determine the surface area of Pd blacks. The Pd black was cathodically polarized to 0 V in 0.5 M H_2SO_4 and then anodically polarized with a constant current (1–3×10^{-7} Å cm^{-2} real surface area). Charge passed in the potential range 0–0.08 V was attributed to the oxidation of absorbed H_2, and charge due to H_2 oxidation from 0.09–0.28 V was assumed to correspond to H_2 chemisorbed on Pd. The Pd surface area calculated assuming 210 $\mu C\ cm^{-2}$ Pd yielded good agreement with BET measurements on Pd blacks with surface areas 2–100 $m^2\ g^{-1}$ Pd. Gromyko[275] reported that H_2 deposited on Pd by the potentiodynamic sweep technique in 0.5 M H_2SO_4 can also be applied to determine the Pd surface area if experimental conditions are chosen such that the quantity of absorbed H_2 in Pd can be neglected. The amount of hydrogen adsorbed on Pd, ${}_{H}Q_{ads}$, does not depend on the potential sweep rate v. On the other hand, the amount of charge ${}_{H}Q_{abs}$ for the absorption of hydrogen in Pd depends on the rate of H_2 diffusion in Pd, so ${}_{H}Q_{abs} = kv^{-1/2}$. The intercept ($v \rightarrow \infty$) of linear plots of Q_0 (total charge) vs. $v^{-1/2}$, gives values for ${}_{H}Q_{ads}$ from which the Pd surface area is calculated. Gromyko applied this technique for determining the surface areas of smooth Pd electrodes. It is doubtful whether potential sweep measurements at high scan rates on highly dispersed supported Pd electrocatalysts will yield meaningful data, since double-layer charging currents will become very significant. Instead of relying on the measurement of chemisorbed H_2 on Pd as a means of determining Pd surface area, Rand and Woods[276] proposed using the O_2 adsorption coverage. This procedure avoids the problems associated with separating H_2 adsorption from absorption, but depends on anodizing Pd in 1 *M* H_2SO_4 under conditions in which monolayer O_2 coverage is achieved. The monolayer oxide is reduced by a cathodic potential sweep, and the integrated oxide reduction charge is used to calculate the Pd surface area (1 real cm^2 Pd = 424 μC for $2e^-$ transfer). Rand and Woods also demonstrated the feasibility of this method for surface area measurements of Rh and Au. With highly dispersed electrocatalytic materials, however, phase oxide formation and anomalous metal–oxygen stoichiometries observed on small metal crystallites may lead to erroneous surface-area measurements.

The adsorption of hydrogen atoms on Pt at potentials positive to the reversible hydrogen electrode potential is an example of "underpotential deposition." Numerous other examples involving cations depositing at an underpotential on Pt can be cited: Cu,[277–281] Pb,[277] Au,[277] Ag,[277,282–284] Tl,[285] and Bi.[286] A detailed survey of underpotential deposition of metals has been presented by Kolb *et al.*,[287] and the results have been correlated with work function differences between the substrate and the deposit.[288] A detailed analysis of the underpotential deposition of Cu on Pt suggests the possibility of developing new adsorption techniques for measuring surface areas of Pt. Careful control of the potential of Pt electrodes in dilute H_2SO_4 with Cu ions added show that a monolayer of Cu will be adsorbed before bulk Cu deposition occurred. By measuring the charge amounting to a monolayer coverage of Cu on Pt and assuming 490 $\mu C/cm^2$, based on the atomic radius of Cu,[278] the Pt surface area can be determined, with a 2.3-fold increase in accuracy over H_2 chemisorption. This increase in accuracy is an advantage, but the routine application of this method has not been fully established for surface-area measurements on highly dispersed Pt. Nevertheless, the adsorption of metals on noble metal electrocatalysts offers an interesting alternative to H_2 chemisorption as a tool for surface-area measurements, and should be investigated further.

From the metal surface area it is easy to calculate an average particle size. Considering a sphere or cube with a dimension d as its diameter or side length, then the surface areas are πd^2 and $6d^2$, respectively, and the corresponding volumes are $\frac{4}{3}\pi d^3$ and d^3. For a spherical or cubic particle with density ρ, the surface area per unit weight S is

$$S = 6/\rho d \tag{19}$$

The average diameter calculated from the surface area is identical to the surface average diameter d_s, discussed later. Electrocatalyst particles on supports would be expected to have lower active surface areas than suggested by the crystallite discussions due to shielding of the crystallite particle by the support interface. Unfortunately, the distribution of particle sizes within a given electrocatalyst sample and the lack of real knowledge of the crystallite shape and the nature of the electrocatalyst support interface preclude this from being an important consideration at this time.

2. X-Ray Analysis

(i) X-Ray Diffraction Line-Broadening

X-Ray diffraction analysis yields information on both the structure and the composition of crystalline materials. Small crystallite sizes or strain in polycrystalline materials cause broadening of the X-ray diffraction line profiles. The average crystallite size perpendicular to the diffraction plane d_x can be calculated from the Scherrer equation[289]

$$d_x = \frac{k\lambda}{\beta \cos \theta} \tag{20}$$

where k is a constant dependent on the crystallite shape and the way in which β and d_x are defined, λ is the wavelength of the X-ray beam, β is the linewidth free of all broadening due to instrumentation and crystallite strain, and θ is the Bragg angle. The term β has been defined in two ways: (1) the angular width at half-maximum intensity $\beta_{1/2}$ and (2) the integral breadth β_I, which is equal to the integral intensity divided by the maximum intensity ($\beta_I = \int I_{2\theta}\, d\theta / I_{\max}$).

Using the integral breadth and assuming a crystallite with a cubic structure, the following equation was obtained[289]:

$$\beta_I = \frac{\lambda}{\cos\theta} \frac{V}{\int T dV} \tag{21}$$

where V is the volume of the crystal and T is the thickness of the crystal measured normal to the reflecting plane. This equation states that the volume average of the thickness of the crystal is obtained from the integral breadth in Eq. (21), using $k = 1$, and the crystallite diameter is usually considered as the cube root of the crystallite volume. The alternative form of β, i.e., $\beta_{1/2}$, does not lend itself to the simple theoretical interpretation as β_I, but is experimentally easier to obtain. Theoretical calculations by Patterson[290] indicate that β_I is consistently higher than $\beta_{1/2}$ by amounts varying from 4 to 30%.

Warren[291] has suggested that the geometric mean value of β corrected for instrument broadening, β_{instr}, and using the measured $\beta_{1/2}$, should be applied for the determination of d_x. The geometric mean is defined as

$$\beta = [(\beta_{1/2} - \beta_{\text{instr}})(\beta_{1/2}^2 - \beta_{\text{instr}}^2)^{1/2}]^{1/2} \tag{22}$$

Other equations have been considered to correct for β_{instr}, but will not be considered here. The value of k can vary from as little as 0.7 to as much as 1.70, depending on the crystallite shape, the indexes of the reflecting plane, the definition of d_x, and the β that is adopted. Due to these uncertainties, the absolute accuracy of the average crystallite diameter does not exceed 25–50% in most studies. Klug and Alexander[289] suggest the use of $k = 0.9$ with the definition $\beta = \beta_{1/2}$ and $k = 1.05$ when $\beta = \beta_I$ is adopted. Experimentally, $\beta_{1/2}$ is generally more accurately measured and has more practical significance.

The relationship between Pt crystallite sizes and the angular widths at half-maximum for the (111) diffraction plane are plotted in Fig. 9 using Eqs. (22) and (20) to calculate β and d_x, respectively. The Pt crystallite size can be calculated using measurements from other diffraction planes [e.g., (200), (220)] if the intensities can be resolved from the background. This is not possible for supported Pt catalysts with small crystallites, in which only the (111) diffraction plane is usually resolved from the background intensity. The range of average crystallite sizes generally detected by X-ray diffraction line-broadening is of the order of 50–1000 Å in diameter for supported catalysts. For very small crystallites, the width of the diffraction line profile due to the crystallite size is very broad and is not discernible above the background, whereas at the other extreme, the line profile of large crystallites is very sharp and insensitive to crystallite size (see Fig. 9). Supported catalysts have a further complication that may affect the line-broadening measurement: the major diffraction lines of the support may be coincident with those of the catalyst particles, but this does not appear to be a problem for noble metals supported on C blacks.

Adopting the interpretation that the average crystallite size determined by X-ray diffraction line-broadening is a volume average diameter, it can be compared with the volume average diameter [defined in Eq. (49)] determined by electron microscopy. Defining the value of d_x as the cube root of the volume implies cubic particles that can be related to the volume average diameter d_v for spherical crystallites by

$$d_v = d_x\left(\frac{6}{\pi}\right)^{1/3} \tag{23}$$

A detailed study of the particle size of Pt supported on silica gel showed

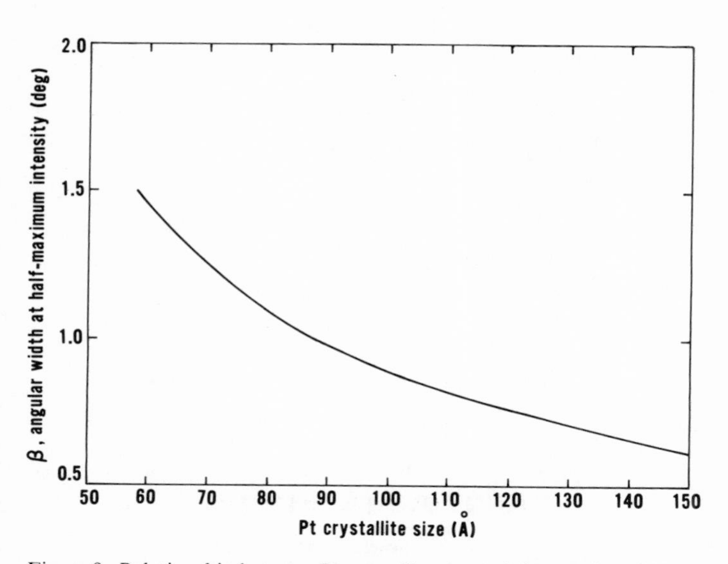

Figure 9. Relationship between Pt crystallite size and the angular width at half-maximum intensity for the (111) diffraction plane.

very good agreement with the volume average diameter of Pt calculated from X-ray diffraction and electron microscopy.[74] A similar analysis for supported and unsupported Pt in this laboratory yielded good agreement for the calculated volume average diameters (see Table 7).

Binary alloys have been investigated extensively as alternative catalysts for Pt in electrochemical systems. X-ray diffraction techniques have yielded definitive information on the phase composition and crystallite size of these binary alloy systems. The presence of single-phase alloys can be determined by the shape of the diffraction peak profiles, with asymmetrical peak profiles indicative of phase separation.[197,292,293] The X-ray diffraction profiles for (111) planes of Pd–Rh alloys[293] illustrate this point very clearly in Fig. 10. The asymmetrical peak profiles are clearly evident at alloy

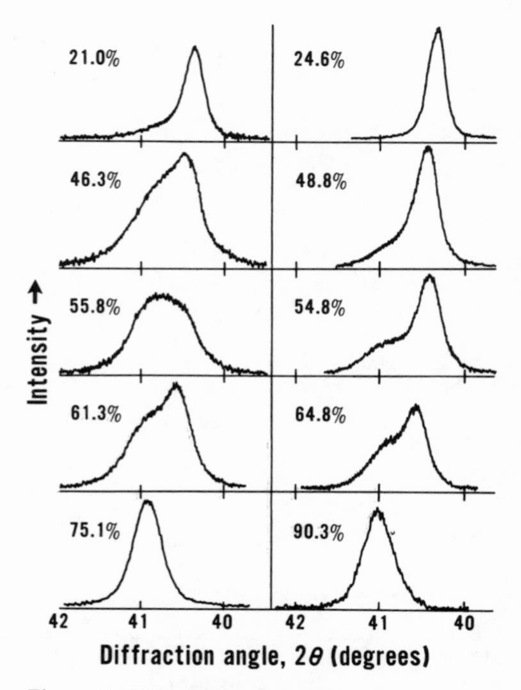

Figure 10. X-Ray diffraction profiles from the (111) diffraction planes of Pd–Rh alloy films. The composition is in at % Rh.[293]

compositions between 25 and 75 % Rh, indicative of phase separation.

The compositions of binary alloys can be obtained from the lattice constants determined by X-ray diffraction (the centroids of the individual X-ray diffraction peak profiles are used to determine the lattice constants). To calculate the lattice constant *a* accurately, there are various mathematical methods for correcting errors. One such method is the Nelson–Riley function[294]

$$a \text{ vs. } \frac{1}{2}\left(\frac{\cos^2\theta}{\sin\theta} + \frac{\cos^2\theta}{\theta}\right) \tag{24}$$

which when extrapolated to $\theta = 90°$, yields the lattice parameter. This procedure was used by Moss *et al.*[293] and Ross *et al.*[198] to determine the composition of Pd–Rh and Pt–Rh alloys, respectively.

(ii) Small-Angle X-Ray Scattering (SAXS)

At small angles from the primary X-ray beam, X-ray scattering is caused by scattering centers in a matrix that arise from differences in electron densities. The intensity of X-ray scattering depends on the size and shape of the heterogeneous electron density centers. Consider the case of a supported metal catalyst. The microporous holes in the catalyst support itself give rise to intense X-ray scattering; furthermore, these scattering centers are approximately the same size as the catalyst particles. Interference by the X-ray scattering signal from the support must be eliminated before the SAXS characteristics of the metal catalyst can be observed.

There are some characteristic features that are different for the two X-ray techniques, which are worth mentioning. X-Ray diffraction line-broadening yields an average crystallite size determined from specific crystallographic planes, whereas SAXS makes no distinction between different crystal orientations. Lattice distortions can therefore be observed by X-ray diffraction, but not in SAXS. Furthermore, since only specific oriented crystals contribute to an X-ray diffraction peak signal, the intensity is much less than the SAXS intensity, to which all particles contribute.

The theory of SAXS will be considered only briefly, since there are very detailed analyses of this technique in the literature.[295,296] The scattered intensity $I(h)$ from a particle depends on the number, distribution, and type of atoms in the particle. The intensity can be expressed[295] as

$$I(h) = I_e(h)F^2(h) \tag{25}$$

where h is the scattering angle 2θ divided by the X-ray wavelength (Å):

$$h = \frac{2\theta}{\lambda} \approx \sin\frac{2\theta}{\lambda} \tag{26}$$

$I_e(h)$ is the intensity of scattering by one electron and $F(h)$ is the "structure factor" of the particle, which is defined as the ratio of the total scattered amplitude to the amplitude, scattered on a single electron under the same conditions, i.e.,

$$F(h) = \frac{A(h)}{A_e(h)} \tag{27}$$

For N noninteracting particles, the scattering intensity is

$$I(h) = I_e(h)NF^2(h) \tag{28}$$

At small angles, the structure factor was approximated by Guinier and Fournet[295] to be

$$F^2(h) = n^2 \exp\left(-\frac{h^2R_g^2}{3}\right) \tag{29}$$

where n is the total number of electrons in the particle and R_g is the radius of gyration of the particle. Substituting into Eq. (28) yields

$$I(h) = I_e(h)Nn^2 \exp\left(-\frac{h^2R_g^2}{3}\right) \tag{30}$$

or

$$\log_{10} I(h) = -\frac{h^2R_g^2}{6.9} + \text{const} \tag{31}$$

Thus, the radius of gyration is obtained from the slope of the $\log I(h)$ vs. h^2 relationship at small angles. The particle diameters that can be observed by SAXS must be in the range of 20–500 Å.[297] This can be more readily visualized by some typical R_g values tabulated at small angles θ calculated in Table 6 for Cu K_α X-rays.[298] At increasing angles, the Guinier and Fournet approximation, Eq. (29), is not valid, and so the initial slope (low angles) must be used to calculate R_g.

Table 6

Radius of Gyration for Various Small Angles Using CuK_{α} X-Rays[298]

θ (deg)	R_g (Å)
0.05	202
0.1	101
0.5	50.5
1	25.3
2	12.1

SAXS was used by Somorjai[298] to follow the growth of Pt particles supported on alumina during heat treatment at elevated temperatures. Since the pores in the alumina matrix are scattering centers that contribute to the scattering intensity, this X-ray signal had to be eliminated before meaningful measurements of Pt particle sizes were possible. This was achieved by applying high pressure ($>$100,000 atm) to the supported catalyst and thereby collapsing the alumina pore structure to a size that did not contribute to the SAXS intensity.

Plavnik *et al.*[299] measured the average Pt particle size of Pt supported on C by SAXS. The effect of X-ray scattering from the porous C on the Pt particle size measurement was compensated by the difference in X-ray scattering intensity of the support with and without Pt. It was assumed that the total intensity was a summation of the intensities of the scattered X-rays by the Pt and the C micropores, and interference effects were neglected. Somorjai[298] and others[300] have pointed out that interference from X-ray scattering due to micropores in the support does not necessarily allow the simple correction applied by Plavnik *et al.* to be a valid assumption.

The problem associated with X-ray scattering due to micropores in silica–alumina cracking catalysts was solved by Gunn,[301] who used high-electron-density liquids as pore maskants to eliminate X-ray scattering from pores. This technique was more recently applied by Whyte *et al.*[300] to study the SAXS of Pt metal dispersions on alumina catalysts. Organic liquids such as CH_2I_2 and C_4H_9I were added to Pt supported on alumina catalysts to eliminate the background X-ray scattering from the micropores of the support. Whyte *et al.*[300] have demonstrated the usefulness of SAXS as a tool for characterizing supported catalysts where the average surface area $\langle S \rangle$ of Pt particles supported on an alumina matrix can be derived from the scattered intensity produced with a line-shaped X-ray beam $I_1(h)$ by the equation

$$\lim_{h \to \infty} I_1(h) = \frac{\Delta\rho^2}{16\pi^2 h^3} \langle S \rangle \tag{32}$$

where $\Delta\rho$ is the difference in the electron density of Pt and Al_2O_3. The average Pt volume $\langle V \rangle$ for a distribution of particle radii can also be determined by use of the relationship

$$\int_0^\infty 2\pi h I_1(h)\, dh = \Delta\rho^2 \langle V \rangle \tag{33}$$

Combining Eqs. (32) and (33) yields another radius known as the Porod radius R_p:

$$R_p = \frac{3 \int_0^\infty 2\pi h I_1(h)\, dh}{\lim_{h \to \infty} 16\pi^2 h^3 I(h)} = \frac{3\Delta\rho^2 \langle V \rangle}{\Delta\rho^2 \langle S \rangle} \tag{34}$$

For a distribution of spherical particles of average radius $\langle r \rangle$

$$\langle S \rangle = 4\pi \langle r^2 \rangle \tag{35}$$

$$\langle V \rangle = \frac{4\pi}{3} \langle r^3 \rangle \tag{36}$$

and substituting Eqs. (35) and (36) into (34) gives

$$R_p = \frac{\langle r^3 \rangle}{\langle r^2 \rangle} \tag{37}$$

For spherical particles, the intensity can be approximated by

$$I_1(h) \approx \frac{8(5)^{1/2}\pi}{9} \Delta\rho^2 \langle r^5 \rangle \exp\left(-\frac{4\pi^2 R_g^2 h^2}{5} \right) \tag{38}$$

where R_g is the spherical radius of gyration or the Guinier radius. R_g can then be determined from the square root of the slope of log $I_1(h)$ vs. h^2 at small values of h, and is given by[302]

$$R_g = \left(\frac{\langle r^7 \rangle}{\langle r^5 \rangle} \right)^{1/2} \tag{39}$$

The two parameters obtained from SAXS, R_p and R_g, can be used to obtain information on the distribution of particle sizes. Whyte *et al.*[300] assumed that the particle-size distribution for $Pt\text{–}Al_2O_3$ catalysts could be represented by the log–normal distribution function

$$P(r) = \frac{1}{(2\pi)^{1/2} r \ln \sigma} \exp\left(-\frac{\ln r - \ln \mu}{(2)^{1/2} \ln \sigma}\right)^2 \tag{40}$$

where μ is the geometric mean of the distribution and σ is the square root of the variance of the distribution. This distribution function has the property that if the variable r is distributed log–normally, then so are its moments r^n; i.e.,

$$\langle r^n \rangle = \int_0^\infty r^n P(r)\, dr \tag{41}$$

and using Eq. (40)

$$\langle r^n \rangle = \exp(n \ln \mu + \tfrac{1}{2} n^2 \ln^2 \sigma) \tag{42}$$

Now the moments of the log–normal distribution can be expressed by the two parameters μ and σ. Since in Eqs. (37) and (38), the moments can be described in terms of experimentally derived SAXS parameters R_p and R_g, respectively, we can write

$$R_g = \exp(\ln \mu + 6 \ln^2 \sigma) \tag{43}$$

$$R_p = \exp(\ln \mu + 2.5 \ln^2 \sigma) \tag{44}$$

or by solving for μ and σ

$$\ln \mu = \ln R_g - 1.714 \ln R_g/R_p \tag{45}$$

$$\ln^2 \sigma = 0.286 \ln R_g/R_p \tag{46}$$

Thus, the log–normal distribution is defined in terms of experimentally determined SAXS parameters, and a Pt particle-size distribution curve can be realized. Using this procedure, Whyte *et al.*[300] showed that the mean Pt particle diameter and the particle size distribution of an aged $Pt\text{-}Al_2O_3$ catalyst shifted to larger sizes relative to a fresh catalyst. Renouprez *et al.*[303] have also examined supported Pt catalysts ($Pt\text{-}Al_2O_3$) and obtained good agreement between Pt surface areas determined by SAXS and gas-phase chemisorption. As a final example, size-distribution measurements obtained on a colloidal ThO_2 sol by electron microscopy and SAXS showed excellent agreement[304] for SAXS data converted to a log–normal distribution.

3. Electron Microscopy

The electron microscope is a powerful tool to characterize electrocatalysts, and has frequently been used to complement other analytical techniques. Electron microscopy allows one to observe the size, shape, and distribution of electrocatalyst particles in a short period of time, but measurements of crystallite parameters from electron micrographs are tedious and time-consuming. It has been reported[233,239,247] that Pt crystallites smaller than 10 Å can be resolved by electron microscopy, and from these observations, average Pt crystallite diameters of less than 20 Å have been obtained. More recent work[305] concluded that for samples with a major fraction of the Pt particles less than 25 Å, the Pt particle-size distributions obtained from electron microscopy become increasingly subject to error. The accuracies of particle-size distributions from electron microscopy were examined from both an experimental and a theoretical point of view. Flynn *et al.*[305] found that the particle sizes and detectabilities depended on defocus conditions, and hence on the elevation of the particle in the supported sample with respect to the plane of focus.

Both bright-field (BFEM) and dark-field (DFEM) electron-microscopy techniques have proved useful for characterizing electrocatalysts. The electron optical arrangements necessary to obtain bright- and dark-field images are shown schematically in Fig. 11. In bright-field operation, the incident electron beam is transmitted through the specimen with a portion of the beam diffracted. The transmitted electron beam passes through the objective aperture and contributes to the bright-field image. On the other hand, the electrons diffracted by the specimen are intercepted by the objective aperture, do not contribute to the final image observed in the electron microscope, and will appear as a dark area on the observation screen. High-resolution dark-field images can be obtained by filtering the incident electron beam so that a chosen diffraction spot moves to the position formerly occupied by the undeviated beam. Now the diffracted beam passes down the axis of the objective lens, and the transmitted beam is intercepted by the objective aperture. The image obtained from the diffracted electron beam is the so-called "dark-field image."

DFEM imaging techniques were discussed by Hall[306] and have been applied to study the microstructures of C blacks[307,308] and

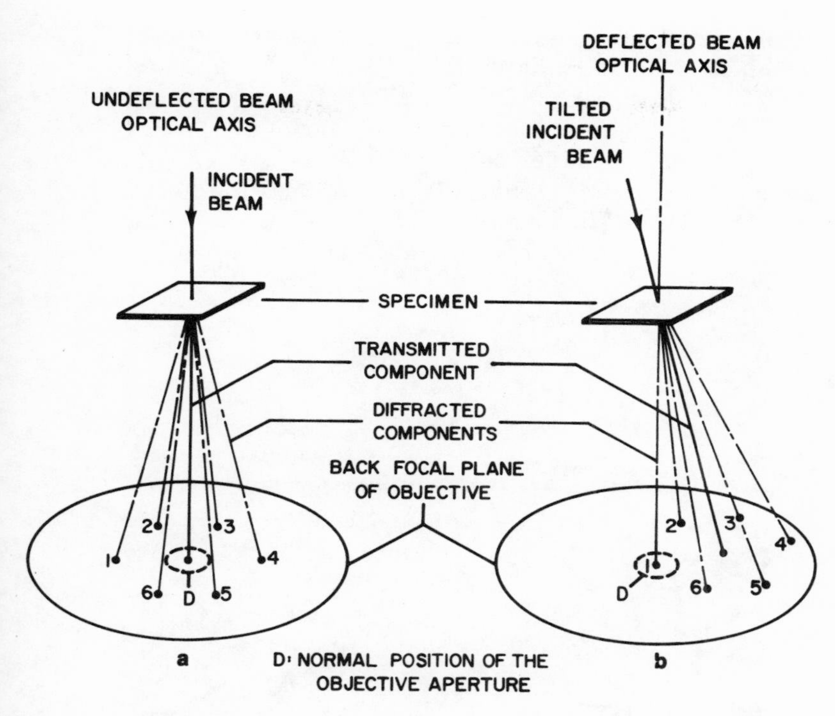

Figure 11. Schematic representation illustrating the mechanics for the formation of bright-field (a) and dark-field (b) images in the electron microscope. (From *Fine Particles Second International Conference*, Ed. by W. E. Kuhn and J. Ehretsmann, 1974, p. 199. Reprinted by permission of The Electrochemical Society, Inc.).

natural graphite.[309] BFEM can also yield information on the microstructure and morphology of electrocatalyst supports such as C blacks. Two such examples are presented in Fig. 12 for a furnace black (Vulcan XC-72R) before and after heat treatment at 2700°C. High-resolution electron micrographs indicated a "turbostratic" nature to the C layer planes, which on heat treatment became more ordered and assumed a polygonal morphological structure. For unsupported electrocatalysts, such as Pt blacks (Fig. 13A), only the outline of the Pt black agglomerate is visible, and the overlap of Pt particles does not allow access to Pt particle-size measurements. Using the DFEM technique, it has been shown by Kinoshita and co-workers[70,218] that Pt-particle-size distributions can be obtained for unsupported Pt

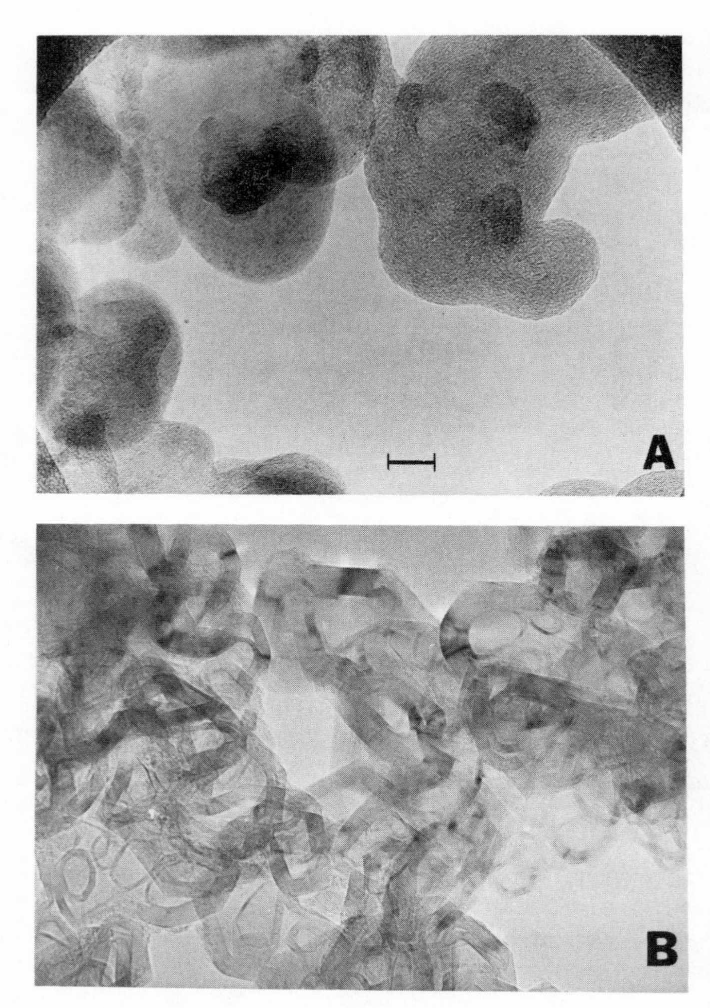

Figure 12. High-resolution electron micrographs of Vulcan XC-72 (A) and graphitized Vulcan XC-72 (B) carbon blacks. Calibration bar equals 100 Å.

blacks. Dark-field conditions were set such that diffraction for the Pt (111) crystal plane was observed. Those crystallites that satisfied the Bragg condition for diffraction appeared uniformly bright against a dark background (Fig. 13B).

Similarly, DFEM can be applied to measure the Pt-crystallite-size distribution of supported Pt electrocatalysts on C. It is not necessary, however, to use DFEM if the Pt crystallites can be distinguished from the support. Figure 14A shows a bright-field electron micrograph of an 18.5% Pt on graphitized Vulcan XC-72R catalyst.[87] Dark spots corresponding to Pt crystallites are evident, but cannot be distinguished clearly from the C at positions where the C support becomes too thick. The dark-field electron micrographs, on the other

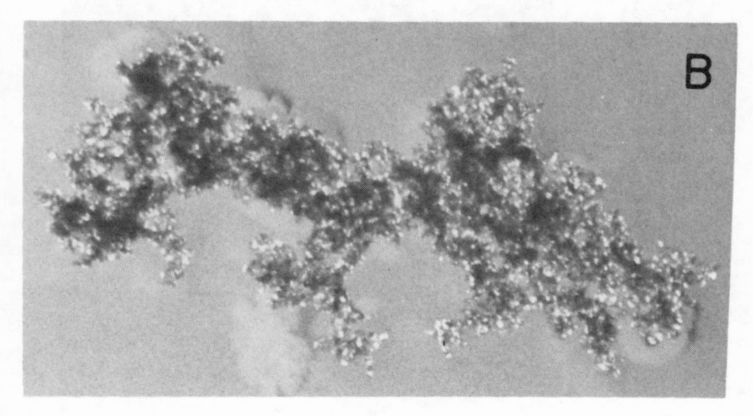

Figure 13. Electron micrographs of unsupported Pt black (30.5 m^2/g Pt): (A) bright-field image and (B) dark-field image from (111) diffraction plane of Pt. Calibration bar equals 1000 Å.

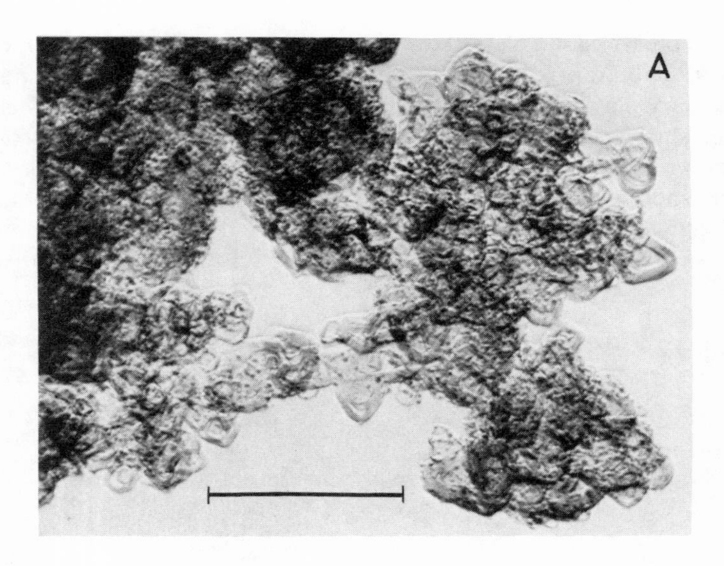

Figure 14. Electron micrographs of 18.5% Pt supported on graphitized Vulcan XC-72 (110 m^2/g Pt): (A) bright-field image and (B) dark-field image from (111) diffraction plane of Pt. Calibration bar equals 1000 Å.

hand, allow a more unambiguous measurement of the Pt crystallite sizes, since the Pt particles appear as white spots on a darker background (Fig. 14B). It should be reiterated that only the Pt crystallites having the correct orientation to the incident beam will be resolved. Our experience has been that less than 10% of the noble metal particles on the support surface have this preferred orientation, so the real noble metal crystallite density was an order of magnitude greater than that implied by the dark-field images in Fig. 14B.

Pt-crystallite-size distributions of the electrocatalysts shown in Figs. 14 and 15 are presented in the histograms in Fig. 16. A total of between 800 and 1000 individual Pt crystallites were measured to obtain the histograms and also to obtain a reasonable statistical set of data to evaluate average Pt crystallite diameters. Since the specimens observed in an electron microscope are so small, numerous electron micrographs and a large number of Pt-crystallite-size measurements must be made to arrive at a statistical average. Adams *et al.*[74] estimated that their electron-microscopy observations of a 2.5 < wt% Pt on silica gel were made on a sample size containing approximately 10^{-17} g Pt. Based on the measurement of 942 Pt particles, the probable error in

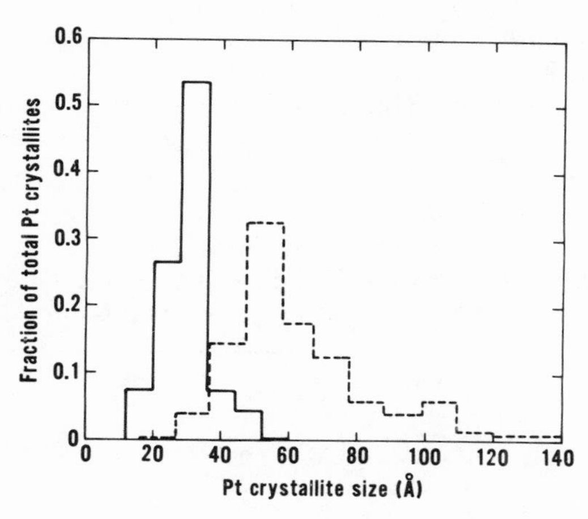

Figure 15. Pt crystallite size distributions obtained from dark-field electron micrographs. Solid line: 18.5% Pt supported on graphitized Vulcan XC-72 (110 m^2/g Pt); dashed line: Pt black (30.5 m^2/g Pt).

the calculation of the average Pt diameters amounted to about 10% among eight observers.

Average Pt-crystallite diameters can be obtained by the use of the following equations.[74] For crystallites of diameter d_i and number m_i, the number average diameter from a Pt size-distribution histogram is given by

$$d_n = \frac{\sum m_i d_i}{\sum m_i} \tag{47}$$

the surface average diameter by

$$d_s = \frac{\sum m_i d_i^3}{\sum m_i d_i^2} \tag{48}$$

and the volume average diameter by

$$d_v = \frac{\sum m_i d_i^4}{\sum m_i d_i^3} \tag{49}$$

Average diameters calculated from electron microscopy can be compared directly with results obtained from surface-area measurements and X-ray diffraction analysis.[74] Since the surface-area measurement yields a surface average diameter and X-ray diffraction line-broadening yields a volume average diameter, these values can be compared with d_s and d_v, respectively. There is, however, one caveat that must be considered in these comparisons. Strictly speaking, X-ray diffraction yields a measure of the crystallite size d_x, which may not

Table 7
Comparison of Average Crystallite Size of Pt Obtained from Surface Area, X-Ray, and Electron Microscopy

Electrocatalyst	Surface area ($m^2\,g^{-1}$ Pt)	d (Å)	d_s (Å)	d_v (Å)	d_x (Å)
Pt black[70]	30.5[a]	92	84	91	108
18.5% Pt–graphitized Vulcan XC–72[87]	110[b]	26	36	38	—
5% Pt—Graphon[70]	53[b]	53	64	76	67

[a]BET measurement.
[b]Electrochemical measurement.

necessarily be the same as the particle size if the crystallite contains a grain boundary. For supported Pt catalysts,[70,74] it would appear from comparisons of d_x and d_v that the crystallite and particle sizes are synonymous.

Calculated Pt diameters of Pt electrocatalysts obtained by the various experimental techniques discussed here are summarized in Table 7. In general, comparisons of d with d_s and d_v with d_x yielded good agreement, which illustrates the applicability and interrelationship of surface-area measurements, X-ray analysis, and electron microscopy to the characterization of electrocatalysts.

4. Reaction-Rate Measurements

The previous discussions on electrocatalyst characterizations have described techniques for evaluating the physical dimensions of the particles or surface dimensions from adsorption phenomena. One other technique that has not been applied generally is to use a well-characterized reaction (in terms of specific activity) to define the electrocatalyst. This technique is commonly used in gas-phase catalysis, but in the electrocatalytic area, there has been little agreement on the kinetic parameters for a specific electrode reaction, even to within a factor of 2 in the reaction rate. It is important, however, that the question of an electrochemical evaluation for highly dispersed electrocatalysts be addressed.

It is an unfortunate truism that only electrocatalysts of high activity are of practical interest. Low-activity electrocatalyst materials can be fabricated into an electrode structure and readily evaluated for electrochemical properties in the absence of diffusion parameters. On the other hand, high-activity electrocatalytic materials must be formed into porous electrode structures in such a way that the electrode structure itself can be described, so that diffusional parameters within the electrode can be accounted for.

If an electrode structure can be formed such that no diffusional effects can be observed, then the effectiveness factor ε, for the electrocatalyst is unity. In practice, with highly active materials, the effectiveness factor is usually less than unity which means that some of the electrocatalyst material is not utilized and is starved of reactant due to diffusion limitations. It is not important that such a situation exists, since electrocatalytic materials can be characterized kinetically, provided the diffusional effects can be described adequately.

A useful electrode structure for such a purpose is the PTFE-bonded flooded layer structure described by Vogel *et al.*,[314] which is a practical realization of a porous electrode structure first described by Austin and Lerner.[315] The PTFE serves only to bond the electrocatalyst particles together and has no other purpose. It is to be expected that small amounts of other stable thermoplastic materials could be used in place of the PTFE.

A porous electrocatalyst layer operating in the absence of mass transport ($\varepsilon = 1.0$) with Tafel kinetics has a current-density–potential relationship of the form

$$I = i_0 L\gamma \exp(-2.303\eta/b) \tag{50}$$

where i_0 is the usual standard exchange current, L is the thickness of the electrode structure, and γ is the surface area of active electrocatalyst per volume of electrode structure (real cm^2 electrocatalyst/cm^3 electrode structure). I, η, and b (Tafel slope) have their usual electrochemical significance.

When the effectiveness factor of the electrocatalyst layer is significantly less than unity, the majority of the electrocatalyst is not being used, and the current-density–potential relationship has the form

$$I = (nFDC_0 i_0 \gamma)^{1/2} \exp(-2.303\eta/2b) \tag{51}$$

where C_0 is the concentration and D the diffusivity of the reactant.

Forms of this equation have been derived several times,[314–316] although the original was due to Thiele[317] for diffusional effects complicating gas-phase reaction rates in porous catalyst pellets. The analogy between temperature increase in gas-phase catalysis and overpotential increase in electrocatalysis is obvious. Vogel *et al.*[314] calculated the current–voltage parameters expected for O_2 reduction in 85% H_3PO_4 at 120°C on porous Pt black layer electrodes shown in Fig. 16. Two different porosities were considered and are given in the caption.

Due to a general inability to make an ideal electrode structure with $\varepsilon = 1.0$, it is preferable to extrapolate measurements from the double Tafel region to zero overpotential. Then the exchange current

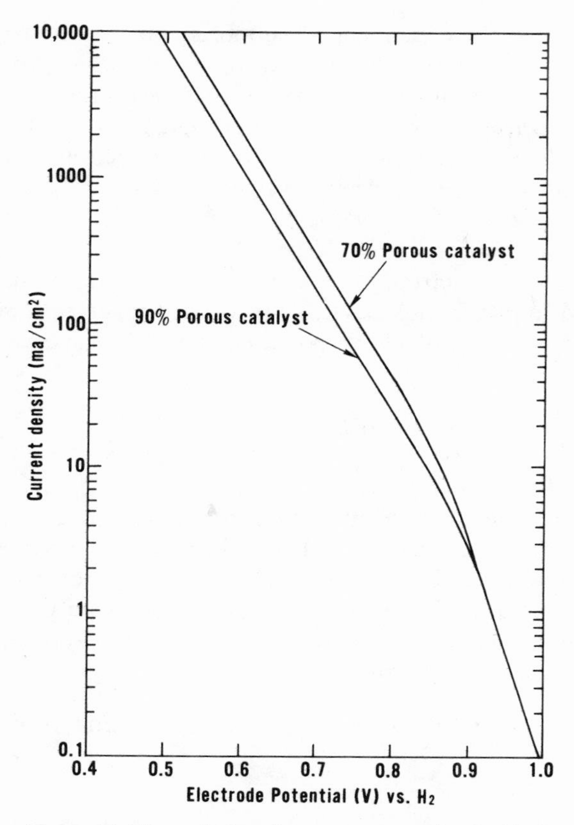

Figure 16. Theoretical performance curves for Pt black layer electrodes.[314] 1 mg Pt/cm^2, 24 m^2/g Pt. 1 atm O_2, 85% H_3PO_4 at 120°C, according to Eqs. (50) and (51). $n = 4$, $b = 60$ mV, $i_0 = 4 \times 10^{-10}$ A cm^{-2}, $DC_0 = 4 \times 10^{-12}$ mol cm^{-1} sec^{-1}. 90% catalyst porosity: $L = 4.77 \times 10^{-4}$ cm and $\gamma = 5.04 \times 10^5$ cm^{-1}; 70% catalyst porosity: $L = 1.59 \times 10^{-4}$ cm and $\gamma = 1.51 \times 10^6$ cm^{-1}.

i_0 is obtained from

$$i_0 = \frac{I^2}{A_i^2 nFDC_0 \gamma} \tag{52}$$

where A_i is a small experimental interfacial roughness factor. This technique was used by Stonehart and Ross[318] to obtain kinetic parameters for O_2 reduction at Pt supported on C in 96 wt% H_3PO_4 between 125 and 160°C. Since it is important to know the regime ($\varepsilon = 1.0$ or $\varepsilon \ll 1.0$) in which the electrocatalytic results are taken, it is necessary to *test* a number of electrode structures and to vary the electrode thicknesses. When $\varepsilon = 1.0$, the current density will increase with the thickness of the electrode structure, whereas when $\varepsilon \ll 1.0$, the current density is independent of the electrode thickness (electrocatalyst loading/cm^2 electrode).

The critical electrode-layer thickness for the transition between single and double Tafel slopes L_{crit} is related to the current density, and is defined as a function of overpotential and exchange current in the manner

$$L_{crit} = \left(\frac{nFD}{i_0\gamma}\right)^{1/2} \exp(-2.303\eta/2b) \tag{53}$$

Flooded layer electrodes have been used to determine the mechanism of the reversible hydrogen reaction on dispersed Pt and Rh electrocatalysts in acid electrolytes.[69,318] The rate-controlling step was shown to be the Tafel reaction (dissociative chemisorption of the hydrogen molecule). For this mechanism, the experimental anodic current density is given by

$$I = (nFDC_0)[1 - \exp(-2F\eta/RT)](\sigma)^{1/2} \tan hL(\sigma)^{1/2} \tag{54}$$

where

$$\sigma = \frac{\gamma i_0}{nFDC_0}\left(\frac{\exp(F\eta/RT)}{\theta_0 + (1-\theta_0)\exp(F\eta/RT)}\right)^2 \tag{55}$$

and θ_0 is the coverage of the electrocatalyst surface by adsorbed hydrogen at the equilibrium potential.

The potential region of greatest kinetic interest is that region in which there is no adsorbed hydrogen (which acts as a poison) on the electrocatalyst surface, so $\eta \to \infty$ and the current density reaches a limit I_{lim}.

At $\varepsilon \ll 1.0$ and $\eta \to \infty$, then

$$I_{lim} = A_i(2FDC_0\gamma i_0)^{1/2} \frac{1}{(1-\theta_0)} \tag{56}$$

At $\varepsilon = 1.0$, $\eta \to \infty$, then

$$I_{\text{lim}} = A_i \gamma L i_0 \left(\frac{1}{1 - \theta_0} \right)^2 \tag{57}$$

Polarization curves for Pt on C and Rh on C electrocatalysts operating as flooded layer electrodes are shown in Fig. 17, and the exchange currents were determined according to Eq. (56). Stonehart and Ross[318] further used this reaction-rate information to deduce the Pt electrocatalyst active surface area when operating as a H_2 electrode in concentrated phosphoric acid in the presence of various partial pressures of CO. In this way, adsorption isotherms for the CO were obtained (Fig. 18).

Hydrogen oxidation kinetics in the presence of CO for dispersed electrocatalyst alloys of Pt–Rh and Pt–Ru on C supports using flooded

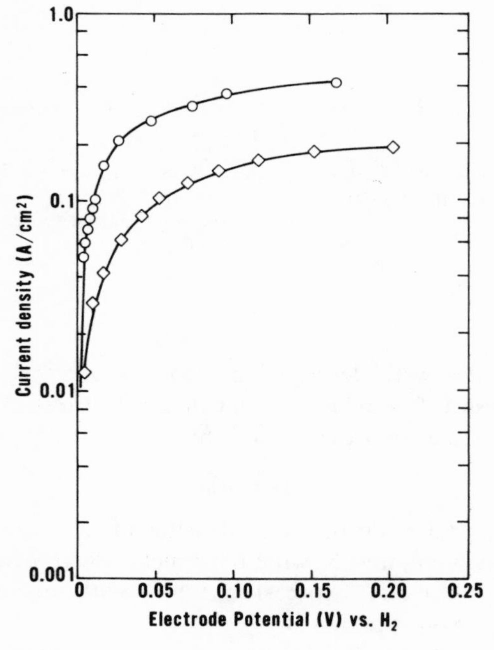

Figure 17. Anodic polarization curves at 160°C (96 wt.% H_3PO_4) for Pt (○) and Rh (◇) with 40% H_2/N_2.[318] $\gamma = 4.5 \times 10^{-4}$.

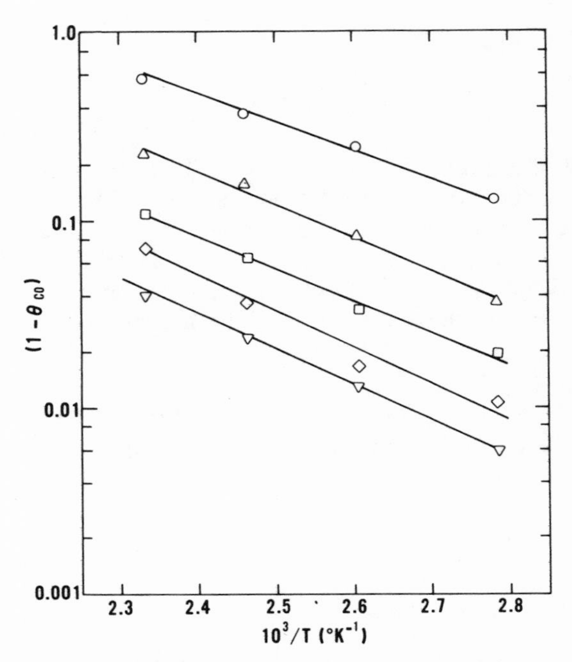

Figure 18. Carbon monoxide isobars on Pt[318]: (○) 10^{-4} atm; (△) 10^{-3} atm; (□) 10^{-2} atm; (◇) 2×10^{-2} atm; (▽) 5×10^{-2} atm carbon monoxide partial pressure. 96 wt.% H_3PO_4.

layer electrodes were described by Ross *et al.*[198,319] The active electrocatalyst surface areas per unit volume of porous layer electrode, γ, were calculated from the relationship

$$\gamma = (1 - \pi)\rho W S \tag{58}$$

where π and ρ are the porosities and densities of the C support, W is the wt% metal content, and S is the total metal electrocatalyst surface areas. A more exact expression for γ with dispersed metal electrocatalysts on supports is

$$\gamma = \frac{(1 - \pi)\rho W S}{(1 - W)} \tag{59}$$

although for low-metal-content electrocatalysts (high dispersion), the correction is not important.

Diagnosis of electrocatalyst activities and the derivation of electrokinetic parameters in difficult environments is as important as the ability to prepare the electrocatalytic material. It is to be expected that there will be further uses of flooded porous layer electrode structures as analytical tools for understanding the operation (and poisoning) of electrocatalyst surfaces.

V. SUMMARY

Although electrocatalytic processes have been studied under ideal conditions at smooth electrode surfaces, the real world must work with highly dispersed electrocatalysts in hostile environments. In order to utilize electrocatalysts in preparative electroorganic chemistry and electrochemical power production, it is fundamentally important to be able to characterize the electrocatalytic material adequately. Without this capability, all speculations with regard to electrocatalytic reaction rates, catalyst–support interaction effects, diffusion effects in porous structures, etc., are negated. Lack of adequate physical analysis of the electrocatalyst material must inevitably cause some confusion in the interpretation of the electrode kinetics for a given process.

There is a considerable body of literature extant describing the preparations of highly dispersed catalysts and electrocatalysts. We have tried to place these techniques at the disposal of future workers in this area, so that they may progress more easily toward the development of the field. It may well be that some favorite references have been excluded from this review. The responsibility for this oversight rests solely with the authors. It is expected that new preparative techniques will evolve as the field matures, and that established techniques will be refined and developed.

One area of electrocatalyst characterization has been excluded from this review by choice. This concerns the use of high-vacuum surface-analysis techniques (LEED, SIMS, Auger, and ESCA) for diagnosing the structure of binary alloy electrocatalyst particles. The gas-phase literature teaches that the surface composition of a binary alloy is not the same as the bulk composition, due to surface segregation by one of the alloy components. A recent review by

Ponec[320] discusses the surface composition considerations for alloys as well as the effects of small particle sizes and substrate effects in gas-phase catalysis. This is sufficient evidence that the earlier, but relatively crude, attempts to correlate bulk compositions of binary alloy[2-4] electrodes (wires and flat sheet) with electrocatalytic processes could be only a first approximation to the realities of electrocatalysis. More sophisticated determinations of electrocatalysis on binary alloy particle surfaces will surely follow. The use of small particles provides another dimension for controlling electrocatalyst behavior. We believe that it is in this dimension that future work in the field will evolve.

REFERENCES

[1] N. Kobosev and W. Monblanova, *Acta Physicochim. URSS* **1** (1934) 611.
[2] J. O'M. Bockris and H. Wroblowa, *J. Electroanal. Chem.* **7** (1964) 428.
[3] S. Srinivasan, H. Wroblowa, and J. O'M. Bockris, in *Advances in Catalysis*, Ed. by D. D. Eley, H. Pines, and P. B. Weisz, Academic Press, New York, 1967, Vol. 17, p. 351.
[4] A. J. Appelby, *Catal. Rev.* **4** (1970) 221.
[5] Y. Okamoto and W. Brenner, *Organic Semiconductors*, Reinhold Publishing, New York, 1964, p. 82.
[6] M. H. Polley and B. B. S. T. Boonstra, *Rubber Chem. Technol.* **30** (1957) 170.
[7] W. R. Smith, in *Encyclopedia of Chemical Technology*, 2nd Ed., Interscience, New York, 1964, Vol. 4, p. 243.
[8] H. P. Boehm, in *Advances in Catalysis*, Ed. by D. D. Eley, H. Pines, and P. B. Weisz, Academic Press, New York, 1966, Vol. 16, p. 179.
[9] J. B. Donnet, *Carbon* **6** (1968) 161.
[10] D. Rivin, *Rubber Chem. Technol.* **44** (1971) 307.
[11] H. B. Puri, in *Chemistry and Physics of Carbon*, Ed. by P. L. Walker, Marcel Dekker, New York, 1970, Vol. 6, p. 191.
[12] B. A. Puri and R. C. Bansal, *Carbon* **1** (1964) 457.
[13] K. Kinoshita and J. A. S. Bett, *Carbon* **11** (1973) 237.
[14] K. Kinoshita and J. A. S. Bett, in *Corrosion Problems in Energy Conversion and Generation*, Ed. by C. S. Tedmon, Electrochemical Society, Princeton, N.J., 1974, p. 43.
[15] H. Binder, A. Kohling, K. Richter, and G. Sandstede, *Electrochim. Acta* **9** (1964) 255.
[16] C. A. Hampel, Ed., *Encyclopedia of Electrochemistry*, Reinhold Publishing, New York, 1964, p. 151.
[17] *Thorpe's Dictionary of Applied Chemistry*, Longmans Green, 1957, Vol. 6, p. 284.
[18] K. Kinoshita and J. A. S. Bett, *Carbon* **11** (1973) 403.
[19] W. G. Berl, *Trans. Electrochem. Soc.* **83** (1943) 253.
[20] M. O. Davies, M. Clark, E. Yeager, and F. Hovorka, *J. Electrochem. Soc.* **106** (1959) 56.
[21] P. Stonehart, Ph.D. thesis, University of Cambridge, England, 1962.
[22] R. J. Bowen and H. B. Urbach, *J. Chem. Phys.* **49** (1968) 1206.
[23] H. B. Urbach and R. J. Bowen, *Electrochim. Acta* **14** (1969) 927.

[24]E. Yeager and A. Kozawa, Tech. Rept. 17, Office of Naval Res. Contr. Nonr. 2391 (00) Project NR 359–277, Jan. 15, 1974.
[25]J. Mrha, *Collect Czech. Chem. Commun.* **32** (1967) 708.
[26]M. Brezina, J. Jinda, and J. Mrha, *Collect. Czech. Chem. Commun.* **33** (1968) 2363.
[27]J. P. Hoare, *Energy Convers.* **8** (1968) 155.
[28]J. Mrha, W. Vielstich, and U. Vogel, *Z. Phys. Chem. (Frankfurt)* **52** (1967) 215.
[29]G. Luft, K. Mund, G. Richter, R. Schulte, and F. von Sturm, *Siemens Forsch. Entwicklungs ber.* **3** (1974) 177.
[30]S. G. Meilbuhr, *Nature, London* **210** (1966) 409.
[31]L. W. Niedrach and D. W. McKee, in *Proceedings of the 21st Annual Power Sources Conference*, May 16–18, 1967, PSC Publications Committee, Red Bank, N.J. p. 6.
[32]W. T. Grubb and D. W. McKee, *Nature, London* **210** (1966) 192.
[33]*Hydrocarbon–Air Fuel Cells*, Semi-Annual Technical Summary, Report No. 12, July 1–December 31, 1967, Contr. DAAK-02-67-C-0080, General Electric Co., Lynn, Mass.
[34]H. A. Laitinen, C. A. Vincent and T. M. Bednarski, *J. Electrochem. Soc.* **115** (1968) 1024; D. Elliott, D. L. Zellmer, and H. A. Laitinen, *J. Electrochem. Soc.* **117** (1970) 1343.
[35]A. C. C. Tseung and S. C. Dhara, *Electrochim. Acta* **19** (1974) 845.
[36]C. A. Vincent and D. G. C. Weston, *J. Electrochem. Soc.* **119** (1972) 518.
[37]R. D. Armstrong, A. F. Douglas, and D. E. Williams, *Energy Convers.* **11** (1971) 7.
[38]J. McHardy and P. Stonehart, Electrochemistry of transition metal oxide bronzes and related compounds in *International Review of Science, Physical Chemistry*, Series 2, Ed. by J. O'M. Bockris, Butterworths, London, 1975 p. 171.
[39]L. W. Niedrach and H. I. Zeliger, *J. Electrochem. Soc.* **116** (1969) 152.
[40]B. Broyde, *J. Catal.* **10** (1968) 13.
[41]A. C. C. Tseung and H. L. Bevan, *J. Electroanal. Chem.* **45** (1973) 429.
[42]D. B. Meadowcroft, *Nature, London* **266** (1970) 847.
[43]C. Pinnington, International Symposium on Fuel Cells, Antwerp, Belgium, October 2–3, 1972, fourth paper.
[44]ESSO Research and Engineering Company, *Hydrocarbon–Air Fuel Cell Report No. 7*, Semi-Annual Report No. 3 (January 1, 1965–June 30, 1965), Contract No. DA 36-039 AMC-03743(E).
[45]L. J. Hillenbrand and J. W. Lacksonen, *J. Electrochem. Soc.* **112** (1965) 249.
[46]C. S. Nicolau, H. G. Thom, and E. Pobitschka, *Trans. Faraday Soc.* **55** (1959) 1430.
[47]C. S. Nicolau and H. G. Thom, *Z. Anorg. Allgem. Chem.* **303** (1960) 133.
[48]P. N. Ross, K. Kinoshita, and P. Stonehart, *J. Catal.* **32** (1974) 163.
[49]V. A. Garten and D. E. Weiss, *Aust. J. Chem.* **8** (1955) 68.
[50]F. Solymosi, *Catal. Rev.* **1** (1967) 233.
[51]N. H. Sagert and R. M. L. Pouteau, *Platinum Met. Rev.* **19** (1975) 16.
[52]B. R. Arora, *Chem. Age India* **25** (1974) 375.
[53]B. S. Hobbs and A. C. C. Tseung, *Nature, London* **222** (1969) 556.
[54]B. S. Hobbs and A. C. C. Tseung, *J. Electrochem. Soc.* **119** (1972) 580.
[55]B. S. Hobbs and A. C. C. Tseung, *J. Electrochem. Soc.* **120** (1973) 766.
[56]A. C. C. Tseung and B. S. Hobbs, *Platinum Met. Rev.* **13** (1969) 146.
[57]J. E. Benson, H. W. Kohn, and M. Boudart, *J. Catal.* **5** (1966) 307.
[58]R. B. Levy and M. Boudart, *J. Catal.* **32** (1974) 304.
[59]P. A. Sermon and G. C. Bond, *Catal. Rev.* **8** (1973) 211.
[60]J. O'M. Bockris and J. McHardy, *J. Electrochem. Soc.* **120** (1973) 61.
[61]A. J. Robell, E. V. Ballou, and M. Boudart, *J. Phys. Chem.* **68** (1964) 2748.
[62]M. Boudart, A. W. Aldag, and M. A. Vannice, *J. Catal.* **18** (1970) 46.
[63]V. S. Bagotzky, L. D. Kanevsky, and V. S. Palanker, *Electrochim. Acta* **18** (1973) 473.

[64]T. V. Balashova, V. S. Bagotskii, A. M. Skundin, and L. G. Gindin, *Electrokhimiya* **8** (1972) 1600.
[65]L. S. Kanevskii, V. V. Emel'yanenko, and V. S. Bagotskii, *Electrokhimiya* **8** (1972) 288.
[66]L. S. Kanevskii, V. S. Palanker, and V. S. Bagotskii, *Electrokhimiya* **6** (1970) 271.
[67]J. Bett, K. Kinoshita, K. Routsis, and P. Stonehart, *J. Catal.* **29** (1973) 160.
[68]K. Kinoshita, J. Lundquist, and P. Stonehart, *J. Catal.* **31** (1973) 325.
[69]W. Vogel, J. Lundquist, P. Ross, and P. Stonehart, *Electrochim Acta* **20** (1975) 79.
[70]D. E. Fornwalt and K. Kinoshita, *Micron 4* (1973) 99.
[71]J. Sinfelt, in *Annual Reviews of Material Science*, Ed. by R. Huggins, R. H. Bebe, and R. W. Roberts, Addison-Wesley, Palo Alto, Calif., 1972, Vol. 2, p. 641.
[72]K. Morikawa, T. Shirasaki, and M. Okada, in *Advances in Catalysis*, Ed. by D. D. Eley, H. Pines. and P. B. Weisz, Academic Press, New York, 1960, Vol. 20, p. 97.
[73]F. G. Ciapetta and C. J. Plank, in *Catalysis, Vol. I, Fundamental Principles* (*Part* 1), Ed. by P. H. Emmett, Reinhold Publishing, New York, 1954, p. 315.
[74]C. R. Adams, H. A. Benesi, R. M. Curtis, and R. G. Meisenheimer, *J. Catal.* **1** (1962) 336.
[75]H. A. Benesi, R. M. Curtis, and H. P. Studer, *J. Catal.* **10** (1968) 328.
[76]T. A. Dorling, B. W. J. Lynch, and R. L. Moss, *J. Catal.* **20** (1971) 190.
[77]T. A. Dorling and R. L. Moss, *J. Catal.* **5** (1966) 111.
[78]T. A. Dorling and R. L. Moss, *J. Catal.* **7** (1967) 378.
[79]O. M. Poltorak and V. S. Boronin, *Zh. Fiz. Khim. SSSR* **39** (1965) 1476.
[80]R. A. Herrmann, S. F. Adler, M. S. Goldstein, and R. M. DeBaun, *J. Phys. Chem.* **65** (1961) 2189.
[81]N. M. Zaidman, V. S. Dzis'ko, A. P. Karnaukhov, N. P. Kradilenko, N. G. Koroleva, and G. P. Vishnyakova, *Kinet. Katal.* **9** (1968) 863.
[82]N. M. Zaidman, V. S. Dzis'ko, A. P. Karnaukhov, L. M. Kefeli, N. P. Krasilenko, N. G. Koroleva, and I. D. Ratner, *Kinet. Katal.* **10** (1969) 386.
[83]N. M. Zaidman, N. P. Krasilenko, L. M. Kefeli, and I. D. Ratner, *Kinet. Katal.* **11** (1970) 741.
[84]L. J. Hillenbrand and J. Lacksonen, *J. Electrochem. Soc.* **112** (1965) 245.
[85]D. M. Drazic and R. R. Adzic, *Electrochim. Acta* **14** (1969) 405.
[86]R. W. Maatman and C. D. Prater, *Ind. Eng. Chem.* **49** (1957) 253.
[87]K. Kinoshita, unpublished results.
[88]F. S. Kemp and M. A. George, U.S. Patent 3,857,737, December 31, 1974.
[89]W. T. Grubb, H. A. Liebhafsky, and E. J. Cairns, in *Fuel Cells and Fuel Batteries, A Guide to Their Research and Development*, John Wiley, New York, 1968, p. 392.
[90]G. C. Bond, P. A. Sermon, G. Webb, and D. A. Buchanan, *J. Chem. Soc. Chem. Commun.* (1973) 444.
[91]R. B. Clarkson and A. C. Cirillo, *J. Catal.* **33** (1974) 392.
[92]J. M. Cece and R. D. Gonzalez, *J. Catal.* **28** (1973) 254.
[93]J. H. Sinfelt and D. J. C. Yates, *J. Catal.* **8** (1967) 82.
[94]G. W. Higginson, *Chem. Eng.* (1974) 98.
[95]Z. Sulcek, M. Vasak, and J. Dolezal, *Microchem. J.* **16** (1971) 210.
[96]I. Furuoya, T. Shirasaki, E. Echigoya, and K. Morikawa, *Kagyo Kagaku Zasshi.* **72** (1969) 1431.
[97]A. Tomita and Y. Tamai, *J. Phys. Chem.* **75** (1971) 649.
[98]V. A. Alekseenko, N. M. Vasil'ev, M. M. Senyairn, D. N. Strazhesko, and I. A. Tarsovskaya, *Zh. Fiz. Khim. SSSR* **45** (1971) 57.
[99]V. A. Garten and D. E. Weiss, Proceedings of the Third Conference on Carbon, held at the University of Buffalo, New York, June 17–21, 1957.
[100]H. P. Boehm, E. Diehl, W. Heck, and R. Sappok, *Angew. Chem. Int. Ed. Engl.* **3** (1964) 669.

[101]K. F. Blurton, *Carbon* **10** (1972) 305.
[102]B. J. Cooper, U.S. Patent 3,793,224 (February 19, 1974).
[103]G. J. K. Acres and B. J. Cooper, *J. Appl. Chem. Biotechnol.* **22** (1972) 769.
[104]B. J. Cooper and D. L. Trimm, Paper presented at the Third International Carbon and Graphite Conference (S.C.I.), London, 1970.
[105]J. L. Schmitt and P. L. Walker, *Carbon* **9** (1971) 791.
[106]J. L. Schmitt and P. L. Walker, *Carbon* **10** (1972) 87.
[107]J. L. Schmitt, Ph.D. thesis, Pennsylvania State University, 1970.
[108]L. G. Christner, Ph.D. thesis, Pennsylvania State University, 1972.
[109]G. C. Bond, *Trans. Faraday Soc.* **52** (1956) 1235.
[110]R. M. Wilenzick, D. C. Russell, R. H. Morriss, and S. W. Marshall, *J. Chem. Phys.* **47** (1967) 533.
[111]K. V. Kordesch and R. F. Scarr, Low Cost Oxygen Electrodes, Contract No. DAAK-02-71-C-0297, July 1972, Union Carbide Corp., Cleveland, Ohio.
[112]J. Giner, J. M. Parry, and S. M. Smith, in *American Chemical Society Symposium on Fuel Cells*, Chicago (September, 1967), *Advances in Chemistry Series 90*, American Chemical Society, Washington, D. C., 1969, p. 151.
[113]N. M. Kagan, G. F. Muchnik, Y. N. Pisarev, Y. A. Kaller, and V. A. Panchenko, *Elektrokhimiya* **9** (1973) 1498.
[114]German Patent 2,264,592 (May 9, 1974); British Patent 1,357,494 (June 19, 1974).
[115]J. Turkevich and G. Kim, *Science* **169** (1970) 873.
[116]Y. Chiang and J. Turkevich, *J. Colloid Sci.* **18** (1973) 772.
[117]B. V. Enustun and J. Turkevich, *J. Am. Chem. Soc.* **85** (1965) 3317.
[118]G. Frens, *Nature London Phys. Sci.* **241** (1973) 20; see also J. Turkevich, P. C. Stevenson, and J. Hiller, *Discuss. Faraday Soc.* **11** (1951) 55.
[119]D. J. Booth, D. Bryce-Smith, and A. Gilbert, *Chem. Ind.* **17** (1972) 688.
[120]P. F. Casella, *Chem. Technol.* **4** (1974) 64.
[121]J. A. Cusamano, *Nature, London* **247** (1974) 456.
[122]H. I. Zeliger, *J. Catal.* **7** (1967) 198.
[123]R. Adams and R. L. Shriner, *J. Am. Chem. Soc.* **45** (1923) 2171.
[124]S. Nichimura, T. Onoda, and A. Nakamura, *Bull. Chem. Soc. Jpn.* **33** (1960) 1356.
[125]D. W. McKee, *J. Catal.* **8** (1967) 240.
[126]D. W. McKee, L. W. Niedrach, J. Paynter, and I. F. Danzig, *Electrochem. Technol.* **5** (1967) 419.
[127]L. W. Niedrach, D. W. McKee, J. Paynter, and I. F. Danzig, *Electrochem. Technol.* **5** (1967) 318.
[128]M. Kobayashi and T. Shirasaki, *J. Catal.* **28** (1973) 289.
[129]A. P. Gorokhov, A. B. Fasman, and D. V. Sokol'skii, *Kinet. Katal.* **12** (1971) 690.
[130]D. W. McKee and M. S. Pak, *J. Electrochem. Soc.* **116** (1969) 516.
[131]C. W. Keenan, B. W. Giesemann, and H. A. Smith, *J. Am. Chem. Soc.* **76** (1954) 229.
[132]D. Cahan and J. G. Ibers, *J. Catal.* **31** (1973) 369.
[133]M. Raney, U.S. Patents 1,563,787 (December, 1925), 1,628,191 (May, 1927), 1,915,473 (June, 1933).
[134]E. Lieber and F. L. Morritz, in *Advances in Catalysis*, Ed. by W. G. Frankenburg, V. I. Komarewsky, and E. K. Rideal, Academic Press, New York, Vol. 5, 1953, p. 417.
[135]A. R. Despic, D. M. Drazic, C. B. Petrovic, and V. L. Vujcic, *J. Electrochem. Soc.* **111** (1964) 1109.
[136]M. Jung and H. H. Von Dohren, in *Power Sources 1966*, Ed. by D. H. Collins, Pergamon, Oxford, 1967, p. 497.
[137]A. D. Semenova, N. V. Kropotova, and G. D. Vovchenko, *Dokl. Akad. Nauk SSSR* (*Phys. Chem.*) **212** (1973) 1162.

[138]T. M. Grishina, L. I. Logacheva, V. I. Fedeeva, V. I. Strat'ev, and G. D. Vovchenko, *Vestn. Mosk. Univ. Khim.* **14** (1973) 586; *Chem. Abstr.* **80** (1974) 52654n.
[139]H. Binder, A. Kohling, and G. Sandstede, in *From Electrocatalysis to Fuel Cells*, Ed. by G. Sandstede, University of Washington Press, Seattle, 1972, p. 43.
[140]H. Binder, A. Kohling, and G. Sandstede, in *Hydrocarbon Fuel Cell Technology*, Ed. by B. S. Baker, Academic Press, New York, 1965, p. 91.
[141]K. I. Yankovskii, A. D. Sevenova, G. D. Vovchenko, and S. V. Smykova, *Vestn. Mosk. Univ. Khim.* **14** (1973) 83; *Chem. Abstr.* **78** (1973) 164508h.
[142]H. Krupp, H. Rabenhorst, G. Sandstede, G. Walter, and R. McJones, *J. Electrochem. Soc.* **109** (1962) 533.
[143]H. A. Smith, W. C. Bedoit, and J. F. Fuzek, *J. Am. Chem. Soc.* **71** (1949) 3769.
[144]R. C. Burshtein, A. G. Pshenichnikov, and F. Z. Sabirov, in *Fuel Cell Systems II, Advances in Chemistry Series 90*, Ed. by R. E. Gould, American Chemical Society, 1969, p. 71.
[145]H. Binder, A. Kohling, and G. Sandstede, in *From Electrocatalysis to Fuel Cells*, Ed. by G. Sandstede, University of Washington Press, Seattle, 1972, p. 15.
[146]M. R. Tarasevich, K. A. Radyushkina, and R. K. Bushtein, *Elektrokhimiya* **3** (1967) 455.
[147]A. E. Newkirk and D. W. McKee, *J. Catal.* **11** (1968) 370.
[148]K. Kinoshita, K. Routsis, and J. A. S. Bett, *Thermochim. Acta* **10** (1974) 109.
[149]S. D. Robertson, B. D. McNicol, J. H. deBaas, S. C. Kloet, and J. W. Jenkins, *J. Catal.* **37** (1975) 424.
[150]M. Yamano and H. Ikeda, in *Hydrocarbon Fuel Cell Technology*, Ed. by B. S. Baker, Academic Press, New York, 1965, p. 169.
[151]J. E. Schroeder, D. Pouli, and H. J. Seim, in *Fuel Cell Systems II, Advances in Chemistry Series 90*, Ed. by R. E. Gould, American Chemical Society, 1969, p. 93.
[152]D. W. McKee, *J. Phys. Chem.* **67** (1963) 841.
[153]A. C. C. Tseung, B. S. Hobbs, and A. D. S. Tantram, Electrochim. Acta **15** (1970) 473.
[154]A. C. C. Tseung and H. L. Bevan, *J. Mater Sci.* **5** (1970) 604.
[155]L. W. Niedrach and I. B. Weinstock, *Electrochem. Technol.* **3** (1965) 270.
[156]P. G. Dickens and M. S. Whittingham, *Trans. Faraday Soc.* **61** (1965) 1226.
[157]K. Mund, G. Richter, and M. Wenzel, German Patent 2,108,396 (September 7, 1962).
[158]G. Schulz-Elkoff, D. Baresel, and W. Sarholz, preprint of paper in Proceedings of the Fifth International Congress on Catalysis, Miami Beach, 1972.
[159]K. Von Benda, H. Binder, A. Kohling, and G. Sandstede, in *From Electrocatalysis to Fuel Cells*, Ed. by G. Sandstede, University of Washington Press, Seattle, 1972, p. 87.
[160]H. Boehm, *Electrochim. Acta* **15** (1970) 1273.
[161]M. Svata and Z. Zabransky, *Collect Czech. Chem. Commun.* **39** (1974) 1015.
[162]F. K. Roehrig and T. R. Smith, *J. Am. Ceram. Soc.* **55** (1972) 58.
[163]S. F. Exell, R. Roggen, J. Gillot, and B. Lux, in *Fine Particles Second International Conference*, Ed. by W. E. Kuhn and J. Ehretsmann, Electrochemical Soc., Princeton, N.J., 1974, p. 165.
[164]P. N. Ross and P. Stonehart, *J. Catal.* **39** (1975) 298.
[165]D. R. Rhodes, U.S. Patent 3,449,169 (June 10, 1969).
[166]Y. S. Kim and F. R. Monforte, *Ceram. Bull.* **50** (1971) 532.
[167]D. W. Johnson and P. K. Gallagher, *J. Am. Ceram. Soc.* **54** (1971) 461.
[168]V. V. Mirkovich and T. A. Wheat, *Ceram. Bull.* **49** (1970) 724.
[169]S. H. Gelles and F. K. Roehrig, *J. Met.* (1972) 23.
[170]*New Scientist*, July 6, 1972, p. 6.
[171]A. Landsberg and T. T. Campbell, *J. Met.* (1965) 856.

[172]R. Kusay, *Metall. Met. Form.* (1972) 377.
[173]M. J. Ferrante, R. R. Lowery, and G. B. Robidart, *Report of Investigations 7485*, U.S. Bureau of Mines, 1971.
[174]J. Rowbottom, British Patent 992,655 (September 16, 1963).
[175]Y. Trambouze, *Bull. Inst. Int. Froid, Annexe* (1969) 13.
[176]A. C. C. Tseung, German Patent 2,119,702 (November 4, 1971).
[177]T. Kudo, M. Yeshida, and O. Okamoto, U.S. Patent 3,804,674 (April 16, 1974).
[178]F. J. Schnettler, F. R. Monforte, and W. W. Rhodes, in *Science of Ceramics*, Ed. by G. H. Stewart, British Ceramic Soc., Stoke-on-Trent, U.K., 1968, Vol. 4, p. 179.
[179]D. W. Johnson and F. J. Schnettler, *J. Am. Ceram. Soc.* **53** (1970) 440.
[180]H. L. Bevan and A. C. C. Tseung, *Electrochim. Acta* **9** (1974) 201.
[181]W. J. King and A. C. C. Tseung, *Electrochim. Acta* **19** (1974) 485.
[182]A. C. C. Tseung and L. L. Wong, *J. Mater. Sci.* **2** (1972) 211.
[183]P. N. Rylander, L. Hasbrouck, S. G. Hindin, I. Karpenko, G. Pond, and S. Stanick, paper presented at 153rd National Meeting of the American Chemical Society, Miami Beach, April 13, 1962.
[184]P. N. Rylander, L. Hasbrouck, S. G. Hindin, R. Iverson, I. Karpenko, and G. Pond, *Engelhard Ind. Tech. Bull.* **8** (1967) 93.
[185]T. M. Grishina, L. I. Logacheva, and G. D. Vovchenko, *Elektrokhimiya* **9** (1973) 1247.
[186]E. L. Holt, *Nature, London* **203** (1964) 857.
[187]D. W. McKee, *J. Catal.* **14** (1969) 355.
[188]J. Forten and E. F. Rissman, U.S. Patent 3,467,554 (September 16, 1969).
[189]J. H. Fishman and M. Yarish, *Electrochim. Acta* **12** (1967) 579.
[190]L. C. Hoffman, British Patent 1,091,347 (November 15, 1967).
[191]D. W. McKee and F. J. Norton, *J. Catal.* **3** (1964) 252.
[192]K. A. Radyushkina and M. R. Tarasevich, *Elektrokhimiya* **6** (1970) 1703.
[193]A. Metcalfe and M. W. Rowden, *J. Catal.* **22** (1971) 30.
[194]D. W. McKee and F. J. Norton, *J. Catal.* **4** (1965) 510.
[195]J. H. Anderson, P. J. Conn, and S. G. Brandenberger, *J. Catal.* **16** (1970) 404.
[196]C. H. Bartholomew and M. Boudart, *J. Catal.* **25** (1972) 173.
[197]D. Cormack, D. H. Thomas, and R. L. Moss, *J. Catal.* **32** (1974), 492.
[198]P. N. Ross, K. Kinoshita, A. J. Scarpellino, and P. Stonehart, *J. Electroanal. Chem.* **59** (1975) 177.
[199]J. H. Sinfelt, *J. Catal.* **29** (1973) 308.
[200]D. W. McKee and A. J. Scarpellino, *Electrochem. Technol.* **6** (1963) 101.
[201]G. T. Pott, British Patent Application 7429/74 (see also British Patent 1,106,708; U.S. Patent 3,340,097).
[202]P. H. Emmett, in *Advances in Catalysis*, Ed. by W. G. Frankenburg, V. I. Komarewsky, and E. K. Rideal, Academic Press, New York, 1948, Vol. 1, p. 65.
[203]J. M. Thomas and W. J. Thomas, *Introduction to the Principles of Heterogeneous Catalysis*, Academic Press, New York, 1967, Chapt. 4, p. 180.
[204]W. B. Innes, in *Experimental Methods in Catalytic Research*, Ed. by R. B. Anderson, Academic Press, New York, 1968, p. 44.
[205]T. W. Whyte, *Catal. Rev.* **8** (1973) 117.
[206]R. J. Farranto, presented at the 77th National Meeting of the A.I.Ch.E., June 2–5, 1974, Pittsburgh, Pa., Paper No. 12C; *Chem. Eng. Prog.* **71** (1975) 37.
[207]S. Brunauer, P. H. Emmett, and E. Teller, *J. Am. Chem. Soc.* **60** (1938) 309.
[208]I. Langmuir, *J. Am. Chem. Soc.* **40** (1918) 1361.
[209]P. H. Emmett and S. Brunauer, *J. Am. Chem. Soc.* **59** (1937) 1553.
[210]A. L. McClellan and H. F. Harnsberger, *J. Colloid Sci.* **23** (1967) 577.

[211]B. Genot, *J. Colloid Sci.* **50** (1975) 413.
[212]D H. Everett, G. D. Parfitt, K. S. W. Sing, and R. Wilson, *J. Appl. Chem. Biotechnol.* **24** (1974) 199.
[213]P. H. Emmett and M. Cines, *J. Phys. Colloid Chem.* **51** (1974) 1329.
[214]P. A. Sermon, *J. Catal.* **24** (1972) 467.
[215]M. A. Vannice, J. E. Benson, and M. Boudart, *J. Catal.* **16** (1970) 348.
[216]F. F. Nelson and F. T. Eggertsen, *Anal. Chem.* **30** (1958) 1387.
[217]H. W. Daeschner and F. H. Stross, *Anal. Chem.* **34** (1962) 1150.
[218]K. Kinoshita, K. Routsis, J. A. S. Bett, and C. S. Brooks, *Electrochim. Acta* **18** (1973) 953.
[219]H. P. Boehm, U. Hofmann, and A. Clauss, in *Proceedings of the Third Conference on Carbon*, University of Buffalo, N. Y., June 17–21 1957, Pergamon Press, New York, 1959, p. 241.
[220]C. H. Giles and H. P. D'Silva, *Trans. Faraday Soc.* **65** (1969) 1943.
[221]H. A. Smith and R. B. Hurley, *J. Phys. Colloid Chem.* **53** (1948) 1409.
[222]J. M. Dalla Valle, P. T. Bankston, and C. Orr, paper presented at the Southwest Chemical Conference held under the joint auspices of the American Chemical Society, local sections in the Southeast and the Southern Association of Science and Industry, Wilson Dam, Ala., October 20, 1951.
[223]J. H. deBoer, B. C. Lippens, B. G. Linsen, J. C. P. Broekhoff, A. van den Heuvel, and T. J. Osinga, *J. Colloid Sci.* **21** (1966) 405.
[224]W. R. Smith and G. A. Kasten, *Rubber Chem. Technol.* **43** (1970) 960.
[225]E. P. Barrett, L. G. Joyner, and P. P. Halenda, *J. Am. Chem. Soc.* **73** (1951) 373.
[226]J. H. Atkins, *Carbon* **3** (1965) 299.
[227]B. Viswanathan and M. V. C. Sastri, *J. Catal.* **8** (1967) 312.
[228]E. W. Lard and S. M. Brown, *J. Catal.* **25** (1972) 451; see also G. R. Lester, *J. Catal.* **8** (1967) 243.
[229]J. Medema and A. Compagner, *J. Catal.* **8** (1967) 120.
[230]A. A. Adamson, *Physical Chemistry of Surfaces*, Wiley-Interscience, New York, 1964, p. 449.
[231]R. R. Paxton, J. F. Demendi, G. J. Young, and R. B. Rozelle, *J. Electrochem. Soc.* **110** (1963) 932.
[232]H. A. Benesi, R. U. Bonner, and C. F. Lee, *Anal. Chem.* **27** (1955) 1963.
[233]J. Freel, *J. Catal.* **25** (1972) 139.
[234]C. E. Hunt, *J. Catal.* **23** (1971) 93.
[235]H. A. Benesi, L. T. Atkins, and R. B. Mosely, *J. Catal.* **23** (1971) 211.
[236]H. L. Gruber, *Abal. Chem.* **34** (1962) 1828.
[237]H. L. Gruber, *J. Phys. Chem.* **66** (1962) 48.
[238]J. A. Cusamano, G. W. Dembinskii, and J. H. Sinfelt, *J. Catal.* **5** (1966) 471.
[239]G. R. Wilson and W. K. Hall, *J. Catal.* **17** (1970) 190.
[240]D. J. C. Yates and J. H. Sinfelt, *J. Catal.* **8** (1967) 348.
[241]C. S. Brooks, *J. Colloid Sci.* **34** (1970) 419.
[242]R. A. Dalla Betta, *J. Catal.* **34** (1974) 57.
[243]K. C. Taylor, R. M. Sinkevitch, and R. L. Klimisch, *J. Catal.* **35** (1974) 34.
[244]P. C. Aben, *J. Catal.* **10** (1968) 224.
[245]S. E. Wanke and N. A. Dougharty, *J. Catal.* **24** (1972) 367.
[246]K. C. Taylor, *J. Catal.* **38** (1975) 299.
[247]G. R. Wilson and W. K. Hall, *J. Catal.* **24** (1972) 306.
[248]E. Kikuchi, P. Flynn, and S. E. Wanke, *J. Catal.* **34** (1974) 132.
[249]D. E. Mears and R. C. Hansford, *J. Catal.* **9** (1967) 125.
[250]P. Wentrcek, K. Kimoto, and H. Wise, *J. Catal.* **33** (1973) 273.
[251]J. E. Benson, H. S. Hwang, and M. Boudart, *J. Catal.* **30** (1973) 146.

[252]P. A. Sermon, *J. Catal.* **24** (1972) 460.
[253]G. W. Sears, *Anal. Chem.* **28** (1956) 1981.
[254]A. Kozawa, *J. Inorg. Nucl. Chem.* **21** (1961) 315.
[255]W. O'Grady, C. Iwakura, J. Huang, and E. Yeager, in *Proceedings of the Symposium on Electrocatalysis*, Ed. by M. W. Breiter, Electrochemical Soc., Princeton, N.J., 1974, p. 286.
[256]A. Kozawa, *J. Electrochem. Soc.* **106** (1959) 552.
[257]M. L. Kronenberg, *J. Electroanal. Chem.* **24** (1970) 357.
[258]A. J. Groszek and G. I. Andrews, in *Third Conference on Industrial Carbons and Graphite*, Society of Chemical Industry, London, 1971, p. 156.
[259]F. G. Will and C. A. Knorr, *Z. Electrochem.* **64** (1960) 258.
[260]S. Srinivasan and E. Gileadi, *Electrochim. Acta* **11** (1966) 321.
[261]S. Gilman, *J. Phys. Chem.* **67** (1963) 78.
[262]S. Gilman, *J. Electroanal. Chem.* **7** (1964) 382.
[263]F. G. Will, *J. Electrochem. Soc.* **112** (1965) 451.
[264]R. Woods, *J. Electroanal. Chem.* **49** (1974) 217.
[265]A. N. Frumkin, in *Advances in Electrochemistry and Electrochemical Engineering*, Ed. by P. Delahay, Wiley-Interscience, New York, 1963, Vol. 3, p. 287.
[266]P. Stonehart, in *Power Sources 1966*, Ed. by D. H. Collins, Oriel Press, Newcastle-upon-Tyne, 1967, p. 509.
[267]K. F. Blurton, P. Greenberg, H. G. Oswin, and D. R. Rutt, *J. Electrochem. Soc.* **119** (1972) 559.
[268]J. F. Connolly, R. J. Flannery, and G. Aronowitz, *J. Electrochem. Soc.* **113** (1966) 577.
[269]J. F. Connolly, R. J. Flannery, and B. L. Meyers, *J. Electrochem. Soc.* **114** (1967) 241.
[270]M. W. Breiter, *J. Phys. Chem.* **72** (1968) 1305.
[271]N. A. Urisson, L. N. Mokrousov, G. V. Shteinberg, Z. I. Kudryavtseva, I. I. Askakhov, and V. S. Bagotskii, *Kinet. Katal.* **15** (1974) 1000.
[272]A. P. Zabotina, T. I. Yakovleva, and D. V. Sokol'skii, *Dokl. Akad. Nauk SSSR* **197** (1971) 621.
[273]M. W. Breiter, *Electrochemical Processes in Fuel Cells*, Springer-Verlag, New York, 1969, p. 51.
[274]S. K. Burshtein, M. R. Tarasevich, and V. S. Vilinskaya, *Elektrokhimiya* **3** (1967) 349.
[275]V. A. Gromyko, *Elektrokhimiya* **7** (1971) 885.
[276]D. A. J. Rand and R. Woods, *J. Electroanal. Chem.* **31** (1971) 29.
[277]S. H. Cadle and S. Bruckenstein, *Anal. Chem.* **43** (1971) 1858.
[278]G. W. Tindall and S. Bruckenstein, *Anal. Chem.* **40** (1968) 1051.
[279]G. W. Tindall and S. Bruckenstein, *Anal. Chem.* **40** (1968) 1637.
[280]M. W. Breiter, *J. Electroanal. Chem.* **23** (1969) 173; B. J. Bowles, *Electrochim. Acta* **15** (1970) 589.
[281]S. B. Brummer and M. J. Turner, in *Proceedings Twenty-Third Annual Power Sources Conference*, PSC Publications Committee, 1969, PSC Publications Committee, Red Bank, N.J. p. 26.
[282]G. W. Tindall and S. Bruckenstein, *Electrochim. Acta* **16** (1971) 245.
[283]G. Raspi and F. Malatesta, *J. Electroanal. Chem.* **27** (1970) 283; F. Malatesta and G. Raspi, *J. Electroanal. Chem.* **27** (1970) 295.
[284]D. Sandroz, R. M. Peekema, H. Freund, and C. F. Morrison, *J. Electroanal. Chem.* **24** (1970) 165.
[285]B. J. Bowles, *Electrochim. Acta* **10** (1965) 717; **10** (1965) 731.
[286]B. J. Bowles, *Electrochim. Acta* **15** (1970) 737.
[287]D. M. Kolb, M. Przasnyski, and H. Gerischer, *J. Electroanal. Chem.* **54** (1974) 25.
[288]H. Gerischer, D. M. Kolb, and M. Przasnyski, *Surface Sci.* **43** (1974) 662; see also D. M. Kolb and H. Gerischer, *Surface Sci.* **51** (1975) 323.

[289]H. Klug and L. Alexander, *X-Ray Diffraction Procedures*, Wiley, New York, 1962, p. 491.
[290]A. L. Patterson, *Phys. Rev.* **56** (1939) 972.
[291]L. V. Azaroff, *Elements of X-ray Crystallography*, McGraw-Hill, New York, 1968, p. 556.
[292]R. L. Moss and H. R. Gibbens, *J. Catal.* **24** (1972) 48.
[293]R. L. Moss, H. R. Gibbens, and D. H. Thomas, *J. Catal.* **16** (1970) 117.
[294]L.V. Azaroff, *Elements of X-ray Crystallography*, McGraw-Hill, New York, 1968, p. 480.
[295]A. Guinier and G. Fournet, *Small Angle Scattering of X-rays*, Wiley, New York, 1955.
[296]H. Brumberger, *Small Angle X-ray Scattering*, Gordon and Breach, New York, 1967.
[297]A. Guinier, *X-ray Crystallographic Technology*, Ed. by K. Lonsdale, Hilger and Watts, London, 1952, p. 268.
[298]G. A. Somorjai, Ph.D. thesis, University of California, Berkeley, 1960.
[299]G. M. Plavnik, B. Parlits, and M. M. Dubinin, *Dokl. Akad. Nauk SSSR* **206** (1972) 399.
[300]T. E. Whyte, P. W. Kirklin, R. W. Gould, and H. Heinemann, *J. Catal.* **25** (1972) 407.
[301]E. L. Gunn, *J. Phys. Chem.* **62** (1958) 928.
[302]S. D. Harkness, R. W. Gould, and J. J. Hren, *Philos. Mag.* **19** (1969) 115.
[303]A. Renouprez, C. Hoang-van, and P. A. Compagnon, *J. Catal.* **34** (1974) 411.
[304]G. F. Neilson, *J. Appl. Crystallogr.* **6** (1973) 386.
[305]P. C. Flynn, S. E. Wanke, and P. S. Turner, *J. Catal.* **33** (1974) 233.
[306]C. E. Hall, *J. Appl. Phys.* **19** (1948) 198.
[307]C. E. Hall, *J. Appl. Phys.* **19** (1948) 271.
[308]W. M. Hess and L. L. Ban, *Norelco Reporter* **13** (1966) 102; presented at the Sixth International Congress for Electron Microscopy, Kyoto, Japan, August 28–September 6, 1966.
[309]G. Terriera, A. Oberlin, and J. Mering, *Carbon* **5** (1967) 431.
[310]N. H. Sagert and R. M. L. Pouteau, *Can. J. Chem.* **52** (1974) 2960.
[311]J. S. Mayell and W. A. Barber, *J. Electrochem. Soc.* **116** (1969) 1333.
[312]J. D. Voorhies, J. S. Mayell, and H. P. Landi, in *Hydrocarbon Fuel Cell Technology*, Ed. by B. S. Baker, Academic Press, New York, 1965, p. 455.
[313]J. A. Shropshire, E. H. Okrent, and H. H. Horowitz, in *Hydrocarbon Fuel Cell Technology*, Ed. by B. S. Baker, Academic Press, New York, 1965, p. 539.
[314]W. Vogel, J. Lundquist, and A. Bradford, *Electrochim. Acta* **17** (1972) 1735.
[315]L. G. Austin and H. Lerner, *Electrochim. Acta* **9** (1964) 1469.
[316]J. Giner and C. Hunter, *J. Electrochem. Soc.* **116** (1969) 1124.
[317]E. W. Thiele, *Ind. Eng. Chem.* **31** (1939) 916; see also A. Wheeler, *Adv. Catal.* **3** (1951) 249.
[318]P. Stonehart and P. N. Ross, *Electrochim. Acta* **21** (1976) 441.
[319]P. N. Ross, K. Kinoshita, A. J. Scarpellino, and P. Stonehart, *J. Electroanal. Chem.* **63** (1975) 97.
[320]V. Ponec, *Catal. Rev. Sci. Eng.* **11** (1975) 1.

5

Charge-Transfer Complexes in Electrochemistry

Jean-Pierre Farges

Laboratoire de Biophysique, Université de Nice, France

and

Felix Gutmann

School of Chemistry, Macquarie University
Sydney, Australia

I. THEORETICAL INTRODUCTION

1. Charge-Transfer Complexes

Molecular associations between two or more components not involving true chemical bonding have been studied extensively.[1,2] They play a crucial role in some of the more recent developments in modern biology. The stability of such associations is due to a variety of intermolecular forces, such as van der Waals–London forces or hydrogen bonding, which are weaker, and less well understood, than the forces involved in a simple chemical reaction.

Molecular associations may exhibit, in addition, properties attributable to the transfer of electrical charge from one of the components, which thus acts as an electron donor, to another, which then acts as an acceptor. When this is the case, the molecular interactions include some extra contribution from electrical-charge-transfer forces. Consequently, such molecular associations have been named charge-transfer complexes (CTC) or electron donor–acceptor complexes.[2]

The relative importance of charge-transfer forces, when compared with the other intermolecular forces, is likely to vary within wide limits. For simplicity, it will be convenient to separate CTCs arbitrarily into two main groups,* viz., weak complexes involving but little charge transfer and strong complexes involving considerable transfer of charge from donor to acceptor. We shall omit discussion of yet another group of very weak CTCs, viz., contact CTCs, in which interaction is not permanent, but transitory acting only while the molecules are in contact during collision.

2. Donors, Acceptors, and Complexation

The ability to form CTCs will depend first on the difference in the electrode potentials of the two components likely to promote charge exchange between them. Thus, the lower the ionization energy of the donor and the higher the electron affinity of the acceptor, the easier is the resulting complexation. However, the "strength" of a donor is not determined solely by its ionization potential; especially in solution, it is often more meaningful to express the electron-donating tendency of a given donor in terms of its "donicity," rather than its ionization potential.

Apart from the general distinction between weak and strong CTCs, based on the strength of the charge-transfer interaction, these complexes may be classified according to the molecular orbitals involved in the electron-donating and -accepting processes.

Donors and acceptors may be classified into three groups of interest: (1) π-donors and acceptors, in which the reacting electron comes from or goes into the π-orbital of a conjugated system, so that charge-transfer bonding is delocalized over the π-system of the molecule; (2) σ-donors and -acceptors, in which the electron comes from or goes into a σ-orbital; and (3) n-donors, in which the charge comes from a more or less localized electron; here, the charge-transfer bonding will be localized on the donor atom (e.g., nitrogen) in the molecule. We may thus define six donor–acceptor combinations (the donor is named first): π–π, π–σ, σ–π, σ–σ, n–π, and n–σ.

Although this classification of electron donors and acceptors often proves useful, these terms are only relative. Thus, under

*Slifkin[3] distinguishes a third class of medium-strength complexes in which the ground state is dative to a certain degree, though not to the extent encountered in the strong complexes. For the purposes of this review, we shall assign complexes to either the weak or the strong category.

appropriate conditions, such as when the highest occupied molecular orbital is located between the orbitals of the potential donor and acceptor, any molecule exhibits both electron-donor and electron-acceptor properties. Thus, dimethylalloxazine,[4] for example, acts as a donor to the strong acceptor TCNE (tetracyanoethylene) and as an acceptor from the strong donor pyrene. This dual character applies particularly to π-bonded molecules, and is especially true for large biological molecules[5] such as the purines[6] or cyclophanes.[16] Even the classic electron acceptor, oxygen, has recently been shown[17] to be capable of donating an electron to the magnesium phthalocyanine molecule, which has the high electron affinity of 14 eV, thus matching the ionization potential of O_2, viz., 13 eV. The resulting CTC is of interest biologically.

Self-complexing may occur,[7] with different regions of the same molecular species acting as donor and acceptor. The resulting entities are referred to as *intramolecular complexes.*[5] These effects are important in solution, where they may lead to association. Donor and acceptor properties can be modified by chemical substitution, and are sensitive to steric hindrances.

Moreover, the donor, having donated, may act as an acceptor and vice versa; reverse charge-transfer complexes[5] are known in which, for example, the π-electron system of an aromatic compound acts as an electron acceptor rather than as a donor, such as the ferrocenes[18] or benzoporophyrazines.[19]

Complex formation is facilitated by a polar environment, though strong CTCs in solvents of sufficiently high permittivity may dissociate into the component ions.

Charge-transfer complexes have also been obtained in the gas phase[10] as well as in melts.[11] The former, often weak to very weak contact CTCs, are outside the scope of this review, while the electrochemistry of fused CTCs has received little attention as yet.

The enthalpy of formation of CTCs is usually[39] of the order of a few kilocalories per mole, in contradistinction to the tens of kilocalories per mole involved in conventional chemical reactions.

The donor–acceptor spacing in weak complexes is only about 0.1–0.2 Å below the sum of the molecular radii, while in strong complexes, the coulombic interaction causes a much closer approach, typically 3–3.5 Å.[20] Even this distance between charge centers, however, is greater than what is generally encountered in true chemical compounds. Complex formation usually involves a lowering of the entropy of the system, because of the increased order, though in solution this may be masked by clustering, association, and solvation effects.

The formation of a CTC is sometimes followed by a true chemical reaction forming a new chemical compound; this occurs, for example, in the reaction between amino acids and iodine,[12] or between indole and chloranil,[13] or between chlorpromazine and 6-hydroxydopamine.

Charge transfer, in this discussion, means electron transfer, and generally we will omit reference to complexes due entirely or in part to proton transfer,[15] although such contributions are known to contribute to the stabilization of the resulting entity.

Some cases of hydrogen bonding,[42] though, exhibit a definite charge-transfer band and behave like weak CTCs.

3. Weak Charge-Transfer Complexes

Many complexes of this type are known; many are of interest in their biological aspects.

The mainly nonionic, nondative ground state characteristic of these complexes may be associated with a high ionization potential of the donor, a low electron affinity of the acceptor, or both these effects. The stability of the complex results from van der Waals forces. The intermolecular distances are only slightly less than the sum of the van der Waals radii, and the complexes do not appear as chemical entities, but possess chemical properties close to the sum of the components. They may constitute an intermediate step to a true chemical reaction.

There is little charge-transfer stabilization of the ground state. An almost complete electron transfer results from the transition to the first excited state, which may be due to the absorption of light, giving the complex the highly colored appearance often observed. Thus, for many so-called CTCs,[3] the charge transfer is significant only in the excited state, and it would be more correct to speak of "complexes showing charge-transfer absorption." It is, however, sometimes difficult to demonstrate the presence of the supposed absorption experimentally.

The theory, applicable only to weak CTCs, is due to Mulliken.[1,2] The ground state ψ_N and the excited state ψ_E wave functions are chosen as hybrids of a "no-bond" function $\psi_{(DA)}$ and a "dative" function $\psi_{(D^+A^-)}$, which describe, respectively, molecular associations with formally zero charge transfer (DA) and total charge transfer (D^+A^-) from the donor D to the acceptor A. In the no-bond structure (DA) the interactions result only from van der Waals–London forces or hydrogen bonding or both:

$$\psi_N = a\psi_{(DA)} + b\psi_{(D^+A^-)} \tag{1}$$

and

$$\psi_E = b^*\psi_{(D^+A)} - a^*\psi_{(DA)} \tag{2}$$

where $a \ll b$ for the ground state, which is essentially nonionic, and $b^* \ll a^*$ for the excited state, which is essentially ionic.

An energy-level diagram of a a weak CTC, as obtained from simple perturbation theory, is shown in Fig. 1. W_0 and W_1 are the energies of the no-bond and of the dative structures, respectively, while W_N and W_E are the energies of the ground state and of the (first) excited state of the complex, respectively. $R_N = W_0 - W_N$ and $R_E = W_E - W_1$ are two resonance energies representing the contribution of the dative structure to that of the ground state and that of the no-bond structure to that of the excited state, respectively.

Within the approximations of the Mulliken theory, then, $W_1 - W_0 = I_D - E_A - U$, where I_D is the ionization energy of the donor and E_A is the electron affinity of the acceptor. Most of the energy U is due to the coulombic interaction between donor and acceptor species

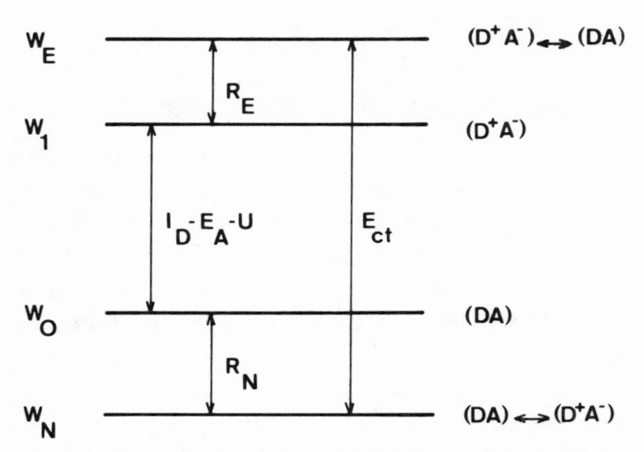

Figure 1. An energy-level diagram for the formation of a weak CTC.

in the (nearly ionic) excited state of the complex. The two resonance energies R_N and R_E are related to this quantity as follows:

$$R_N = C_N/I_D - E_A - U \qquad \text{and} \qquad R_E = C_E/I_D - E_A - U \quad (3)$$

where C_N and C_E are (approximately) constants.

The energy $E_{ct} = W_E - W_N$ of the electronic charge-transfer absorption characteristic of the complex can then be expressed in the approximate form

$$E_{ct} \simeq I_D - E_A - U + C/I_D - E_A - U \quad (4)$$

where $C = C_N + C_E \simeq$ const.

This result predicts a parabolic relationship between I_D and E_{ct} for a given acceptor, and between E_A and E_{ct} for a given donor. When the resonance energies R_N and R_E are small enough to be neglected (very weak complexes), a simple linear relationship will result:

$$E_{ct} \simeq I_D - E_A - U \quad (5)$$

Thus, the Mulliken theory predicts the appearance of a new, charge-transfer, band in the optical absorption spectrum of the complexes, which would otherwise be similar to the superposition of the spectra of the separate components. Of course, this band will depend strongly on the relative character of the donor and acceptor species forming the complex.

Formally, the charge-transfer terms, as well as the common polarization and dispersion terms, should be included within the so-called van der Waals–London molecular interactions.[1]

4. Strong Charge-Transfer Complexes

The formation of such complexes involves both strong donors and strong acceptors: the ionization potential of the donor must be low and the electron affinity of the acceptor must be high. The resulting molecular associations exhibit a marked ionic character in their ground state. In this case, the charge-transfer forces, which contribute to the molecular interaction, are strong enough to produce an appreciable stabilization of the ground state of the complex; other contributions arise from electrostatic, induction, and dispersion forces. The perturbation theory of Mulliken is less applicable under these circumstances, and other methods, such as the variation method, should be used to evaluate the resonance energies and the energy of the charge-transfer band of the resulting molecular associations exhibiting marked ionic character in their ground state.

The electronic absorption spectrum of the complex resembles that corresponding to the sum of the spectra of the two separate and ionized components. Intermolecular distances between the two components are significantly less[20] than the sum of the van der Waals radii. In the limit, formation of two molecular ions D^+ and A^- may occur.

In solution, solvation and association effects are likely; sometimes, these effects can be detected by means of the concomitant NMR chemical shifts. The association constants of CTCs have thus been shown[93] to vary exponentially with the energy of the charge-transfer transition.

While no ESR absorption occurs in the case of weak complexes, broad bands are often observed with strong CTCs; the g value is usually below 2.0023, indicating that the electrons are not completely free.

Strong CTCs in which the ground state is predominantly ionic exhibit very high dipole moments and therefore high permittivities; e.g., the dipole moment of the triethylamine–iodine complex has been experimentally determined as 11.3 Debye (D) units,[41] though other workers report only 5.23–6.55 D, depending on the solvent.[43] The pyridine–iodine complex, for example, has a moment of 4.44 D.[36] However, their generally low resistivity renders the measurements difficult and imprecise. Weaker complexes, especially of the π–π type, formed between nonpolar donors and acceptors show high dipole moments because of the resulting mutual polarization of the constituents[22]; the mean dipole moment of pyrene–naphthalene, for example, is 2.0 ± 0.3.[23] In CTCs formed between one polar and one nonpolar constituent, these same induced polarizations result in a diminution of the net polarization; e.g., nitrobenzene has a dipole moment of 4.22–4.28 D, while the resulting CTC exhibits a value of 3.78 D.[24]

II. SOLID CHARGE-TRANSFER COMPLEXES[21]

1. Crystal Structures

A property of CTCs is their ability to form crystals having simple stoichiometries. The charge-transfer interactions are usually enhanced in the solid state because the molecules are held in close and stable proximity, whereas in solutions, the close approach of the two components forming the complex is hindered by Brownian motion.

Detailed X-ray data are now available for many CTCs. The crystal structures often consist of infinite stacks[21] of alternate donor and acceptor.

Since it is the requirement of maximum overlap of the positive and negative regions of the donor and acceptor wave functions that determines the structures of the CTC, these stacks are frequently skewed; a typical structure is shown in Fig. 2. This skewing holds especially for π-complexes formed between aromatic donors and acceptors. In n–π complexes, however, the requirement means maximizing that the overlap between the donor lone pair orbital wave function and that of the unfilled orbital of the acceptor must be maximized, which often results in the donor and acceptor planes becoming parallel to each other.

Layered structures in which planes of donor and acceptor species are alternatively superimposed also exist. In these cases, the molecular interactions are likely to extend over the entire crystal lattice, whereas only discrete molecular associations having the stoichiometry of the complex are likely to exist in solution.

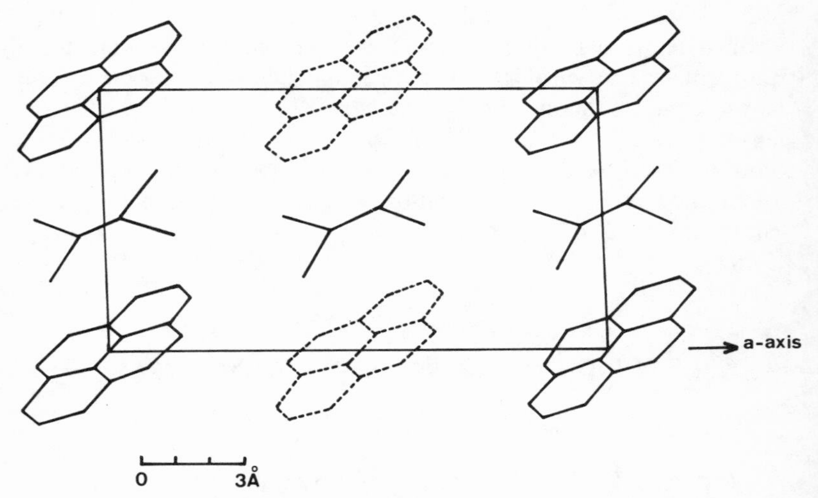

Figure 2. X-Ray structure of the pyrene–TCNE complex (after Rembaum[21]).

Furthermore, in a crystal lattice, CTCs are more closely packed than neutral molecules in molecular crystals, and the solid-state properties are affected by steric changes.

2. Electrical and Magnetic Properties

Many solid-state CTCs are good or even excellent organic semiconductors. As a class, they tend to have higher conductivities than their constituents, although the resulting conductivities are spread over an extended range; e.g., some perylene–fluoranil CTCs,[25] exhibit a conductivity of only about $10^{-15}\,\Omega^{-1}$, while the phenothiazine–iodine complex,[37] for example, conducts about 10^{15} times better.

The anisotropy of the charge-transfer interactions within the crystal lattice is often reflected in a significant anisotropy of the physical properties of the materials.

Some correlation appears to exist between the degree of ionicity of the ground state of the complex and the electrical and magnetic properties in the solid state. Thus, the weak complexes are usually diamagnetic and very poor electrical conductors, whereas the strong complexes often exhibit paramagnetism and are highly conductive. However, an ionic ground state is not a sufficient condition for high conductivity.

Often, 1:1 complexes are found to exhibit conductivities several orders of magnitude below those found in 1:2 complexes formed from the same components. Such complexes usually show two ESR lines, though they are often difficult to resolve. Generally, there is no quantitative relationship between the electrical conductivity and the magnetic properties, especially the ESR data,[35] of the solids, though in some cases of polymeric CTCs, such a relationship has been demonstrated.[26]

The charge-transfer interaction involves the uncoupling of an electron, which is thus left unpaired. The resulting NMR shifts [cf. Section III.4(ii)] have been observed, but are hard to interpret and may be associated with other effects such as hydrogen bonding. In some cases, however, such as certain trinitrobenzene complexes,[49] valuable information has been obtained; thus, in interactions between delocalized π-electron systems, it has been shown[27] that the interaction is indeed highly localized.

As has been mentioned, strong CTCs exhibit high dipole moments and thus high permittivities. This holds *a fortiori* for the solid state. In addition, macromolecular solids, including some CTCs, may show extremely high permittivities, up to perhaps 10^4, by virtue of hyperelectronic polarization.[8] This polarization is due to "excitons, in the form of thermally generated, separated charge pairs lying on the very long electronic orbital domains present in such highly conjugated molecules."[9] Thus, an externally applied electric field causes the displacement of an electron within a polymeric region exhibiting something akin to local superconductivity extending over several hundred or even 1000 Å. This effect is not confined to CTCs.[8]

3. Modes of Charge Transfer in the Solid State[21]

Over most of the temperature ranges usually encountered, a common feature of the electric behavior of organic semiconductors, including the CTCs, is the exponential rise of conductivity with increasing temperature; the activation energy involved in CTCs may have values of up to more than 2 eV for very weak complexes to nearly zero for strong CTCs featuring a near-degenerate electron distribution (see below).

Charge may be transported by (1) electrons moving in an energy band; (2) electrons transiting from site to site by thermally activated hopping; (3) electrons "percolating" by diffusion from site to site, via localized extended states in the energy gap of the solid; (4) ionic transfer; and (5) excitons.[21]

Motion in energy bands is familiar from covalent semiconductors such as Si, Ge, GaAs, InSb, and many others. The extent, if any, to which this model is applicable to poorly conducting organic semiconductors, including weak CTCs exhibiting energy gaps of the order of 2–3 eV and more, is still not clear, though the model is applicable to the highly conductive CTCs with energy gaps of less than, say, 1 eV. In common with organic semiconductors,[21] however, the mobility of the carrier—be it an electron or a positive hole, viz., an electron vacancy—is low because the conduction bands are generally narrow, and the effective mass of the carrier is high. Moreover, since organic materials cannot be purified to the extent possible in the case of Si or Ge, impurities are bound to be important. Even a single crystal is not a perfect crystal, and organic systems can exhibit a far greater concentration of lattice faults than is encountered in a single crystal of, say, Ge.

These considerations, then, further limit the spatial extent of the periodicity and cause further narrowing of the energy bands. Nevertheless, the band model has been successfully applied to a great number of CTCs.[21]

In this model, as the temperature rises, so does the concentration of electrons in the conduction band, until most of or nearly all the valency electrons find themselves in that band. The system then tends to become degenerate and approach, in its conduction behavior, that of a poorly conducting metal. The Arrhenius form of the temperature dependence of conductivity encountered at lower temperatures flattens out and changes into a roughly linear decrease of conductivity with further rise in temperature.[40] This behavior is illustrated in Fig. 3, which refers to one of the tetracyanoquinodimethan (TCNQ) complexes to be discussed in a later section.

Thermally activated hopping may involve[38] an electron transfer over the top of the intermolecular energy barrier in a way similar to a chemical reaction, or the electron may tunnel through the barrier. While tunneling as such is nearly temperature-independent, the classic distribution of accentor levels yields an exponential temperature dependence of the resulting current. The tunneling probability rises exponentially with decreasing width of the barrier, so that rising temperature causes a narrowing of the barrier. The electron may thus tunnel through, provided there is an exactly equal energy level available to it on the other side of the barrier. It is not easy to distinguish between these two modes of transfer.[21]

In very disordered systems such as amorphous solids,[44] the density of states within an isolated energy band can be depicted as in Fig. 4. Instead of the sharp band edges, we find "tails of localized states" caused by the potential energy fluctuations due to the structural disorder.[44] If there is, in addition, disorder of composition, i.e., if the chemical nature of the system varies from region to region, these tails may overlap. The carrier mobility at the band edges, i.e., in the tails, drops from a high value characteristic of motion within an energy band to the low value associated with thermally activated hopping (see Fig. 5). Within the tails, however, where the energy states change from an extended to a localized regime, there prevails a transport mechanism that is akin to Brownian motion; i.e., it is a Markovian process with a mobility intermediate between that of the propagation of an electron wave within an energy band and thermally assisted hopping (see Fig. 6). This mode of charge transfer is sometimes referred to as percolation, and in poorly

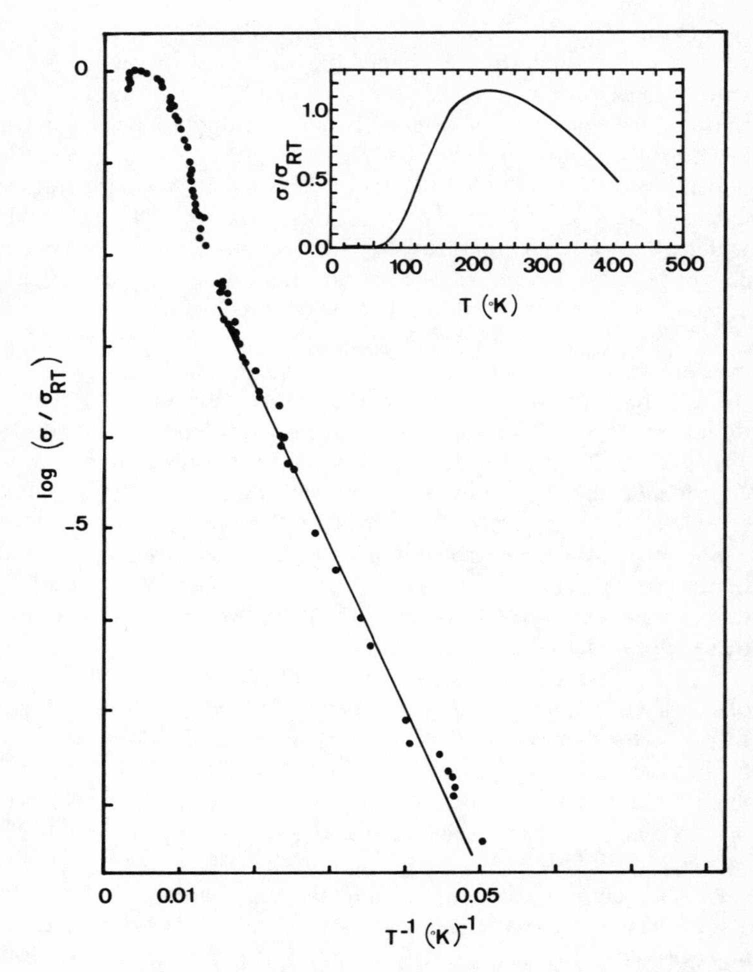

Figure 3. Temperature dependence of the conductivity of the TTF–TCNQ complex (after Coleman *et al.*[47]). (RT) Room temperature.

Ⓐ

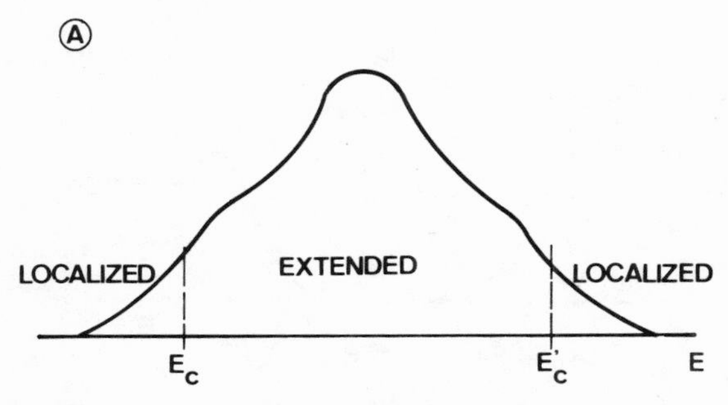

Figure 4A. The simplest model for the density of states of a single isolated energy band in a disordered material. There is a band of extended states inside the energies E_c and E'_c, with tails of localized states outside. These are expected to be universal features of the electronic structures of disordered materials (after Cohen[44]).

Ⓑ

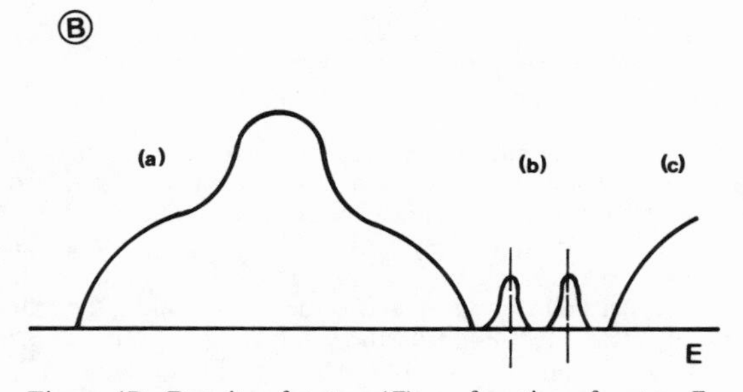

Figure 4B. Density of states $n(E)$ as a function of energy E for (a) a perfect crystal, (b) a crystal containing only one localized imperfection, and (c) a crystal containing a low concentration of localized imperfections. In all three cases, there are continuous bands of energy levels separated by gaps with square root singularities at the band edges. The square root singularities associated with saddle points within the band in (a) and (b) are eliminated by scattering in (c). If the potential change ΔV introduced by the imperfection is strong enough in (b), localized levels split off from the bands (off the bottom for attractive ΔV or the top for repulsive), which broadens into bands in (c) (after Cohen[44]).

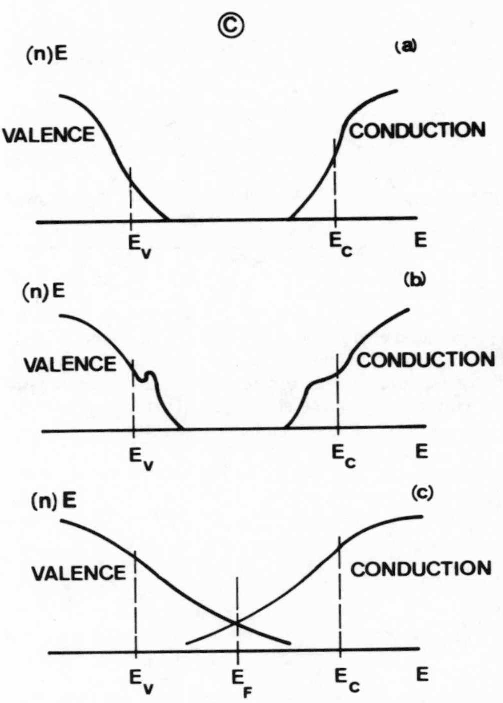

Figure 4C. Band models for amorphous semiconductors. In (a), the simplest possible case, there are valence and conduction bands with extended states for $E < E_v$ and $E > E_c$, and tails of localized states on each band. For elements and compounds (b), the well-defined local order leads to the presence of well-defined structure defects and nonmonotonicity in $n(E)$ in or near the tails. For alloys (c), there is, in addition, compositional disorder and translational disorder enhanced by the requirement that valences be locally satisfied. The tails of the conducting and valence bands therefore overlap, which leads to a finite density of states at the Fermi energy $n(E_F)$ and a finite concentrate of localized charged states near E_F (after Cohen[44]).

conducting CTCs, is likely to be the predominant mode of conduction.[44]

Ionic charge transfer may occur in strong CTCs, in which the ground state is nearly completely ionic, so that the system resembles an ionic crystal.[90] Some electrochemical studies in such systems will be discussed below.

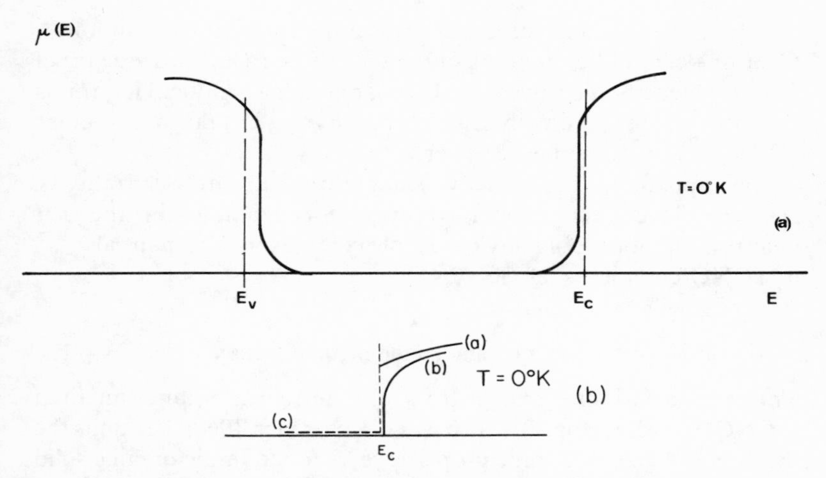

Figure 5. Simple models for the energy dependence of the mobility μ: $T \neq 0°\mathrm{K}$; (b) $T = 0°\mathrm{K}$. In (a), E_v and E_c play the role of mobility edges because the mobility drops there from values characteristic of band conduction to values characteristic of phonon-assisted hopping conducting within the mobility gap ketween E_v and E_c. This drop is termed a *mobility shoulder*. In (b), various possible energy dependencies of μ near E_c are shown for $VT = 0°\mathrm{K}$ that are consistent with: case (a) (solid line) a continuous drop from a finite value to zero, case (b) (solid line) an abrupt but continuous drop toward zero, and case (c) (dotted line) a finite though small value of μ within the mobility gap (after Cohen[44]).

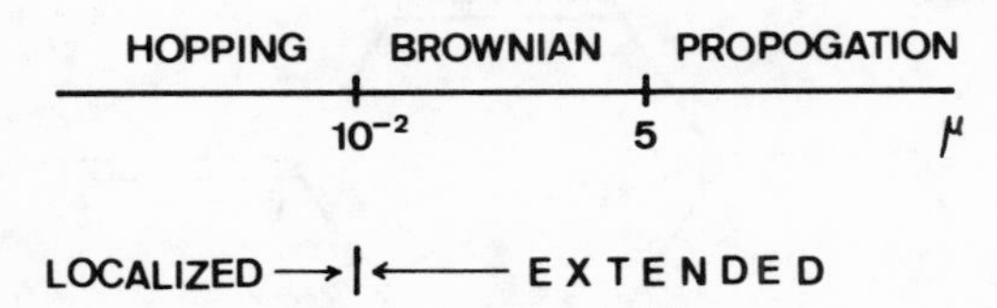

Figure 6. Mobility scale for semiconductors at room temperature in units of $\mathrm{cm}^2/\mathrm{V\,sec}$. The various regimes proposed in the text are indicated above the scale and the character of the wave function below the scale (after Cohen[44]).

Ionic conductance need not involve the translation of an ion. A Grotthus-chain-like "handing on" of charge, i.e., a linked sequence of RedOx reactions, also gives rise to apparent ionic conductivity; this is probably the main contribution to charge transfer in intramolecular CTCs, such as some benzene derivatives.[48]

An apparent A.C. conductivity may also arise from dielectric losses, such as have been shown to arise, for example, in the 1:1 trinitrofluorenone–polyvinylcarbazole complex and perhaps also[152] in TCNQ complexes (cf. Sec. 2.4).[45]

4. Tetracyanoquinodimethan Salts

The π-associations formed from the tetracyanoquinodimethan (TCNQ) molecule (Fig. 7), which was discovered in 1960,[28] exhibits the most remarkable electronic properties ever found in the organic solid state, so that special reference to the important class of compounds comprising complexes formed with this molecule must be made. Strictly speaking, they are not CTCs, since there is in the ground state a complete charge transfer from many donors, organic molecules or metals, to the TCNQ molecule, which is a powerful electron acceptor. These complexes might be more appositely referred to as charge-transfer *salts*, or as free radical salts, in view of the extra electron unpaired on the $TCNQ^-$ ion.

Their crystal structures differ from those of the common CTCs,[29] the donor and acceptor molecules forming distinct stacks, parallel and separated. Furthermore, as shown in Fig. 7, TCNQ is a planar

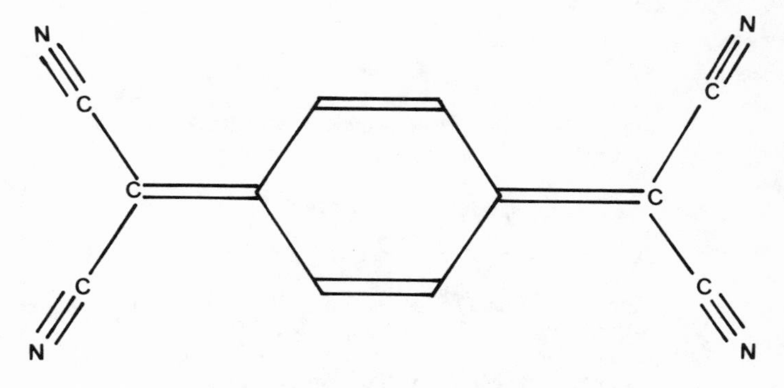

Figure 7. Structure of the TCNQ molecule.

molecule that allows a close-packed arrangement along the TCNQ stacks, with relatively large π-overlap compared with that encountered in usual molecular crystals.

There are two immediate consequences of these particular structural features. First, the close-packed arrangement leads to exceptionally high values for the electrical conductivity. Several salts are known that exhibit values at room temperature higher than $100\,\Omega^{-1}\,cm^{-1}$. Although the conductivity is activated at low temperatures, they may exhibit electrical and magnetic properties suggestive of a "metal" above some characteristic temperature (see Fig. 3).[46,47] Second, the columnar structure of TCNQ molecules, as shown in Fig. 8 for the TEA–TCNQ entity,[29] leads to a strong electrical anisotropy, which is illustrated in Fig. 9 for the same entity. This entity exhibits conductivities at room temperature of 7.4, 4.5×10^{-2}, and $2.6 \times 10^{-3}\,\Omega^{-1}$ for the three principal crystal directions.[46] The activation energy, too, is anisotropic. Other charge transfer complexes of that type, but not involving TCNQ, have recently been reported.

Those TCNQ salts that exhibit nearly perfect one-dimensional order are of considerable interest; the unpaired electrons, which are responsible for the conductivity, are exchange-coupled, so that the ground magnetic state is always diamagnetic. The narrowness of the one-electron energy bands and the large density of unpaired electrons may enhance the effect of the coulombic interactions, which are consequently a possible limiting process for electrical conduction. Thus usually, but not always, a 1:2 ratio with only one electron per two TCNQ molecules is more favorable to a high conductivity than a 1:1 ratio. Thus, for example, it can be seen from Fig. 9 that phenazinium TCNQ exhibits a highly anisotropic conductivity of the order of $10^{-6}\,\Omega^{-1}$ at room temperature, while the $TCNQ_2$ adduct, which contains one neutral TCNQ molecule in addition to the TCNQ radical anion,* has a conductivity about 10^4 times higher. The activation energies are 0.2 and 0.03 eV, respectively. Further characterization of these salts is illustrated by Table 1, which lists the conductivities of several TCNQ salts as functions of the cation–anion ratio. Polyiodine and polysulfurnitride "canal" type complexes also are essentially one-dimensional "metals".[151]

Up to now, however, the importance of their one-electron band

*The existence of $TCNQ^-$ ions has now been spectroscopically confirmed.[149]

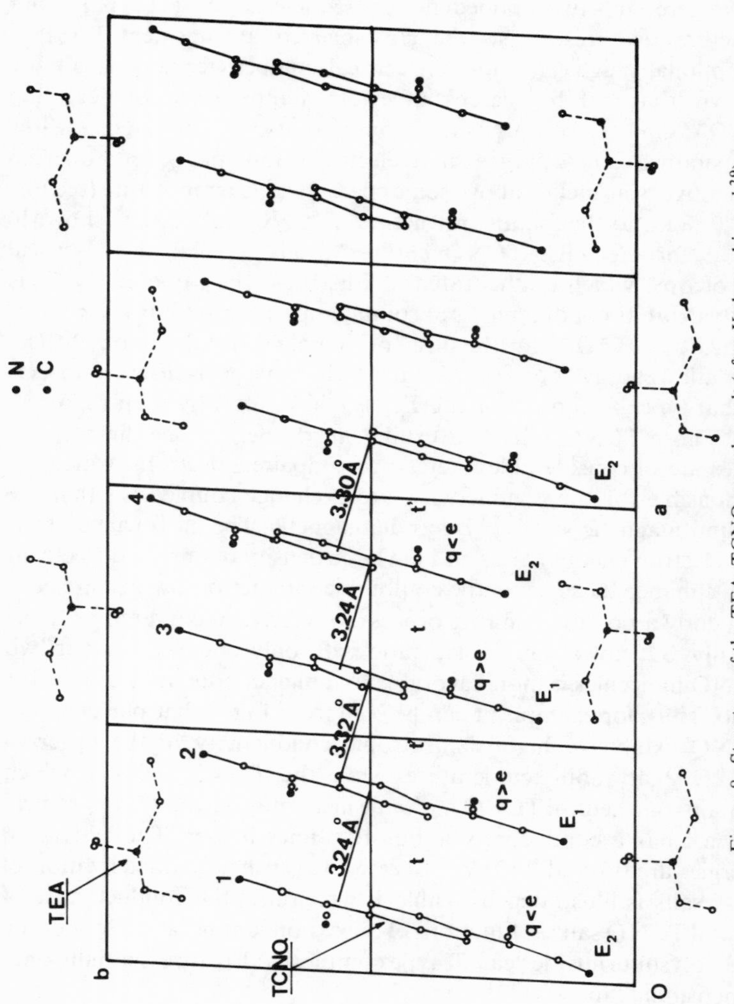

Figure 8. Structure of the TEA–TCNQ complex salt (after Kobayashi *et al.*[29]).

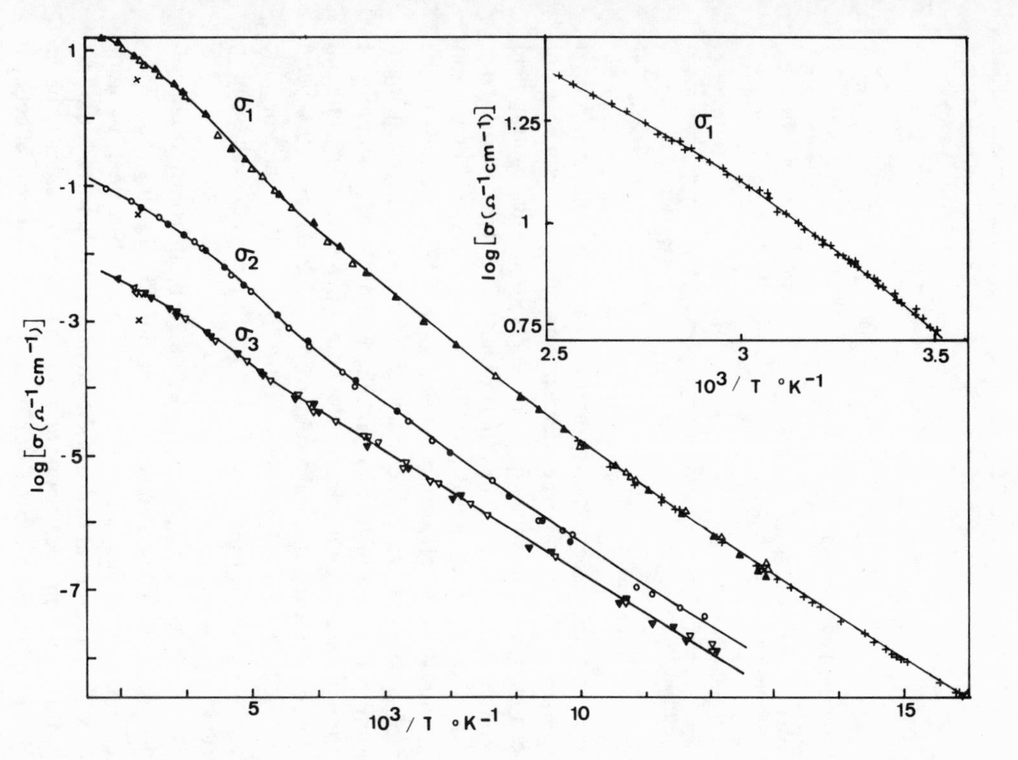

Figure 9. Temperature dependence and anisotropy of the electrical conductivity of the TEA–$TCNQ_2$ salt, measured along the three principal crystal axes (after Brau and Farges[46]).

Table 1
Electrical Conductivities of Polycrystalline Samples of TCNQ Salts with Multiple (TCNQ)–Cation Ratios[a]

TCNQ salts	Electrical conductivity at 23°C ($\Omega^{-1}\,cm^{-1}$)	
	$n = 1$	$n = 2$
$Cs(TCNQ)_n$	5×10^{-4}	1×10^{-5b}
Triphenylmethylphosphonium $(TCNQ)_n$	2×10^{-11}	2×10^{-2c}
Triethylammonium $(TCNQ)_n$	1×10^{-9}	5×10^{-2}
4-Cyano-*N*-methylquinolinium $(TCNQ)_n$	7×10^{-6}	2×10^{-2}
N-Methylquinolinium $(TCNQ)_n$	1×10^{-7}	1×10^{-1}

[a]After Le Blanc.[21]
[b]$n = 1.5$.
[c]Single-crystal measurements.

structure, of the electron interactions, and of the disordered arrangement of the cations as evidenced from the X-ray data have not yet been clarified. The most spectacular results in this exciting field concern the report of an extraordinarily high value of the conductivity, more than $10^6\,\Omega^{-1}$, obtained at 58°K on some crystals of the salt tetrathiofulvaline–TCNQ.[30] Such behavior has been ascribed to the onset of superconducting fluctuations, the material experiencing a quantum mechanical instability, viz., the Kohn–Peierls instability: it transforms back into an insulator before the superconducting state can be stabilized. The experimental results are reproduced in Fig. 10. Several reviews on TCNQ salts have been published recently,[31] and organic "metals" with donors other than TTF, e.g., dipyranylidene, have been reported.[150]

Some other layered compounds that exhibit a partial charge-transfer interaction are known[32] to be superconducting, e.g., TaS_2 intercalation compounds such as the TaS_2-substituted pyridine complex, the structure of which is shown in Fig. 11. In this compound, charge transfers from the organic donor to a metallic layer result in the formation of "evanescent Cooper pairs." Precursor effects to superconductivity have been reported[32] at temperatures as high as 35°K, e.g., polysulfurnitride becomes superconducting at 0.26°K.[151]

Yet another way to attain superconductivity might result from the observation that a negative value of the permittivity ε appears[33] to be

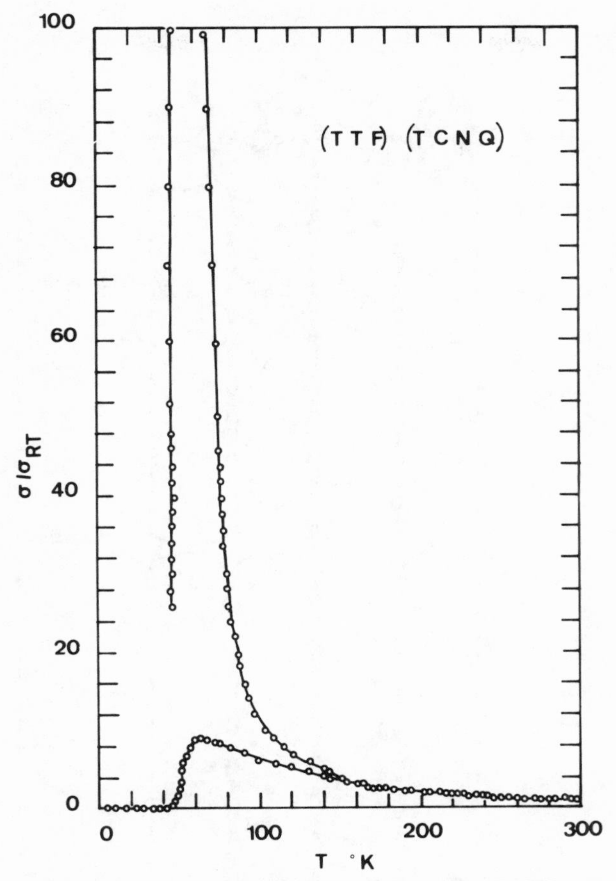

Figure 10. Temperature dependence of the TTF–TCNQ complex, showing the approach to a superconductive region (after Coleman *et al.*[30]).

a precursor for the onset of superconductivity. It may be possible to devise a CTC between a strong donor and a weak acceptor, allowing the injection of so much space charge that the quantity $dC = dQ/dV$ becomes negative; dC here stands for the apparent capacitance change resulting from a charge increment dQ causing an increment dV in the voltage across the capacitor. But for a geometric factor, C is a measure of the permittivity. The theory of these thought experiments is not as yet clear.[34]

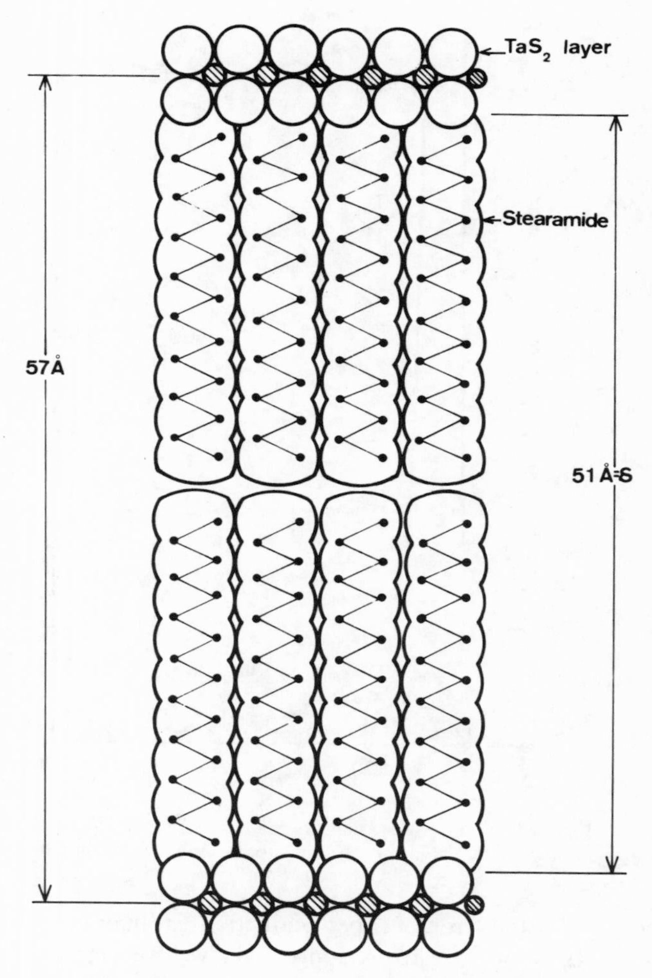

Figure 11. Schematic representation of the structure of the TaS_2–stearamide complex. The large open circles represent the sulfur atoms. The small shaded circles represent the tantalum atoms (after Fernandez-Moran[32]).

III. IDENTIFICATION OF CHARGE-TRANSFER COMPLEXES

1. Spectroscopy

There is a very large body of literature[71] on this subject. Several spectroscopic methods can be employed to determine the σ presence or absence of a charge-transfer interaction: absorption spectroscopy in the visible, IR, and UV regions, fluorescence and phosphorescence spectroscopy, and flash photolysis. Polarized single-crystal spectroscopic studies, which may permit the simultaneous examination of charge-transfer and "molecular" transitions, as well as the resolution of and distinction between these transitions through their different polarizations, hold the potential of providing important information in regard to this problem.[72]

For a solid complex, reflectance spectra[89] can give a great deal of information; the diffuse reflectance has been shown to be related[84] to the excitation coefficient of the crystal as measured by absorption spectroscopy. In some of the more highly conducting CTCs, a nearly metallic reflectance is sometimes noted; this type of reflectance would indicate the presence of an electron plasma[85] in the surface.

Many of the poorly conducting solid CTCs are photoconducting—the effect is masked by the high dark conductance in the case of the better-conducting complexes—allowing the photocurrent spectra to be studied.[86] In general, these spectra resemble the absorption spectra.

In donor–acceptor systems that show a charge-transfer band, the threshold of the intrinsic photoconduction[73] is regularly about 0.2 or 0.3 eV above the peak of the charge-transfer band because of the dissociation of an ion pair. A second and smaller threshold at lower energies (e.g., 0.7 eV) than the first may be attributable to defects.

2. Conductimetric Titrations[74]

The probability of charge transfer in solution increases with increasing permittivity[58] ε of the medium. If ε is sufficiently high, the complex may dissociate into ions, giving rise to appreciable ionic conductivity:

$$\begin{aligned} D + A &\rightharpoondown DA \\ DA &\rightharpoondown D^{+} + A^{-} \end{aligned} \tag{6}$$

The formation of CTCs can be followed by measuring changes in the electrical conductivity of a solution of, say, the donor in an inert solvent of sufficiently high permittivity, consequent on additions of a solution of the acceptor to the same solvent, or vice versa. In effect, this amounts to a conductimetric titration; the donor may be titrated with the acceptor, or vice versa.

The conductivity σ of a given system of noninteracting carriers present in concentrations n_i carrying z_i electronic charges e with a mobility μ_i is given by

$$\sigma = e\sum n_i \mu_i z_i \tag{7}$$

where the summation has to extend over all the i species of carriers present.

In the absence of interaction, chemical or other, the addition of a solution of conductivity σ_A to another solution of conductivity σ_D yields a conductivity σ given by

$$\begin{aligned} \sigma &= \sigma_D + \sigma_A \\ &= Ze[C_D(\mu_D^+ + \mu_D^-) + C_A(\mu_A^+ + \mu_A^-)] \end{aligned} \tag{8}$$

where C_D and C_A stand for the concentrations of the solutes A and D, respectively, and the μs refer to the mobilities of the carriers contributed by the D and by the A solution. We shall assume that $z = 1$ for all carriers. Concentrations will be used instead of activities, and the conductivity of the solvent, or medium, itself will be neglected.

In a titration, where the volume changes on addition of titrant, donor and acceptor are present in concentrations C_D^0 and C_A^0, respectively, so that

$$\begin{aligned} C_D^0 &= V_0^D C_0^D / V_0^D + V_0^A \\ C_A^0 &= V_0^A C_0^A / V_0^A + V_0^D \end{aligned} \tag{9}$$

where V_0^D and V_0^A refer to the volumes of stock solutions of D and A, with concentrations C_0^D and C_0^A, respectively.

In the absence of interaction, C_D^0 and $C_A = C_A^0$. Since these conductivities are additive, it follows that σ, in the absence of interaction, should be linearly related to the concentration of titrant.

When charge transfer occurs, following Eq. (6), it gives rise to an increase in conductivity. Let the concentration of the complex DA be given by C_{DA}, and the concentrations of D^+ and A^- by C_D^+ and C_A^-, respectively. C_D and C_A stand for the concentrations of unreacted donor and acceptor, as before.

We then can write the following equations for the equilibrium state of the system:

Complex formation: $D + A \rightleftarrows DA$

$$\frac{C_{DA}}{C_D C_A} = K_1 = f(\varepsilon, T\ldots) \tag{10}$$

Complex dissociation: $DA \rightleftarrows D^+ + A^-$

$$\frac{C_D^+ C_A^-}{C_{DA}} = K_2 = f'(\varepsilon, T\ldots) \tag{11}$$

Direct transfer and
recombination: $D^+ + A^- \rightleftarrows D + A$

$$\frac{C_D C_A}{C_D^+ C_A^-} = K_3 = f''(\varepsilon, T\ldots) \tag{12}$$

The solution must be electrically neutral, so that

$$C_D^+ = C_A^- = n \tag{13}$$

which is the carrier concentration required.

The conductivity then may be written

$$\sigma = en(\mu^-) = \sigma_0 n \tag{14}$$

Thus, now $C_D \neq C_D^0$ and $C_A \neq C_A^0$, and in the solution there will be present carriers D^+ and A^- in concentrations C_D^+ and C_A^-, respectively; undissociated complex molecules DA in concentration C_{DA}; unreacted donor and acceptor molecules in concentrations C_D and C_A, respectively. Therefore,

$$\begin{aligned} C_D &= C_D^0 - n - C_{DA} \\ C_A &= C_A^0 - n - C_{DA} \end{aligned} \tag{15}$$

From the above, it follows that

$$[C_D^0 - (n + K_2/n^2)][C_A^0 - (n + n^2/K_2)] = n^2 K_3 = n^2(K_1 K_2)^{-1} \tag{16}$$

The conductivity σ, which is proportional to n, is a maximum if $C_D^0 = C_A^0$, for a 1:1 complex. Equating dn/dC_D^0 to zero yields

$$-C_A^0 + n + n^2/K_2 = 0 \tag{17}$$

Since from Eq. (17) $n + n^2/K_2 = C_D^0 - C_D$, a σ peak will appear if

$$C_D^0 - C_D = C_A^0 \tag{18}$$

so that

$$n + C_{DA} = C_A^0 \tag{19}$$

or in terms of the titrating solutions,

$$\begin{aligned} V_0^D C_0^D &= V_0^A C_0^A \\ V_0^D / V_0^A &= C_0^A / C_0^D \end{aligned} \tag{20}$$

holds for the conductivity peak. The stoichiometry of the complex may thus be determined.

In practice, conductivity peaks occur slightly off the stoichiometric, say, 1:1 ratio. They thus indicate nonideal behavior.

A typical example[57] of such a conductivity titration is shown in Fig. 12. The data are consistent with two stoichiometries, viz., 1:2 and 2:3, which the method at present cannot resolve, though more sophisticated instrumentation may be able to do so.

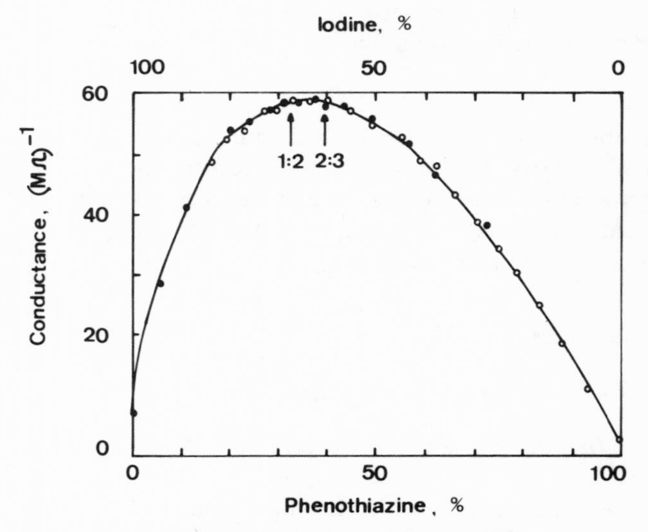

Figure 12. Conductimetric titration of the phenothiazine–iodine complex formed in acetonitrile as solvent. Two stoichiometries are evident (after Gutmann and Keyzer[87]).

Conversely, Fig. 13 illustrates[74] the linear relationship obtained if there is no complex formation or the complex fails to dissociate appreciably. This result is generally due to the permittivity of the solvent being too low. A well-developed conductivity maximum (see Fig. 12) is not always obtained. A steeply sloping base line caused by serious mismatch of the conductivities of the pure donor and acceptor solutions tends to yield ill-defined or obscured conductivity peaks. This situation is illustrated in Fig. 14, which refers to the iodine–naphthalene complex formed in acetonitrile.[74] The decidedly nonlinear plot indicates complexing. The stoichiometry of this complex can be determined by correcting for the background conductivity obtained by linear interpolation, i.e., by plotting $\sigma - \sigma_0$ relative concentration. This plot is shown in Fig. 15. A peak becomes evidence at a stoichiometric ratio of 0.57 and indicates a 1:1 complex.[74]

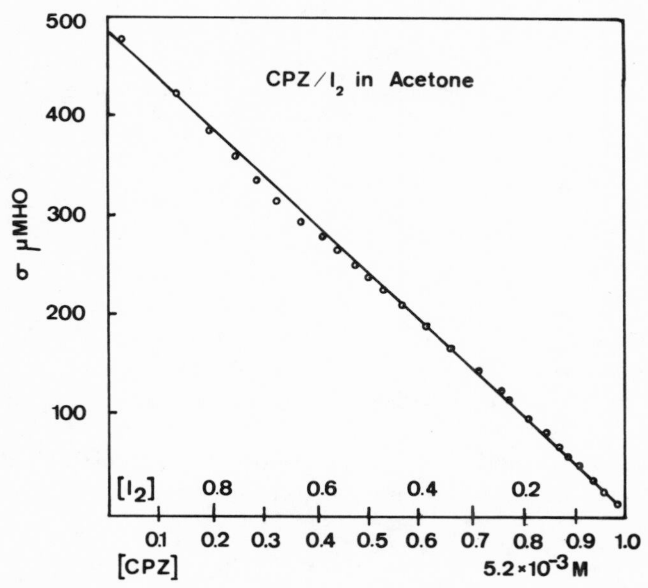

Figure 13. Conductimetric titration of the chlorpromazine (CPZ)–iodine complex formed in acetone as solvent. This is a well-known complex that, however, fails to yield a conductivity peak in acetone because the permittivity of the medium is too low to dissociate the complex (after McGlynn[24]).

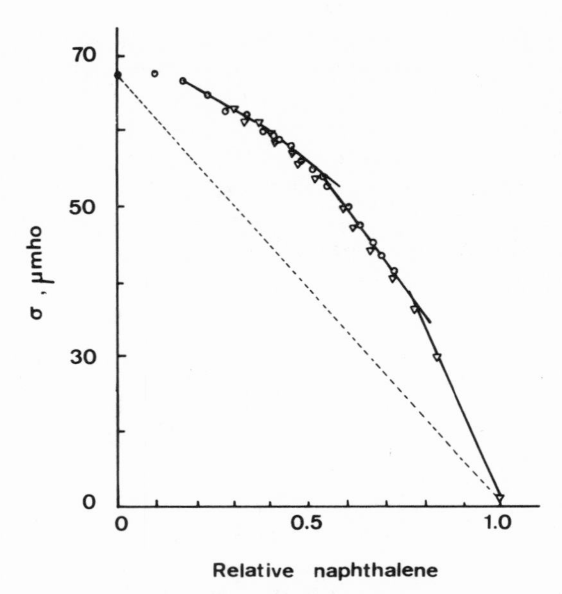

Figure 14. Conductimetric titration of the naphthalene–iodine CTC, formed in acetonitrile as solvent. While complexing is evident from the departures above the base line, the steep slope makes the determination of the stoichiometry of the complex uncertain (after Gutmann and Keyzer[74]).

The method has been used to identify other CTCs, e.g., chloropromazine complexes,[87] amaranth,[56] pyridine,[97] neomycin,[59] acetylcholine,[60] 6-hydroxydopamine,[60] serotonine,[60] and several proteins.[91] It supplements spectroscopy because it is at its best for highly ionized complexes that do not yield a charge-transfer band and are difficult to identify by spectroscopy only.

3. Dielectric Properties

(i) Dipole Moment[99]

The formation of a CTC is generally accompanied by an increase in dipole moment, though complexes between a polar and a nonpolar component result in a reduction of the molecular polarization (cf. Section I.4). Thus, measurement of changes in the molecular polarization following complexation provides an indication, though by no means an infallible proof, of the formation of CTCs. The changes[77] are small differences between comparable quantities, if both donor and acceptor are polar, and are also small if both are apolar. It is important that the changes in the molecular dipole moment be shown to be temperature-reversible in order to establish a charge-transfer interaction.

However, the polarizations being nonadditive indicates, though it falls short of proving, a charge-transfer interaction; e.g., the *m*-cresol–pyridine and –quinoline complexes were initially studied in this way, the interaction then being confirmed by means of conductimetric titrations.[98] For strong complexes, the measured permittivity is linearly related to the weight fraction of the dissolved complex, using the same solvent[99]; the specific polarization of the complex solution has been shown to follow[61] an equation derived from the

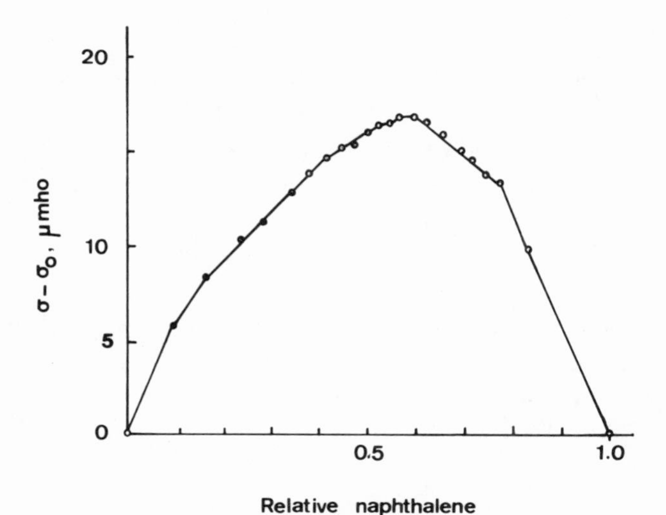

Figure 15. Replot of the data displayed in Figure 14. The excess conductances above the base line, i.e., $\sigma - \sigma_0$, are used in this plot in place of the conductances σ. A clear conductance peak results, indicating a 1:1 complex stoichiometry (after Gutmann and Keyzer[74]).

Clausius–Mosotti equation:

$$\frac{\varepsilon - 1}{\varepsilon + 2} \cdot \frac{1}{d} = P \tag{21}$$

having the form[61]

$$vP_3 = P_{12}\left(1 - \frac{B}{P_{12}} + \frac{3A}{(\varepsilon_{12} + 2)(\varepsilon_{12} - 1)}\right) - (P_{1,2} - P_1)(P_2 - P_1) \times \left(w_2 + \frac{M_2}{M_3}(1 - w_w)\right) \tag{22}$$

Here, P stands for the density, ε for the permittivity, M for the molecular weight, P for the specific polarization, and w for the weight fraction; the subscript 1 refers to the solvent, 2 to the donor, and 3 to the complex. A and B are proportionality constants derived from the linear relationships[61]

$$\varepsilon_{123} = \varepsilon_{12} + Aw_3 \qquad d_{123} = d_{12} + Bw_3 \tag{23}$$

Excitation leads to a significant change in the dipole moments.[62] It is often accompanied by intramolecular charge transfer.

For the π-electron components of the dipole moments in the ground (μ^0), fluorescent (μ^V), and photophorescent (μ^T) states in all compounds, the following inequality is valid:

$$|\mu^V - \mu^0| = |\Delta\mu^V| > |\Delta\mu^T| = |\mu^T - \mu^0| \tag{24}$$

That is, the change of the dipole moment in the phosphorescent state is smaller than its change in the fluorescent state.

In almost all compounds, the following inequality is fulfilled:

$$\mu^V > \mu^T \tag{25}$$

Excitation to higher singlet and triplet states is accompanied by intramolecular charge transfer and changes in dipole moments, though there is no relationship between the extent of this transfer and the energy of the transitions.[62]

(ii) Permittivity Titrations

The conductivity titrations discussed in Section III.2 involve, in practice, a series of conductance measurements using one of the well-known bridge methods. Balancing an alternating current bridge requires a quadrature balance, i.e., a measurement of the equivalent capacitance of the system. It has thus been noticed[100] that at least in many systems involving the formation of known CTCs, such as the chlorpromazine–iodine in acetonitrile system, there occurred large capacitance changes that showed a capacitance maximum at the same stoichiometry, which indicates complex formation from the conductivity changes, as well as from other evidence. Similar observations have been made[100] for several phenothiazines.

No capacitance change was observed[100] using DMSO as a solvent; the effect appears to be confined to relatively inert solvents of not too high a permittivity. The value of the capacitance peak depends on the direction of the titration; e.g. in the titration of acetylcholine chloride vs. chlorpromazine–HCl in water, the capacitance rise was roughly 10-fold when adding acetylcholine, while the converse yielded only roughly a 2-fold rise. Both peaks occurred at the same, correct, stoichiometry, which indicated complexation.

Such large capacitance changes cannot be associated with a bulk effect, since the concentrations employed were of the order of $10^{-3}\,M$; they must be due to processes occurring within the double layer at the electrode surface(s). Preferential adsorption appears to play a major role; e.g., the initial capacitances for a chlorpromazine[60] HCl solution in water was $2.96 \times 10^{-4}\,\mu$F, while that for an acetylcholine–Cl solution, also in water, at the same $10^{-3}\,M$ concentration and under identical conditions, was 7.13 μF, which approached the expected double-layer capacitance of the electrode system. The peak was flat and hard to resolve. Chlorpromazine is known to be surface active,[101] and thus appears to be preferentially adsorbed at the electrode surface.

Phenothiazine does not form a CTC with acetylcholine; the capacitance changes linearly during the titration, increasing from about 1 pF for the phenothiazine in acetonitrile solution to 60 pF for pure acetylcholine chloride in a linear fashion.[100]

Capacitance changes are expressed by means of a plot of the effective capacitance of the system against relative donor to acceptor concentrations.[75] Slifkin[71] has shown that any such property will peak at the complex stoichiometry. Considering the complex $(D_m : A_n)$ formed between a donor D and an acceptor A, we can write[71]

$$mD + nA \overset{K_c}{\rightleftarrows} (D_m : A_n)$$

$$K_c = \frac{[(D_m : A_n)]}{[D - m(D_m : A_n)]^m [A - n(D_m : A_n)]^n} \tag{26}$$

This function has a maximum for complex stoichiometry, where $nD - nA$, given that the total concentration $D + A$ remains constant.

(iii) Other Dielectric Methods

Other dielectric properties have been employed to aid identification of CTCs. Their formation in a dielectric solvent causes the appearance of space charges, which may set up an electric field, causing changes[50] in the refractive properties of the solution via the "electrooptic" or Kerr effect[101]; e.g., chlorobiphenyl complexes have been studied by this method[50].

Circular dichroism and rotational dispersion (i.e., the Cotton effect), which involve spectroscopy, have been used to identify several CTCs, e.g., those between optically active alcohols and TCNE.[51] The method is applicable to intermolecular as well as to intramolecular CTCs exhibiting sufficient asymmetry to render them optically active.[76]

There have been few studies of dielectric relaxation effects, though they may facilitate the identification of CTCs.[52] Intramolecular CTCs of thiazine derivatives have been identified by this method.[53] Contact CTCs may give rise to dielectric absorption peaks[52] because of collision-induced dipole moments; so do ion pairs.[54] Dielectric relaxation measurements identify[63] stable CTCs in contradistinction to contact CTCs. The stable complex exhibits a relaxation time characteristic of a rigid molecule, while the relaxation time(s) obtained from a contact CTC depends on its lifetime and will be much shorter than for a rigid molecule,[63] viz., a few times 10^{-11} to 10^{-12} sec. If the lifetime of the complex is shorter than the relaxation time of the entity itself, then the faster process will determine the experimental result.[63]

4. Other Electrical and Magnetic Methods

In this section, we will briefly discuss the results of electron spin resonance (ESR), nuclear magnetic resonance (NMR, nuclear quadrupole resonance, and related methods).

(i) Electron Spin Resonance

The following information can be derived from ESR spectra[102]:

1. The presence of more than 10^{10} free radicals or triplet states

can be detected. The sensitivity of detection depends on the width of the line observed: the wider the line, the less the sensitivity. Ingram[102] gives a table of linewidths.

A low spin concentration generally indicates a nonbonding ground state. The ESR absorption in such complexes in most cases obeys a Curie type law.

2. The paramagnetic species present can be identified in some instances. Each unpaired electron interacts with its environment; as a result, its gyromagnetic ratio, i.e., its *g* value, changes from that characteristic of an electron in free space, 2.0023. Consequently, the measured *g* value can identify the species.

In a molecular complex D^+A^-, each of the two ions has an unpaired electron, and therefore two lines can, in general, be expected. These lines are not always resolved,[102] especially if a low temperature has stopped that rotation of groups within the molecules that enables a hyperfine structure to be seen at higher temperatures. Examples of spectra in which two *g* values are attributed to donor and acceptor ions are given by Matsunaga and McDowell[80] and by Bose and Labes.[80]

The occurrence of hyperfine structure in an ESR line enables useful identification of radicals present. Hyperfine splitting results from interactions between the electron and such nearby nuclei as have magnetic moments (e.g., ^1H, ^{14}N).

3. The onset of tumbling of the molecules can be observed as the removal of some, but not all, of the hyperfine structure.

4. A change in a solid from a crystalline to an amorphous form can be observed as a smearing out of the hyperfine structure. This phenomenon has been useful in the study of some polymers.

5. The rotation of parts of a molecule can be detected as the temperature is changed. At sufficiently low temperatures, the averaging effect caused by this rotation can be removed, and thus hyperfine structure is eliminated from observation. Such an effort has been found, for example, with the *p*-phenylenediamine–chloranil complex.

6. Spin–lattice as well as spin–spin interactions can be studied through linewidth; in a CTC, the linewidth should decrease with rising temperature.[103]

Spin paramagnetism in solid CTCs is often complicated by spin exchange correlations between the unpaired electrons in the columnar and alternating stacks of donor and acceptor species. Amino acid–chloranil,[64] melanin,[65] phenothiazines,[66] purine and pyridine complexes with iodine,[67] and other CTCs have thus been studied by this method.[68]

(ii) Nuclear Magnetic Resonance

The method is based on observing the interaction of the magnetic moment of the proton with externally applied magnetic fields; for details of the methodology, consult the literature.[69] Of interest here is the shift in the proton resonance frequency as a function of its chemical environment, i.e., the chemical shift.[69]

To identify a CTC, one looks for the chemical shift associated with an unpaired electron; however, this evidence is by no means decisive by itself. The method is best for short-lived, weak complexes; the observed chemical shift δ is given by[70]

$$\delta = \frac{[DA]}{[A]}(\delta_{\text{complex}} - \delta_{\text{free}}) + \delta_{\text{free}} \tag{27}$$

where we consider the formation of a CTC from a donor D and an acceptor A with, say, the donor present in excess to ensure that all molecules of the acceptor do interact. Then, molecule of A interacts with a certain number of molecules of D within the characteristic time (period) of the method. The problem consists in finding the portions of this time that the molecule spends free and as part of the complex; the δ values above refer to these two situations.[70] The solvent should be inert, though this is difficult to achieve. Equilibrium constants may be thus evaluated.[88] Application of these methods to CTCs is still very limited.

(iii) Nuclear Quadrupole Resonance

This effect[70] involves an interaction between the quadrupole moment, if any, of a nucleus with the gradient of the electric field that it experiences; the resonance frequency is affected by the character of the bonds, and sometimes allows identification of a CTC as such. Only asymmetrical nuclei have a quadrupole moment—typically,[35] Cl and other halogen nuclei. The method has been applied to a few complexes such as those containing the acceptor chloranil[78] and those between chloroform and several donors.[79] However, the triiodide ion, I_3^-, has thus been identified in amine complexes,[95] as well in complexes with the donors picoline, pyridine, acridine, phenzine, and hexamethylenetetramine.[96] The quadrupole interaction is related to the ionization potential of the donor,[92] as deduced from studies involving 26 CTCs.[92]

(iv) Magnetic Susceptibilities

The uncoupling of one or more electrons involved in the formation of a CTC causes a lowering of the diamagnetic susceptibility χ_d generally exhibited by most organic compounds. A minimum in χ_d thus results at the complex stoichiometry, though this is rather flat; such a minimum has been found, for example, for the violanthrene–iodine 1:2 entity.[83]

While in CTCs with a nondative ground state the paramagnetic contribution from the unpaired electron(s) causes only a reduction in χ_d, in strong complexes in which the ground state is mainly, or largely, ionic, the resulting complex becomes paramagnetic,[94] characterized by a paramagnetic susceptibility χ_p. A paramagnetic complex may be studied by means of ESR, a method that, of course, does not apply to diamagnetic systems.

The magnetic susceptibilities of several other CTCs have been studied.[86]

IV. BIOLOGICAL APPLICATIONS

1. Introduction

Electrochemical processes of biological systems have recently been reviewed in this series[104]; it suffices to supplement this review by discussing aspects involving CTCs. Several other reviews containing sections dealing with these fields are available,[105] and charge-transfer interactions of biomolecules have been extensively reviewed.[105]

2. Charge-Transfer Interactions on Biological Membranes

The capability of synthetic membranes to act as electrodes has been well substantiated.[110] Lipids such as cholesterol form surface CTCs with acceptors such as iodine; if only one face of the membrane is complexed, then the membrane exhibits a Zener-diode-like rectification characteristic with a rectification efficiency of many orders of magnitude. The application of external electric fields to membranes rendered asymmetrical by different complexation states on both faces may cause precipitation and electrodeposition reactions. Even metallic specularly reflecting mirrors may be deposited.[110] Under irradiation, the membranes show photoconductivity as well as generation of photovoltages.[110] Conduction is due to both ionic and electronic carriers.[110]

This experimental work has received theoretical support that stipulates the existence of multiple and distinct discrete (excited?) states on the surfaces of biological membranes associated with conformational changes and labile structures.[111]

CTCs have been reported on natural biological membranes, e.g., between the coat–(viral) protein and specific regions of exposed bases in RNA.[112] A CTC is also formed with bacteriophage protein, but its formation is inhibited by Mg ions.[112]

A hydrophobic protein fraction derived from *Electrophorus electricus* (electric fish) appears to complex with synthetic lipid membranes, raising their reactivity to acetylcholine.[113] Acetylcholine is usually considered a donor, but has been shown to behave as an acceptor.[114]

The effective permittivity of colloidal particles,[115] as well as that of ion-exchange resins,[116] may attain high values at low frequencies; the ion exchanger Bio-Rex 70 (an acrylic lattice containing weakly acidic groups within large pores) has been shown[116] to exhibit permittivities of the order of 10^6 at 1 kHz, though the resistivity is too low for the dielectric to be of practical utility. Bacteria and their protoplasts also exhibit[116] low-frequency permittivities in excess of 10^4. There is evidence[116] that these effects are associated with the cell wall. They are likely associated with a charge-transfer interaction between the membrane and its environment in a mechanism akin to that proposed by Yomosa.[106] The local field, indicated by the arrow in Fig. 16, is due to anionic and cationic charges in the residues of the amino acids of the proteins plus the permanent dipole moments of the peptide groups, which comprise aromatic moieties of the protein as well as induced dipole moments. All these forces give rise to a local field that acts on the surface of the membrane in a direction from an accepting center A to a donating one D. The ground state of the entity formed by their interaction becomes polarized by the local field. When this polarization reaches a certain critical value, the electronic structure of the ground state changes, and the dipole moment rises suddenly. The resulting electronic structure resembles that of an excited state[106]—and in fact may be considered an excited state, being predominantly of charge-transfer character (Fig. 16).

It has also been suggested[114] that the physiological activity of chlorpromazine (CPZ) and related phenothiazine drugs is due to an electrode reaction involving the formation of a surface CTC with 6-hydroxydopamine on the synaptic membrane.

3. Charge-Transfer Complexes and Drug Activity

The hypothesis[114] described above is based on a theory proposing that the chemical reason for many forms of schizophrenia is that 6-hydroxydopamine, if allowed to accumulate in the synapse, e.g., for genetic reasons, tends to attack the nerve endings.[118] The *in vivo* interaction of CPZ with 6-hydroxydopamine then involves the formation on the surface of the postsynaptic membrane of an initial CTC that acts as an electrode; CPZ is preferentially adsorbed. The

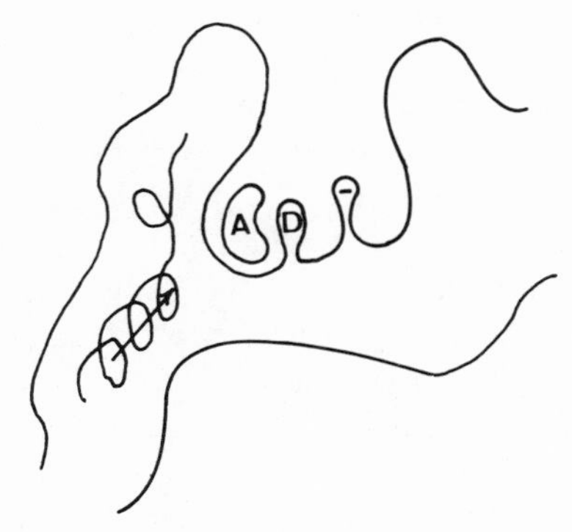

Figure 16. Schematic diagram of a charge-transfer molecular compound formed between a donor *D* and an acceptor *A* under the action of a biological local field, the direction of which is indicated by the arrow, It is seen to be directed from acceptor to donor, polarizing the ground state of the *DA* adduct. If the local field is strong enough, the electronic structure of the adduct resembles that of its excited state and has a strong charge-transfer character (after Yomosa[106]).

complex formation, which is fast though the resulting bond is weak, is followed by a chemical reaction preventing the accumulation of 6-hydroxydopamine and a consequent attack on the nerve ends.*[114]

Other results[122] indicate that phenothiazines are bound to the receptor site on the synaptic membrane by a combination of charge-transfer and electrostatic forces; these interactions occur at five main points, viz., the 2π-electron clouds on the benzene rings, the two electron-rich bridgeheads, and the side-chain nitrogen. Hydrophobic reactions are also often involved.[153] Several studies[121] of the interaction of dopaminergic neurons are at least compatible with this hypothesis.

CTCs of several drugs, especially phenothiazines, with organic acceptors have been shown to yield ions in high-permittivity solvents, while charge-transfer bands are obtained in those of low dielectric constant.[120] However, there appears to be no correlation between donor strength and pharmacological activity of psychotropic drugs.[120]

The antibiotic activity of chloramphenicol and of streptomycine has been suggested to be due to the formation of a CTC on the surface of the bacterial ribosomes.[119]

The drug activity of CTCs has recently been reviewed,[108] and a list of 26 relevant references is given in a paper dealing with cytosine–resorcylic acid 2:1 complexes.[118] There is some evidence that *in vivo* drug activity depends on excited states and, more specifically, on free radicals.[119] UV irradiation of CPZ, as well as its complexation with an acceptor such as SO_2, considerably increased the activity of the drug, as measured by hypothermia induced in mice.[119]

Interactions of heparin are also of special interest. The drug complexes with several dyes such as toluidine blue[117]; this dye has been suggested to act as an indicator that in its complexing behavior models biogenic amines bound in *in vivo* systems.[108]

Heparin also complexes with chlorpromazine[126] and protamine,[126] though there is no evidence for a charge-transfer interaction with amino acids or lignocaine.[126] The chlorpromazine interaction yields a conductivity minimum, which indicates a recombination reaction.[126]

*Charge-transfer complex formation as a precursor of a chemical reaction has also been reported for the anthracene–metal interaction.[154]

4. Charge-Transfer Complexes in Biological Energy Transfer

Since this subject has been recently reviewed,[123] this discussion need do no more than offer a few additional comments.

(i) Aerobic Respiration

This is an energy dissimilation process involving an electron transfer chain terminating in the transfer of an electron to oxygen, which is thereby reduced. Actually, two electrons are required to yield three molecules of adenosine triphosphate, the energy-rich fuel of cell metabolism. This oxidative phosphorylation is coupled to the respiratory process; the overall efficiency of the complete oxidation of one molecule of glucose, for example, is about 52%.

It has been suggested[123] that the flow of electrons involves the formation of CTCs between the prosthetic groups of the redox proteins; thus, the reaction between nicotinamide adenine dinucleotide (NAD) and flavoproteins goes via formation of a CTC as an intermediate step in the oxidation reaction.[124] The dative structure of the complex consists of two free radicals, which then pair off to yield the singlet ground state of the CTC.[124]

In cytochrome redox reactions, CTCs are likely to be involved in, for example, the transfer of an electron from a donor site connected to the heme group of the cytochrome molecule to an acceptor site on the heme group of another, neighboring cytochrome molecule.[123]

CTCs appear to be involved in the charge transfer between mitochondrial cytochromes,[134] as well as in their enzymatic reactions.[135]

Like all porphyrins,[132] hematoporphyrin may act as donor and acceptor,[136] though it is generally thought of as a donor. It is a fact that in electron-transfer centers of biological systems, the redox potentials of the electron donors and acceptors differ.[125]

(ii) Photosynthesis

CTCs involving porphyrins are likely to be involved in photosynthesis.[127] A correlation between the semiconductive properties of the adduct formed by the chlorophyll molecules in the chloroplast and its photosynthetic properties has been demonstrated.[128] Addition of a powerful electron acceptor, such as one of certain nitriles, to chloroplasts has been shown to divert the transfer of electrons from the natural acceptor of the photosystem.[129] A toxic complex between cytochrome oxidase and component(s) of tobacco smoke has been proposed.[130] The energy of the incident radiation appears to travel as an exciton to certain reaction centers involving an intermediate energy storage act[131] likely to involve a CTC. There is evidence that chlorophyll acts as the electron donor and forms a CTC with quinones such as plastochinone, though chlorophyll monolayers have been shown to act also as electron acceptors.[133] The charge transfer within chloroplasts may thus involve a complexation in which triphosphopyridine–nucleotides act as electron acceptors.

5. Charge-Transfer Complexes and Carcinogenesis

The carcinogenic activity of some hydrocarbons has been associated with their activity as electron donors,[137] though later work[138] has suggested that these carcinogens act as acceptors rather than as donors. A protein molecule is thought to exhibit an energy band structure allowing electrons from its valency band to be donated to a vacant energy level of the hydrocarbon, thus forming at least a transient type of CTC.[138] Further support for this hypothesis has been adduced from radiocarbon labeling,[139] as well as from the localization of the carcinogenic activity of the hydrocarbons within certain electron-deficient regions within the hydrocarbons.[140] It has been suggested that the benzpyrenes owe their carcinogenic activity to multiple reactive centers produced by a metabolic charge-transfer interaction with complex ions.[141]

Likewise, charge-transfer forces are likely to be involved in the interaction of amino acridines with nucleic acids; it has been suggested that the amino acridines form a layer-type complex by interposing themselves between two complementary pairs of purine–pyridine bases.[142] It could be that the activity of drugs found active against tumors may be partly due to their competition for electrons—or holes?—with the abnormal metabolite.

Free radicals are likely to play a major role, as would indeed be expected if there is a CTC interaction. ESR studies have shown that iodine complexes of hydrocarbons that are carcinogenic have a higher free-radical concentration than those that are not carcinogenic[143]; moreover, the free radicals of these compounds, such as benzpyrene, are quenched by purine, pyrimidine, and nucleosides,[143] which indicates at least an excitonic transfer if not a charge transfer.

There is evidence that the interaction between the carcinogen and a protein commences as a CTC formation,[147] which is then followed by a chemical reaction; a similar mechanism has been proposed for the action of chlorpromazine on the postsynaptic membrane,[114] as discussed in Section IV.3.

6. Enzymatic Charge-Transfer Complex Reactions

The role of CTCs in enzymatic reactions has been discussed recently in several reviews.[145]

In a way, *all* enzymatic reactions are CTC reactions, because all transfers of atomic groups are started by a charge transfer,[107] the rupture of a peptide bond or of the phosphate bond is just another example in addition to those already mentioned.

The CTCs involved in biological enzymatic reactions fall into two general groups[107]:

1. CTC "interlevel substrates," i.e., ionic complexes due to transitions between molecular levels, i.e., from a substrate donor to a substrate acceptor. Such complexes are involved in the intermolecular transfer of functional groups of low pK; transphosphorylation and decarboxylation are examples of such processes.

2. CTC formation via the 3*d* electronic orbital of the central cation, which results in a bond between the donor and acceptor groups. π-Electron resonance between the substrate donor, a (metallic) cation, and the substrate acceptor may also occur, resulting in a reversible reaction such as isomerization.

In a CTC, we are often concerned[107] with activating (metallic) cations having a maximum coordination number of six. In such an enzyme–metal–substrate complex, one of the following requirements must be fulfilled[107] for two substrates to interest with each other:

1. The substrate–cation bonds must be in the same plane. The bonds linking the enzyme protein with the (metallic) cation are then aligned along an axis perpendicular to this plane; this is the case, for example, in the peptidases or in the hexokinases.

2. Alternatively, the (metal) cation must be linked by four bonds within the same plane with the enzyme protein, so that the bonds between the central cation and the substrate(s) are aligned parallel to an axis perpendicular to that plane containing the four linkage. The

respiratory enzymes are examples of such interactions. In either case, the interaction between the delocalized electrons of the enzyme protein and those of the substrate is quite small.[107]

The organic moiety of the enzyme affects the activity of the central (metal) cation because the latter finds itself in the field of the molecules to which it is coordinated. This effect is important for chelate complexes. If the central cation is paramagnetic, it reacts in the same sense as the field in which it is situated, and, conversely, in the opposite sense if it is diamagnetic.

Enzyme activity is also often related to the redox state of the system; thus, amylase, esterase, lipase, ribonuclease, and urease are activated by reducing agents and inactivated by oxidizing agents.[146] The converse holds for such enzymes as pepsin and pepsinogen.[146] During the growth of *Escherichia coli*, it was found that alcohol dehydrogenase was activated by electron donors such as tryptophan or phenylalanine, while electron acceptors such as catechol inactivate the enzyme, probably by competing for electrons.[146] Other examples are discussed in Ref. 108.

Sometimes it is convenient to use a simpler model compound in place of the enzyme; electron transfer from donors of the indole type to pyridine coenzymes has thus been studied using 1-benzyl-3-carboxamide pyridinium chloride as the model.[144]

ACKNOWLEDGMENTS

We wish to thank Professor D. Vasilescu, Director, Laboratoire de Biophysique de l'Université de Nice, for his encouragement. We are indebted to Professor John O'M. Bockris, The Flinders University of South Australia, for much helpful advice, and for his patience.

REFERENCES

[1] B. Pullman, ed., *Molecular Associations in Biology*, Academic Press, New York, 1968.
[2] R. S. Mulliken and W. B. Person, *Molecular Complexes*, Wiley New York, 1969; R. Foster, *Organic Charge Transfer Complexes*, Academic Press, London, 1969; G. Brieglab, *Elektronen Donator–Akzeptor Komplexe*, Springer-Verlag, Berlin, 1961; M. A. Slifkin, *Charge Transfer Interactions of Biomolecules*, Academic Press, London, 1971.
[3] M. A. Slifkin, *Chem. Phys. Lett.* **7** (1970) 195.
[4] Y. Matsinaga, *Nature* (*London*) **211** (1966) 182.
[5] R. S. Mulliken and W. B. Person, Ref. 2; M. A. Slifkin and R. H. Walmsley, *Photochem. Photobiol.* **13** (1971) 57; M. A. Slifkin, private communication, 1937.

[6]T. Montenay-Garestier and C. Hélène, *Nature* (*London*) **217** (1968) 844.
[7]J. Ferrari *et al.*, *J. Am. Chem. Soc.* **95** (1973) 948; S. Basu and J. H. Greist, *J. Chem. Phys.* (1963) 407; W. M. MacIntyre, *Science* **147** (1965) 507; F. Gutmann and H. Keyzer, *J. Chem. Phys.* **46** (1967) 1969; *Rev. Aggressol.* **9** (1968) 27; cf. also M. A. Slifkin, Ref. 2; T. Tamaki, *Bull. Chem. Soc. Jpn.* **46** (1973) 2527; *J.A.C.S.* **97** (1975) 3209; Y. Matsunaga *et al.*, *Bull. Chem. Soc. Jpn.* **47** (1974) 2926; S. C. Abbi and D. M. Hanson, *J. Chem. Phys.* **60** (1974) 319.
[8]R. D. Hartmann and H. A. Pohl, *J. Polymer Sci. A–1* **6** (1968) 1135; B. Rosen and H. A. Pohl, *J. Polymer Sci.* **4** (1966) 1135.
[9]J. R. Wyhof and H. A. Pohl, *J. Polymer Sci. A*–2 **8** (1970) 1741.
[10]F. T. Lang and R. L. Strong, *J. Am. Chem. Soc.* **87** (1965) 2345; J. Prochrow and A. Tramer, *J. Chem. Phys.* **44** (1966) 4545; M. Tamres and J. M. Goodenow, *J. Phys. Chem.* **71** (1967) 1982; **43** (1965) 3393; M. Kroll and M. L. Ginter, *J. Phys. Chem.* **69** (1965) 3671; W. K. Duerksen and M. Tamres, *J. Am. Chem. Soc.* **90** (1970) 1379; M. Tamres *et al.*, *J. Phys. Chem.* **72** (1968) 966; M. Tamres and J. Grundnes, *J. Am. Chem. Soc.* **93** (1971) 801; J. Aikara *et al.*, *Bull. Chem. Soc. Jpn.* **40** (1967) 2460; **42** (1969) 1842; **43** (1969) 2439; E. A. Chandross and H. T. Thomas, *Chem. Phys. Lett.* **9** (1971), 393.
[11]J. Aikara *et al.*, *Bull. Chem. Soc. Jpn.* **43** (1970) 3323; F. Pellizza *et al.*, *Thermochim. Acta* **4** (1972) 135.
[12]M. A. Slifkin, *Spechtrochim. Acta* **20** (1964) 1543.
[13]R. Foster and P. Hanson, *Tetrahedron* **21** (1965) 755.
[14]F. Gutmann *et al.*, *Adv. Biochem. Psychopharmacol.* **9** (1974) 15.
[15]Y. Matsunaga and G. Saito, *Bull. Chem. Soc. Jpn.* **45** (1972) 963; G. Saito and Y. Matsunaga, *Bull. Chem. Soc. Jpn.* **46** (1973) 1969; I. Agranat and S. C. Cohen, *Bull. Chem. Soc. Jpn.* **47** (1974) 723.
[16]I. Schroff *et al.*, *Tetrahedron Lett.* (1973) 1649.
[17]V. F. Gachkovskii, *Russ. J. Phys. Chem.* **47** (1973) 208.
[18]D. W. Slocum *et al.*, *Tetrahedron Lett.* **46** (1971) 4429.
[19]Z. Witkiewicz, *Biul. Wojsk. Akad. Tech.* **20** (1971) 69.
[20]J. C. A. Boeyens and F. H. Herbstein, *J. Phys. Chem.* **69** (1960) 2160; cf. also Refs. 2, 4, and 5.
[21]F. Gutmann and L. E. Lyons, *Organic Semiconductors*, Wiley, New York, 1967; D. H. LeBlanc, Jr., in *Physics and Chemistry of the Organic Solid State*, Ed. by D. Fox, M. M. Labes, and A. Weissberger, Interscience, New York, 1967, Vol. 3, pp. 133 ff., S. N. Bhat and C. N. R. Rao, *Can. J. Chem.* **47** (1969) 3899; A. Rembaum, *J. Polymer Sci. C* **29** (1970) 157; D. M. Manson, *Crit. Rev. Solid State Sci.* **3** (1973) 243; H. Mohwald and D. Harrer, 4th Internatl. Symp. Org. Solid State, Bordeaux (France), *Proc.* (1975) p. 67.
[22]M. J. Mantione, *Theor. Chim. Acta* (*Berlin*) **11** (1968) 119; D. Bauer *et al.*, *J. Phys. Chem.* **74** (1970) 4594.
[23]H. Kuroda *et al.*, *J. Am. Chem. Soc.* **89** (1967) 6056.
[24]S. P. McGlynn, *Chem. Rev.* **58** (1958) 1130.
[25]F. Gutmann and L. E. Lyons, Ref. 21, p. 725.
[26]P. Ehrlich, in *Electrical Properties of Polymers*, Ed. by K. C. Frisch, Technomic, Westport, Conn., 1972, p. 26.
[27]H. Strehlow, *Bern. Bunsenges. Phys. Chem.* **67** (1963) 250; H. Günther, *Angew. Chem. Intern. Ed. Engl.*, **11** (1972) 861; Lan-Hua, T. Chen, Ph.D. thesis, University of Tennessee, Knoxville, Tenn., March, 1972.
[28]R. G. Kepler *et al.*, *Phys. Rev. Lett.* **5** (1963) 503.
[29]H. Kobayashi *et al.*, *Acta Crystallogr. B* **26** 459 (1970).
[30]L. B. Coleman *et al.*, *Solid State Commun.* **12** (1973) 1125; *Phys. Today* **26** (1973) 17; J. Bardeen, *Solid State Commun.* **13** (1973) 357; *Phys. Today* **26** (1973) 141; L. I.

Buravov *et al.*, *JETP Lett.* **11** (1970), 99; E. H. Halpern and A. A. Wolf, in *Cyrogenic Engineering*, Ed. by K. D. Timmerhaus, Plenum Press, New York, (1972), Vol. 17, p. 109.

[31]I. F. Shchegolev, *Phys. Status Solidi* **12a** (1972) 9; H. R. Zeller, *Advances in Solid State Physics*, Pergamon Press, Oxford, (1973), Vol. 13; T. Ishiguro, *J. Phys. Soc. Jpn.* **41** (1976) 35; A. L. Lefros and B. I. Shklovskii, *Phys. Stat. Solidi.* **B-76** (1976) 475.

[32]H. Fernandez-Moran *et al.*, *Science* **174** (1971) 493; T. H. Geballe *et al.*, *Phys. Rev. Lett.* **27** (1971) 314.

[33]L. V. Keldysh, *Usp. Fiz. Nauk* **86** (1965) 327.

[34]V. L. Ginzburg, *Phys. Rev. Lett.* **13** (1964) 101; D. A. Kirzhints and Yu. V. Kopaev, *Zh. Eksp. Teor. Fiz. Pis'ma Red.* **17** (1973) 379; M. J. Braithwaite, Organic charge transfer complex superconductors, Dissertation, University of Nottingham, 1971; W. A. Little, *Phys. Rev.* **A134** (1964) 1416; *J. Polymer Sci. C*, **17** (1967) 3; L. Z. Kon, *Fiz. Tekh. Puolprovodn*, **7** (1973) 830; H. H. McConnell *et al.*, *Proc. Natl. Acad. Sci. U.S.A.* **57** (1967) 1131.

[35]V. S. Grechishkin, *Opt. Spektroskop.* **20** (1966) 300; L. H. Piette and I. S. Forrest, *Biochim. Biophys. Acta* **57** (1962) 419; W. Caspary *et al.*, *Biochemistry* **12** (1973) 2649; S. A. Jortner and J. Rice, in *Physics and Chemistry of the Organic Solid State*, Ed. by D. Fox, M. M. Labes, and A. Weissberger, Interscience, New York, 1967; R. Foster, *Organic Charge Transfer Complexes*, Academic Press, London, 1969; F. Gutmann and L. E. Lyons, Ref. 21; A. Rembaum, Ref. 21.

[36]W. E. Vaughan, *Dig. Lit. Dielectr.* **35** (1971) 75.

[37]F. Gutmann and H. Keyzer, *J. Chem. Phys.* **50** (1969) 550; **46** (1967) 1969.

[38]F. Gutmann *et al.*, *Nature (London)* **221** (1969) 1237; F. Gutmann, *Jpn. J. Appl. Phys.* **8** (1969) 1417; W. Brenig *et al.*, *Phys. Lett. A* **35** (1971) 77.

[39]I. G. Orlov *et al.*, *Chemistry of Acetylene, Proceedings of the 3rd All-Union Conference*, 1968, A. A. Petrov, ed., Science Publishing House, Moscow, 1972.

[40]A. Rembaum *et al.*, *J. Phys. Chem.* **73** (169) 513; J.-P. Farges *et al.*, *J. Phys. Chem. Solids* **33** (1972) 1723; T. Ishiguro, Ref. 31; A. L. Lefros and B. I. Shklovskii, Ref. 31.

[41]S. Kobinata and S. Nagakura, *J. Am. Chem. Soc.* **88** (1966), 3905; H. Tsubomura and S. Nagakura, *J. Chem. Phys.* **27** (1957) 819.

[42]S. Bratoz, in *Advances in Quantum Chemistry*, Ed. by P.-O. Löwdin, Academic Press, New York, (1967), Vol. 3, p. 221; K. Toyoda, *Bull. Chem. Soc. Jpn.* **42** (1969) 1767.

[43]A. Funatsu and K. Toyoda, *Bull. Chem. Soc. Jpn.* **43** (1970) 279.

[44]H. Fritzsche, *IBM J. Res. Dev.* **13** (1969) 515; *J. Non-Cryst. Solids* **2** (1970) 432; **4** (1970) 391; M. M. Cohen, *J. Non-Cryst. Solids* **4** (1970) 391; E. A. Davis, *Endeavour* **30** (1971) 55; B. Y. Tong, *Phys. Rev. A* **1** (1970) 52; J. Tauc, *Science* **158** (1967) 1543.

[45]K. K. Kanalawa and B. H. Schechtmann, in *Electrets*, Ed. by M. M. Perlmann, Dielectric and Insulator Division of Electrochemical Soc., Princeton, N. J., 1973, p. 405.

[46]A. Brau and J.-P. Farges, *Phys. Status Solidi* **61** (1974) 257; J.-P. Farges *et al.*, *Phys. Status Solidi* **37** (1970) 745; J.-P. Farges *et al.*, *J. Phys. Chem. Solids* **33** (1972) 1723; A. Rembaum *et al.*, *J. Phys. Chem.* **73** (1969) 513.

[47]A. J. Epstein *et al.*, *Phys. Rev. B* **5** (1972) 952; E. Ehrenfreund *et al.*, *Phys. Rev. Lett.* **28** (1972) 873; L. B. Coleman *et al.*, *Phys. Rev.* **7** (1973) 2122; L. I. Buranov *et al.*, *Soviet Phys. JETP Lett.* **12** (1970), 99; *Zh. Eksp. Teor. Fiz. Pis'ma Red.* **8** (1968) 353; A. F. Garito and A. J. Heeger, *Coll. Prop. Phys. Syst. Proc. 24th Nobel Symp.*, B. Lindquist, ed., Nobel Foundation, Stockholm, 1974, p. 129; A. F. Garito and A. J. Heeger, *Accts. Chem. Res.* **7** (1974) 232; D. O. Cowan *et al.*, 4th Internatl. Symp. Org. Solid State Bordeaux (France), *Proc.* (1975) p. 70; H. R. Zeller, *Adv. Solid State Phys.* **13** (1973).

[48]A. E. Lutskii *et al.*, *Russ. J. Phys. Chem.* **47** (1973) 367.

[49]K. K. Deb *et al.*, *Inorg. Chem.* **11** (1972) 2428.
[50]P. Durand and R. Fournié, *Dielectric Materials Measurements and Applications Conference*, IEE Conference Publication No. 67, IEE (London), 1970, p. 142.
[51]A. I. Saitt and A. D. Wrixon, *Chem. Commun.* **1969** 1184; J. Bolard, *J. Chim. Phys. Phys-Chim. Biol.* **66** (1969), 221; T. B. Lakdar *et al.*, *C. R. Acad. Sci.* (*Paris*) **271** (1971) 1201; G. Briegleb *et al.*, *Ber. Bunsenges. Phys. Chem.* **76** (1972) 101.
[52]R. Pottel, *Ber. Bunsenges. Phys. Chem.* **75** (1971) 286.
[53]F. Gutmann and H. Keyzer, *Electrochim. Acta* **13** (1968) 693; H. Keyzer, Ph.D. thesis, University of New South Wales, 1966.
[54]J. E. Anderson, *Ber. Bunsenges. Phys. Chem.* **75** (1971) 294; G. Williams, *Adv. Mol. Relaxation Processes* **1** (1970) 409; G. Schwarz, *J. Phys. Chem.* **74** (1970) 654; cf. also Ref. 52.
[55]F. Gutmann and L. E. Lyons, Ref. 21, p. 504.
[56]D. H. Rodgers, *J. Pharm. Sci.*, **54** (1965) 459; W. A. Harris, *Aust. J. Pharm.* **49** (1968) 587.
[57]M. Brau *et al.*, *Electrochim. Acta* **17** (1972) 1803.
[58]R. L. Ward, *J. Chem. Phys.* **39** (1963) 852; T. N. Misra and B. Rosenberg, *J. Chem. Phys.* **48** (1968) 2096; J. W. Eastman *et al.*, *J. Am. Chem. Soc.* **84** (1962) 1339; H. Kainer and A. Überle, *Chem. Ber.* **88** (1955) 1147; cf. also Ref. 33.
[59]B. C. Gilbert, in *Essays in Chemistry*, Ed. by J. N. Bradley, R. D. Gillard, and B. F. Hudson, Academic Press, New York, 1972, Vol. 4.
[60]F. Gutmann *et al.*, *Adv. Biochem. Psychopharmacol.* **9** (1974) 15.
[61]S. Kobinata and S. Nagakura, *J. Am. Chem. Soc.* **88** (1966) 3905.
[62]N. Tyutyulkov *et al.*, *Theor. Chim. Acta* (*Berlin*) **20** (1971), 385; H. Beens *et al.*, *J. Chem. Phys.* **47** (1967) 1183.
[63]R. A. Crump and A. H. Price, *Trans. Faraday Soc.* **66** (1971) 92.
[64]E. H. Gause *et al.*, *Biochim. Biophys. Acta* **141** (1967) 217.
[65]F. Gutmann *et al.*, *Rev. Aggress.* (*Paris*) **7** (1966) 147.
[66]H. Keyzer, Ph.D. thesis, University of New South Wales, 1966.
[67]J. B. Jones *et al.*, *Nature* (*London*) **211** (1966) 309.
[68]M. A. Slifkin, Ref. 71; H. Inokuchi *et al.*, *Bull. Chem. Soc. Jpn.* **33** (1960) 1622; M. Kinoshita, *Bull. Chem. Soc. Jpn.* 1040, 1137; M. Ohmasa *et al.*, *Bull. Chem. Soc. Jpn.* **42** (1969) 2402; S. Walker and H. Straw, *Spectroscopy*, Chapman and Hall, London, 1966, Vol. 1.
[69]J. A. Pople, W. G. Schneider, and H. J. Bernstein, *High Resolution Nuclear Magnetic Resonance*, McGraw-Hill, New York, 1959; Varian Associates, *NMR and EPR Spectroscopy*, Pergamon Press, Oxford, 1966; K. Higasi, H. Baba, and A. Rembaum, *Quantum Organic Chemistry*, Interscience, New York, 1955; R. Blanc, ed., *Magnetic Resonance and Relaxation*, North-Holland, Publishing Co., Amsterdam, 1967; N. M. Atherton and A. J. Blackhurst, *J. Chem. Soc. Faraday Trans. 2* **68** (1972) 470; E. F. Mooney, ed., *Annual Reports on NMR Spectroscopy*, Academic Press, New York, 1900; R. J. Abraham, *Analysis of High Resolution NMR Spectra*, Elsevier, Amsterdam, 1970; R. Foster, *Ind. Chem. Belg.* **37** (1972) 547.
[70]J. Homer *et al.*, *J. Chem. Soc. Trans. Faraday 2* **68** (1972), 68; for reviews, see *Forschr. Chem. Forsch.* **30** (1972).
[71]M. A. Slifkin, *Charge Transfer Interations of Biomolecules*, Academic Press, London, 1971; R. Foster, *Organic Charge Transfer Complexes*, Academic Press, London, 1969; G. Briegleb, *Elektronen Donator–Akzeptor Komplexe*, Springer-Verlag, Berlin, 1961; F. Gutmann and L. E. Lyons, *Organic Semiconductors*, Wiley, New York, 1967; B. Person, *Rev. Phys. Chem.* **13** (1962) 107; J. G. Heathcote *et al.*, *Spectrochim. Acta* **27A** (1971) 1391; J. Rose, *Molecular Complexes*, Pergamon Press, Oxford, 1967; R. S.

Mulliken and W. B. Person, *Molecular Complexes*, Interscience, New York, 1969; J. B. Birks, *Photo Physics*, Interscience, New York, 1970; E. F. H. Brittain, W. O. George, and C. H. J. Wells, *Introduction to Molecular Spectroscopy*, Academic Press, New York, 1970.

[72]L. J. Parkhurst, Ph.D. thesis, Yale University, 1965; B. J. Anex and E. B. Hill, *J. Am. Chem. Soc.* **88** (1966) 3648; Y. Ihaya *et al.*, *Bull Chem. Soc. Jpn.* **45** (1972) 1004.

[73]H. Akamatu and H. Kuroda, *J. Chem. Phys.* **39** (1963) 3364; cf. also Ref. 86.

[74]F. Gutmann, *J. Sci. Ind. Res. Sect. B* **26** (1967) 19; F. Gutmann and H. Keyzer, *Electrochim. Acta* **11** (1966) 555, 1163; N. E. Wisdom and E. O. Forster, *J. Polymer Sci. C* **17** (1967) 125; L. Libera and H. Bretschneider, *Z. Chem.* **14** (2) (1974) 68.

[75]P. Job, *C. R. Acad. Sci.* (*Paris*) **180** (1925), 928.

[76]J. P. Carrion *et al.*, *Helv. Chim. Acta* **51** (1967) 459; cf. also Shifkin, Ref. 71, p. 34.

[77]J. M. Bonnier and R. Arnard, *J. Chim. Phys.* **68** (1971) 423; W. Wacklawek, *Bull. Acad. Pol. Sci. Ser. Sci. Math. Astron. Phys.* **21** (1973) 189; cf. also Slifkin, Ref. 71, pp. 22 ff, 41.

[78]D. C. Douglas, *J. Chem. Phys.* **32** (1960) 1882; H. Chihara and N. Kakamura, *Bull. Chem. Soc. Jpn.* **44** (1971) 2676; Y. Matsunaga and Y. Suzuki, *Bull. Chem. Soc. Jpn.* **45** (1972) 3375.

[79]R. A. Bennett and H. O. Hooper, *J. Chem. Phys.* **47** (1967) 4855.

[80]Y. Matsunaga and C. A. McDowell, *Nature* (*London*) **185** (1960) 916; M. Bose and M. M. Labes, *J.A.C.S.*, **83** (1961) 4505.

[81]P. W. Selwood, *Magnetochemistry*, Interscience, New York, 1956; A. Pacault *et al.*, *Adv. Chem. Phys.* **3** (1961) 171.

[82]Y. Matsunaga, *Bull. Chem. Soc. Jpn.* **28** (1955) 475.

[83]H. Akamatu and Y. Matsunaga, *Bull. Chem. Soc. Jpn.* **26** (1953) 364.

[84]F. Forster and D. M. L. Goodgame, *Inorg. Cham.* **4** (1965) 823; M. Hashimoto *et al.*, *Bull. Chem. Soc. Jpn.* **44** (1971) 2322.

[85]F. Gutmann and L. E. Lyons, Ref. 71, pp. 621 ff.

[86]M. Ohmasa *et al.*, *Bull. Chem. Soc. Jpn.* **41** (1968) 1998; M. Sano *et al.*, *Bull. Chem. Soc. Jpn.* **41** (1968) 2204; **42** (1969) 548; **43** (1970) 2370.

[87]F. Gutmann and H. Keyzer, *J. Chem. Phys.* **50** (1969), 550; *Electrochim. Acta* **13** (1968) 693; A. Brau *et al.*, *Electrochim. Acta* **17** (1972) 1803.

[88]M. W. Hanna and A. L. Ashbaugh, *J. Phys. Chem.* **68** (1964) 811; R. Foster and C. A. Fyfe, *Trans. Faraday Soc.* **61** (1965) 1626; R. Foster and D. R. Twiselton, *Rec. Trav. Chim. Pays-Bas* **89** (1970) 325.

[89]B. G. Anex and W. T. Simpson, *Rev. Mod. Phys.* **32** (1960) 446; J. J. Eckhart and J. Merski, *Surface Sci.* **37** (1973) 937; Y. Matsunaga and N. Miyajima, *Bull. Chem. Soc. Jpn.* **44** (1971) 361; A. P. Kushelevsky and M. A. Slifkin, *Radiat. Res.* **50** (1972) 56; A. A. Bright *et al.*, *Phys. Rev. Lett.* **34** (1975) 206.

[90]F. Gutmann and L. E. Lyons, Ref. 71, p. 463; G. Briegleb, Ref. 71.

[91]M. A. Slifkin, Ref. 71, p. 73.

[92]M. F. Shostokovskii *et al.*, *Izv. Akad. Nauk SSSR, Ser. Khim.* (1973) 15.

[93]H. J. M. Andriessen *et al.*, *J. Chem. Soc. Perkin Trans.* **2** (1972) 861.

[94]M. Kinoshita *et al.*, *Bull. Chem. Soc. Jpn.* **44** (1971) 2267.

[95]G. A. Bowmaker and S. Hacobian, *Aust. J. Chem.* **21** (1968), 551.

[96]R. Brüggemann *et al.*, *Z. Naturforsch. A* **27** (1972) 1524.

[97]Y. Y. Borovikov, *Izv. Vyssh. Uchebn. Zaved. Khim. Khim. Tekhnol.* **11** (1968) 20.

[98]Y. Y. Borovikov, Ref. 97.

[99]J. Caldorford, *Complex Permittivity*, English University Press, London, 1971; D. Bauer *et al.*, *J. Phys. Chem.* **74** (1970) 4504; W. E. Vaughan, in *Dig. Lit. Dielectr.* **35** (1971) 1973.

[100]F. Gutmann and H. Keyzer, unpublished results.

[101]Z. Croitoru, *Prog. Dielectr.* **6** (1965) 103.

[102]F. Gerson, *High Resolution ESR Spectroscopy*, Wiley, New York, 1970; G. I. Subbotin and R. V. Grechishkina, Ref. 103; E. E. Budzinski *et al.*, *J. Chem. Phys.* **59** (1973) 2899; D. J. Ingram, *Free Radicals as Studied by ESR*, Butterworths, London, 1958; Varian Associates, *NMR and EPR Spectroscopy*, Pergamon Press, Oxford, 1960; P. G. Lykos, in *Advances in Quantum Chemistry*, Ed. by P.-O. Löwdin, Academic Press, New York, 1064, Vol. 1, p. 194; cf. R. Foster, Ref. 71; F. Gutmann and L. E. Lyons, Ref. 71.

[103]G. I. Subbotin and R. V. Grechischkins, in *Magnetic Resonance Phenomena, Proceedings of the 16th Ampére Congress, 1970*, Ed. by I. Ursu, North Holland Publishing Co., Amsterdam, 1971.

[104]L. J. Mandel, in *Modern Aspects of Electrochemistry*, Ed. by J. O.M. Bockris and B. E. Conway, Plenum Press, New York, 1972, p. 239.

[105]F. W. Cope, *Adv. Biol. Med. Phys.* **13** (1970) 1; M. A. Slifkin, *Charge Transfer Interactions of Biomolecules*, Academic Press, London, 1971; cf. Refs. 3 and 4.

[106]S. Yomosa, *Prog. Theor. Phys. Suppl.* **40** (1967) 249.

[107]A. Goudot, *Wave Mechanics Mol. Biol.* **1966** (1966) 28.

[108]A. Popova, *C. R. Acad. Bulg. Sci.* **19** (1966) 77.

[109]A. Goidot and M. Faguet, *C. R. Acad. Sci.* (*Paris*) *D* **265** (1967) 2139.

[110]H. C. Pant and B. Rosenberg, *Chem. Phys. Lipids* **6** (1971) 39; B. Rosenberg and B. B. Bhowmik, *Chem. Phys. Lipids* **3** (1969) 109; B. Rosenberg, *Discuss. Faraday Soc.* **51** (1971) 39; B. B. Bhowmik *et al.*,`*Nature* (*London*) **215** (1967) 842; Y. A. Liberman *et al.*, *Biophysics* **14** (1969) 56; P. Läuger *et al.*, *Biochim. Biophys. Acta* **135** (1967) 20; *Ber. Bunsenges. Phys. Chem.* **71** (1967) 906.

[111]S. V. Konev *et al.*, *Vestsi Akad. Navuk B. SSR, Ser. Biyal. Navuk* (1973) 61; B. Nilsson and A. Ågren, *Acta Pharm. Suec.* **11** (1974) 391.

[112]J. Chroboczek, *J. Virol.* **12** (1973) 230.

[113]T. A. Reader *et al.*, *Biochem. Biophys. Res. Commun*, **53** (1973) 10.

[114]F. Gutmann *et al.*, *Adv. Biochem. Psychopharmacol.* **9** (1974) 15; B. Schreiber, *Bioelectrochem. Bioenergetics* **1** (1974) 355; D. N. Gillbanks, M.Sc. thesis, Victoria University, Wellington, New Zealand, 1973.

[115]G. Schwarz, *J. Phys. Chem.* **66** (1962) 2636.

[116]C. W. Einolf and E. L. Carstensen, *J. Phys. Chem.* **75** (1971) 1091; *Biochem. Biophys. Acta* **148** (1967) 506; E. L. Carstensen *et al.*, *Biophys. J.* **5** (1965) 289; C. W. Einolf, The low frequency dispersion of micro-organisms, Ph.D. thesis, University of Rochester, Rochester, N.Y., 1968.

[117]J. M. Dabney and A. J. Stanley, *Tex. Rep. Biol. Med.* **18** (1960) 194.

[118]C. Tamura, *Bull. Chem. Soc. Jpn.* **46** (1973) 2388.

[119]C. M. Goodley *et al.*, *Nature* (*London*) **223** (1969) 80.

[120]M. Saucin and A. van de Vorst, *Acad. R. Belg. Bull. Cl. Sc. Ser. 5*, **55** (1969) 166; *Biochem. Pharmacol.* **21** (1972) 2673.

[121]B. S. Nunney *et al.*, *Pharmacol. Exp. Theor.* **185** (1973) 560; L. M. Jordan and W. D. Willis, *Pharmacol. Exp. Theor.* **185** (1973) 572.

[122]A. Fulton, Ph.D. thesis, The University of Queensland, St. Lucia, Brisbane; B. Schreiber, Ref. 114.

[123]J. Llopis, *Proceedings of the I.S.E. Conference on Bio-Electrochemistry*, Rome, 1971; *Experienta Suppl.* **18** (1971) 413; I. Pecht and M. Faraggi, *Proc. Natl. Acad. Sci. U.S.A.* **69** (1972) 902.

[124]B. Grabe, *Biochim. Biophys. Acta* **30** (1958) 560; *Ark. Fys.* **17** (1960) 97; I. Fischer-Hjalmar, *Q. Rev. Biophys.* **1** (1969) 311; T. Sakurai and H. Hosoya, *Biochim. Biophys. Acta* **112** (1966) 459.

[125]R. J. P. Williams, 5th Keillin Memorial Lecture, *Biochem. Soc. Trans.* **1** (1973) 1; F. W. Cope, *J. Biol. Phys.* **3** (1975) 1.

[126]G. Eckert and F. Gutmann, unpublished results.
[127]M. Calvin, *J. Theor. Biol.* **1** (1961) 258; F. Gutmann and L. E. Lyons, *Organic Semiconductors*, Wiley, New York, 1967, p. 497; cf. Refs. 105 and 136.
[128]F. F. Litvin and V. I. Zvalins, *Biofizika* **16** (1971) 420.
[129]P. C. Brandon and O. Elgersma, *Biochem. Biophys. Acta* **292** (1973) 753.
[130]M. C. Fu *et al., Proceedings of the 3rd Tobacco Workshop Conference, 1972*, Tobacco Health Research Institute, University of Kentucky, Lexington, Ky., p. 239.
[131]M. Delbrück, *Angew. Chem.* **84** (1972) 1; *Angew. Chem. Intern. Ed. Engl.* **11** (1972) 1.
[132]M. A. Slifkin, in *Physico-Chemical Properties of Nucleic Acids*, Ed. by J. Duchesne, Academic Press, New York, 1973, p. 92.
[133]M. Tomkiewicz and M. Klein, *Proc. Natl. Acad. Sci. U.S.A.* **70** (1973) 143; S. S. Brody, *Z. Naturforsch.* **26** (1971) 134.
[134]D. D. Green, *Discuss. Faraday Soc.* **27** (1959), 206.
[135]Refs. 104 and 105.
[136]J. G. Heathcote *et al., Biochim. Biophys. Acta* **153** (1968) 13; cf. Ref. 105.
[137]G. M. Badger, *Adv. Cancer Res.* **2** (1954) 73.
[138]R. Mason, *Discuss. Faraday Soc.* **27** (1959), 129.
[139]P. M. Bhargava and C. Heidelberger, *J. Am. Chem. Soc.* **78** (1956) 3671.
[140]A. B. Pullman, *Cancerisation par les Substances Chimiques et Structure Moleculaire*, Masson, Paris, 1955.
[141]O. G. Fahmy and M. J. Fahmy, *Cancer Res.* **33** (1973) 302.
[142]G. Lober *et al., Biopolymers* **11** (1972) 2439; B. L. van Duuren *et al., Ann. N. Y. Acad. Sci.* **153** (1969) 744.
[143]C. Hélène, *Photochem. Photobiol.* **18** (1974) 534; W. Caspary *et al., Biochemistry* **12** (1973) 2649.
[144]G. Cilento and P. Giusti, *J. Am. Chem. Soc.* **81** (1959) 3801.
[145]Refs. 104, 105, and 122.
[146]R. Raunio and E.-M. Lilius, *Acta Chem. Scand.* **27** (1973) 990.
[147]S. S. Sung, *C. R. Acad. Sci.* (*Paris*) *D* **275** (1972) 1307.
[148]G. H. Moxon and M. A. Slifkin, *Biochem. Biophys. Acta* **304**(3) (1973) 693.
[149]B. H. Schlechtman, S. F. Lin, and W. E. Spicer, *Phys. Rev. Lett.* **34** (1975) 667; P. Nielsen *et al., Solid State Commun.* **15** (1974) 53.
[150]J. Alizon *et al.*, 4th Internatl. Symp. Org. Solid State, Bordeaux (France), *Proc.* (1975) p. 53.
[151]M. M. Labes, 4th Internatl. Symp. Org. Solid State, Bordeaux (France), *Proc.* (1975) p. 54; P. M. Grant, 4th Internatl. Symp. Org. Solid State, Bordeaux (France), *Proc.* (1975) p. 56; R. L. Greene *et al., Phys. Rev. Lett.* **34** (1975) 577.
[152]W. A. Barlow, 4th Internatl. Symp. Org. Solid State, Bordeaux (France), *Proc.* (1975) p. 63.
[153]C. Appelgren *et al., Acta Pharm. Suec.* **11** (1974) 325; K. A. Johansson *et al., Acta Pharm. Suec.* **11** (1974) 333.
[154]V. Chen-Teh-Bien, Ph.D. thesis, University of Queensland, Australia, 1975.

Index

Abdel Hamid *et al.,* and conductance in nonaqueous solutions, 8
Abnormal conductance, and recent suggestions, 23
  by proton transfer, theories, of proton conductance, 18
Accascina *et al.,* and conductance in nonaqueous solutions, 8
Acceptors, 268
Acetamide, and NMR, 111
Acids, and Gutowsky's function, 72
Activity coefficient, gradient of, 35
Adams' method, 204
Adsorption, and ion exchange, 197
  forming of solution, of powdered sample, 217
Aerobic respiration, 305
Akitt, and dependence of hydrogen shifts upon concentration, 76
  and hydrogen shifts, 70
  and shifts in solution, 72
  and upshield shift of oxyanions, 91
  his extrapolation, 71
Alcohol in water, 59
Alcoholic solutions, 13
Alloys, binary, and platinum-ion catalysts, 213
Alum, in salt solutions, 110
Analysis, of proton transfer in solution, 53
Angell, and structure of water, 51
Anions, and interaction with solution, 97
Anodic processes, and oxidation of carbon blacks, 187
Areas of noble metal catalysts, tabulated, 203
Argade and Gileadi, and work function, 170
Arrhenius, effects in solution, 80
Association of water, 57
Associations, molecular, and electrochemistry, 267
Aukshen and Levin, and potential of zero charge, 169
Average size, of crystallites on platinum, 250

Baker, and spin–rotation, 105
Band model, 280
BDM, and interacting dipoles, 177
BDM theory, 139
Beech and Miller, and protons in solution, 99
Berman and Verhock, and conductance in nonaqueous solutions, 7
Bernal and Fowler, and frequency of proton jumping, 19
  and publication in 1933, 19
  and suggestion of proton transfer, 5
BET and catalysts, 216
BET equation, applied to catalysts, 215
Bett, and potentiodynamic sweeps, 231

Binary alloys, 238
  unsupported, on catalysts, 213
Binder, and platinum surface areas, 206
Biological applications, 301
Biological energy transfer, and charge transfer, 304
Biological materials and hydrogen bond, 2
Bockris, and a two-state model, different from the Watts-Tobin model, 137
  and p.d. across double layer, 139
  and proton conductance theory, 18
  and proton transfer theory, 20
  and theory of maximum adsorption of butanol, 136
  and theory of water at electrodes, 154
Bockris and Argade, absolute potential difference: calculated, 169
  and first absolute calculation of metal solution p.d., 167
Bockris and Habib, a model for mercury-solution at interface, 142
  and the calculation of ion–dipole interaction, 148
  and the orientation of water molecules, 146
  and the water molecule distribution at interfaces, 149
  and three-state water molecule, 140
  calculation of absolute metal-solution p.d., 167
  their theory for water molecules at surface, 141
Bockris and Habib model, and dimers, 144
Bockris and McHardy, and effect of platinum on catalysis of bronzes, 193
Bockris and Potter, and first quantitative paper with water dipoles, 131
Bockris *et al.*, and conductance in non-aqueous solutions, 8
  and proton transfer theory, 28
Bond, and higher crystallite size, 200
Booth, and preparation of unsupported platinum, 202
Borides, 209
Boudart, and the spillover theory, 193
Bragg–Williams' approximation for alloys, 161
Brickmann, and proton transfer theory, 36
Bright field, 244
Broadening, of lines in solution, 109
Bromide shifts, and Rogers' work, 94
Bronsted's principle of specific interaction, 92
Bronzes, 208
  and activity in catalysis, 190
Buckingham, and organic solvents, 53
  frequency of shifts in water, 64
  his equation, 71

Cadmium shifts in solid solutions, 91
Caesium shifts, and NMR in solution, 92
  in concentrated solution, 93
  concentration dependence, 107
Capacitance bridges, 297
Capillary condensation, as function of Kelvin radius, 221
Carbon, and electrocatalysis supports, 187
  and porous electrodes, 185
Carbon black, as basis for catalysis, 195
Carbon monixide, 256
Carbons, pyrolized to 1000°, 189
Carcinogenesis, and charge transfer, 306
Carrington, and isotopic shifts in solution, 85
Catalysis, and high-surface area, 183
  and hydrogen bond, 2
Catalysts, microstructure and porosity, 218
Catalytic materials, dispersed, 183
Cation–anion effects on NMR studies in aqueous solution, 77
CBL theory, explained, 22
Charge transfer, and carcinogenesis, 306
  and complexes in electrochemistry, 267
  and drug activity, 303
Charge transfer, in biological energy transfer, 304
  in solid state, moments of, 276

Charge transfer (*cont'd*)
in strong complexes, 282
Charge transfer complexes, 269, 270
identification, 289
for solids, 273
for solutions, 272
from various substances, 294
Charge transfer interactions, 301
Charge-potential curves, and change of work function, 175
Charging curves and platinum blacks, 230
Charging of double layer, 234
Chemisorption, 222
from gas phase, 224
Chloride shift, in solution, 87
Chlorpromazine, 298
Closed shell ions, observations on NMR spectra, 84
Cluster model, 49
Clusters, and hydrogen bonding, 14
in water, average size, 26
on electrodes, 163
Complex dissociation, 291
Complex of water molecules, 20
Complex reactions, and enzymes, 307
Complexation, 268
Components of entropy at electrodes, 160
Conclusions, concerning structure of water at interface, 173
Conductance, and methanol, 15
Conduction, of polycrystalline samples, 286
Connolly, his hydrogen chemisorption from gas phase, 231
his work on transition times, 231
Contribution, of proton transfer, to proton-transport process, 31
Contributions to proton shifts in water,
Conway, and detailed analysis of proton transfer, 2
Bockris and Linton, and conductance in nonaqueous solutions, 8
and proton-conductance theory, 18
and proton-transfer theory, 20
and review of proton conductance, 18
Conway and Gordon, and libration of solvent dipoles, 151
Conway and Salomon, and upper limit of isotope ratio, 20
Conway *et al.*, and proton-transfer theory, 28
Coordination number, of aluminum, 99
Coupling, mechanism in proton transfer, 46
Covington, and equilibrium constants for formation of solvates, 112
and polarity scale, 94
and studies of NMR theory, 37
and thermodynamic analysis of proton shifts, 121
Creekmore and Reilley, and shifts in water, 66
Crystal structure, effects of charge transfer, 283
Crystallite density, 250
Crystals in charge-transfer complexes, 273

Dalla and Betta, and ruthenium catalysts, 224
Damaskin, and water molecules at interface, 140
Damaskin and Frumkin, and a three-state water model, 149
Danneel, his suggestion of proton jumping, *1905,* 1
Dark field electron micrographs, 247
Darmois, and proton-transfer value, 20
Davies, and the concentration dependence of NMR measurements in solution, 75
his work on diagrammatic shifts, 69
work on hydrogen shifts in solution, 76
Decomposition, thermal, 206
Dempworf, and abnormal mobility of methanoleum ion, 3
Density of states, as function of energy, 279
Dependence of water molecules, 138
Deposition, underpotential, 234

Deverell, and hydrogen-isotope effects, 105
and review of paramagnetic studies in solution, 82
and solvation effects, 104
Deverell and Richards, and bromide shifts, 93
and the concentration of counter ions at magnetically active materials, 85
and the linearity of shifts in solution, 87
Dielectric properties, 294, 298
Dimer molecules, 153
Dimers, and effective surface entropy, 159
and the Bockris and Habib model, 145
Dioxane–water mixtures, and conductance, 12
Dipole capacitance, at metal and solution interface, 148
Dipole contributions, to electrodes, summarized, 174
Dipole model for interfaces, 146
Dipoles, at electrode–solution interface, 131
"Dipping and soaking" technique, 197
Disordered systems, 277
Dispersion of noble metals, 225
Distribution, of particles in pores, 242
of platinum crystallite size, 249
of pore size, 219
Dnieprov and Aggan, and conductance in nonaqueous solutions, 8
Donors, 268
Dorling and Moss, with silica-gel systems, 195
Double layer, charging of, 234
Drost and Hansen, and the three-state water molecule, 175
Drug activity, and charge transfer, 303

Eigen and De Maeyer, 23
Electric and magnetic fields, and electron spin resonance, 298
Electrical and magnetic properties, 275
Electrocatalysis supports, 191
Electrocatalyst supports, 184
Electrocatalysts, characterization of, 214
with colloids, 200
from metal oxides, 208
preparation of, 194
on supports, 257
Electrochemical measures, on surface areas, 233
Electrochemical surface areas, 233
Electrochemistry, and breakthrough by Bockris and Argade, 176
Electrode structure, 251, 252
Electronegativity considerations, 65
Electron micrographs, dark field, 248
of unsupported platinum complex, 247
Electron microscopy, 244
Electron overlap potential, calculated by Bockris and Habib, 168
Electrons, unpaired, in solution, 79
Energy, repulsion, water–water, 144
Energy dependence, of mobility, 281
Energy level, diagram for charge-transfer complex, 271
Enhancement, structural, detected by, 59
effects in NMR, 96
Entropy, and modelistic approaches in the double layer, 150
in the double layer, 133
of electrodes, components of, 161
of formation, and Harrison, 135
of formation of surfaces, and charge, 134
of libration, of water molecules, 157
of solutions, and Harrison, 136
of solvent at electrode, 160
vibrational, at surface, 158
Enzymatic enzymes, and charge transfer, 307
Enzyme activity, and charge-transfer activities, 308
Equations, origin of, for porous electrodes, 252
Equilibrium of water in solution and on surface, 141

Equivalent conductance in nonaqueous solutions, 15
Essin and Markov effect, 178
Ethers, polycyclic, 121
Eucken model, and auto-association structure happening in water, 20
Exchange reactions, 32

Fabricand and Goldberg, linear correlation, 62
Fang, and proton-transfer theory, 36
Florian and Lei, and structure of water, 52
Fluoride shifts, 119
Flux, and proton-transfer theory, 34
Force constants for librating water molecules, 158
Four-state water model, 162
Fraction of hydrogen bonds broken, according to various models, 54
Franck and structure of water, 51
Frank, and water structure, 57
Frank and Wen, their solvation model, 63
Fratiello, and solvation numbers, 68
  his technique, 110
Freeze-drying, 210
  techniques for, 211
Frens, and control of gold particles, 202
Frumkin, and proton-transfer processes, 6
  his criticism of Trasatti, 171
Frumkin and Gorodetskaya, and potential of zero charge, 170

Galvanostatic stripping, 230
Gas phase chemisorption, 223
General principles in proton-transfer theory, 42
Geometric model, and proton transfer, 32
Gierer and Wirtz, and proton transfer in solution, 20
Giner *et al.*, and reduction of platinum complexes, 205
Glycerolic solutions, 13
Glycol and NMR measurements, 108
Gold particles, controlled sizes, 202
Goldammer, and proton-transfer theory, 42
Goldschmidt and Dahl, and conductance in nonaqueous solutions, 7
Graphite, 245
Graphite structures, and layers of carbon atoms, 185
Green and Shepherd, and hydrogen signals in magnesium containing solutions, 112
Greenberg, and sodium shifts in solution, 103
  his work of formamyde, 102
Grubb and McKee, and boron-carbide catalysts, 195
Grubb *et al.*, and silver catalysts, 196
Guinier and Fournet, approximations, 240
Gyration for various small angles, 240

Hamann, and value of pressure coefficient for proton transfer, 24
Hammaker and Clegg, and initial-model shift, 97
Hantzsch and Caldwell, and abnormal mobility of acetate ion, 3
Harned and Dravy, and conductance of protons, 15
Harrison, and entropy of formation of mercury interfaces, 133
Hartree, and proton-transfer theory, 36
Heating of platinum, effect on, 191
Heparin, and charge transfer, 304
High resolution, and shift in proton, 52
  of electron micrographs, 245
Hills and Hsieh, and maximum on entropy-charge curve, 132
Hindmen, and contributions to chemical shift, 64
Hinton and Amis, and solvation review, 81
Horne, and specific conductance as a function of hydrostatic pressure, 24
Huckel, activation energy, 19
  and difference in mobilities, 5

Huggins, and ionic clusters, 19
  and the hydrogen bond, 19
Hydrogen bond, nature of, 56
Hydrogen bonds, and oscillation, 27
  and proton transfer, 2
Hydrogen ions, in solvents other than water, 6
Hydrogen monolayers, for iridium surfaces, 233
Hydrogen oxidation, 256
Hydrogen peroxide–water mixtures, 118
Hydrogen shift, 62
  and dependence upon concentration, 73
  diagramed shifts, 69
  and function of temperature, 63
Hydrogen shifts, diagram of, 63
  for large ions, diagramed, 76
Hydrogen signals from bound water, 113
Hydrogen spectra, in solution, 100, 115
Hydrogen spectrum, in solution, 111
Hydrogen spillover, and platinum or carbon catalysts, 193
Hydroxide, and the isoelectronic fluoride ion, 4
Hydroxyl ions, and resonance shift, 97

Identification, of charge-transfer complexes, 289
Ikenbryy and Dass, and the electron pair model for water, 107
Imaging techniques, 244
Impregnation of catalysts, 194
Interacting dipoles, a characteristic of the BDM model, 177
Interaction, hydrophobic, 58
  lateral, theory due to Bockris *et al.*, 140
Interactions, electronic, on metal catalysts, 192
Ion adsorption, and the four-state water model, 164
Ion collisions, and Rogers' work, 88
Ion-exchange, 197
  adsorption, 198
Ion exchange (*cont'd*)
  and electrocatalysts, 199
  properties, of porous materials, 226
Ions, paramagnetic, 98

Jackson, and ^{17}O in solution, 81

Kemball, and adsorption of water on mercury, 140
Kikuchi, and the Wilson–Hall observations, 225
Kingston and Symons, and proton transfer in solution, 61
  structure of water, 57
Kinoshita and Betta, and examinations in concentrated phosphoric acid, 187
Kinoshita *et al.*, and pore-size distribution, 219
Knight, shift in metals, 78
Kortüm and Wilski, and conductance in nonaqueous solutions, 7
Kramer, and proton-transfer theory, 37
Krichnan and Friedman, 58
Kronenberg, and determination of surface area, 226
Krupp, and the preparation of radium-nickel catalysts, 206
Kuhlmann and Grant, and counter-ion effects for coupling in solution, 95
Kondo–Yamashito formalism, 86

Laidler, Levich, and electron transfer, 32
Latimer and Rodebush, and consideration of entity in water, 19
  and recognition of "free pair of electrons" on water, 2
Law, and adsorption of water on mercury, 140
  and change of work function, 175
Lengyel, and proton-transfer theory, 34
Lengyel *et al.*, and conductance of protons, 15
Lenz, and ion association, 95
Levitch, and proton-transfer processes, 6

Lewis and Dood, their publication of the mobility ratio for protons and deuterium ions, 19
Libration, at electrodes, 155
in solution, and entropy, 156
Lieber and Morritz, and nickel catalysts, 205
Lincoln, and low value for coordination number of gallium, 100
Line width, and relaxation in line width, 45
Literature, on highly dispersed catalysts, 257
Lithium shifts, in chloride-containing solutions, 118
Lorenz, and calculation of mobility, 4
his field, 18
Lown and Thirsk, and structure breaking of water, 25
Lundgren and Williams, and neutron scattering, 3
Lynton, and proton-transfer theory, 20
Lynton *et al.*, and proton-transfer theory, 28

MacDonald and Barlow, and models for water, 137
Maciel, and large solvent shifts, 101
Magnesium ion, its coordination number, 96
Magnetic field, fluctuating, 46
Magnetic properties, and proton-transfer theory, 42
Magnetic studies, in solution, 82
in solution, NMR effects in solution, 78
Magnets, superconducting, used by Richards, 88
Magnetic susceptibility, 301
Malalovsky, and proton shifts in water, 65
Materials, highly dispersed, 237
Maximized solvents and NMR, 108
Mechanics, of electron microscopes, 245
Mechanism, of proton transfer, 27
prototropic, 6
Meiboom, and ion association, 95
Meiboom *et al.*, rate constants, for proton transfer, 16
Metal and gases, on chemisorbed layers, 222
Metal, and solvent properties on, 165
Metal oxide, bronzes, 208
Metal oxides, 189
Metal surfaces, and charge, 135
Methanol, 14
Micrographs, high resolution, 246
Migration, hydrodynamics, 3
Minimum, of orientation of water molecules, at electrodes, 136
Mobilities, hydrodynamic, and Walden's role, 9
Mobility, of ions, abnormal, 3
Model, for temperature rise, due to disc rises, 277
Models, for proton transfer in solution, 47
Monolayer coverage, and sweep rate, 228
Monomeric water, water molecules, monomeric, 25
Mrha, and triangular potential suite, 188
Myoglobin and hydrogen bond, 2

Naberukhin and Shuiskaii, and water aggregates, 26
National Research Council, 122
Nemethy and Scheraga, and structure of water, 54
Nernst, his equation, 4
Nickolau, and ESR studies of platinum, 191
NMR, and studies of electrolytic solutions, 37
reviewed, 122
in solution, 67
NMR studies, and water structure, 49
in electrolytic solutions, 77
of electrolytic solutions, 61
in mixed solutions, 68
NMR theory, 45
in solution, 41
Noble metal areas, tabulated, 203

Nonaqueous conductance, 10
Nonaqueous solutions, conductance with protons, 8
Nuclear magnetic resonance, 300
Nuclear quadrupole resonance, 300
Nuclei bound, relaxation rate, 80

^{17}O studies in solution, 82
Orbitals, *s* and *p*, in bonding in solution, 86
Organic ions, large, and NMR spectra of organic ions, 74
Organic molecules, and interaction in water, 56
Orientational entropy, at electrodes, 153
Oscillation, and water aggregates, 27
Overlap, and structure of water, 49
Overlap theory, in solution, 89
Oxygen, and surface function, 186
Oxygen chemisorption, 224
Oxygen equilibria, 113
Oxygen isotopes and hydrogen shift in water, 55
Oxygen shifts, in diamagnetic electrolytes, 83

Pacemakers, 254
Parsons, and four-state water model, 162
  anomalies facing his four-state water model, 164
  and model for interface, 140
  and parameters needed for four-state water model, 164
Paxton *et al.*, and pore-size distribution, 221
Performance curve, of platinum black, 253
Permativity titrations, 296
Pervoskite complexes, and electrolytic properties, 190
Perovskite structures, 189
Perrault, his lack of proton transfer, 30
pH titrations, for catalysts, 199
Photosynthesis, 305
Physical adsorption on catalysts, 215
Platinum, and tungsten carbide, 209
  crystallite relationship, 237
  unsupported, 202
Platinum complexes, in catalysis, 198
Platinum crystallite size, 249
Platinum crystallites, 250
Platinum supports, in tungstic-oxide catalysts, 192
Platinum surfaces, area of, 232
  impregnated, 196
Platinum–rodium alloys, supported, 212
Polarization, nonadditive, 295
Polarization curve, 256
Polarity scale, and Covington, 94
Poneck, and surface composition, 258
Popel, and free water molecules in solution, 49
Popov, and correlations in solution, 102
Pore size, for carbon blacks, 220
Pore-size distribution, and Shastri, 220
Pore-size measurements, and nitrogen adsorption, 221
Porosity, as function of temperature, 220
Porthaus *et al.*, and effect of halide composition on NMR, 92
Position of maximum, criterion of water molecule models, 178
Potential drop, in double layer, 166
Potential drop across the double layer, 138
Potential region, of greatest interest, 255
Potentials, and carbon-black oxidation in solution, 186
Potentiodynamic stripping of CO on platinum, 229
Pott, his work on binary catalysts, 214
Pretreatment of catalysts, 216
Properties, ferroelectric, and hydrogen bond, 2
Protein fractions and charge transfer, 302
Proton incidence, and shifts in solution, diagramed, 75
Proton NMR, 116

Proton NMR studies, aprotic solvents, 98
Proton studies, in solution, 99
Proton transfer, concluding comments, 36
Proton transfer, and rate determining step in aqueous solutions, 26
  Conway's analysis, 2
  due to field of ion, 22
  and structural changes, 7
Proton-transfer theories, 29
Proton-transfer theory, and mechanism of transfer in solution, 43
  CBL, 21
Proton-tunneling frequency, 29
Prototropic contribution to conductance, 12
Pyridine, and charge-transfer complexes, 294

Quantum mechanical theory, and strong coupling, 36

Radiation, of water molecule, and the field of ion, 22
Radich, and hydrogen shift in solution, 74
Raman spectroscopy, and hydrogen shift, 73
Ramanathan, and cosphere overlap in solution, 78
Raney's method, and preparation of catalysts, 205
Rate constants, second order, for prototropic reactions, tabulated, 17
  second order, for reactions of proton transfer, tabulated, 17
Rate variations, with temperature, 81
Reaction rate measurements, 251
Reduction, of aqueous solutions, 201
Reduction program, by temperature, 207
Reeves, and entropy maximum, 133
Relaxation, NMR, 75
Resonance, cation, 90
  in solutions, 44
Respiration, aerobic, 305
Richards, and quantitative results on hydrogen shifts, 106
  and the problem of shifts with cesium, 88
Roberts and Northey, and fluidity in aqueous solutions, 26
  and the diffusion constant of protons, 6
Rode, and effect of alkaline metals on solution, 98
Ruff, and proton-transfer theory, 33
Ruff and Friedrick, and contribution of exchange reactions to proton transfer, 32
Russian work, its uniqueness, 193

Sagert and Pouteau, and effect of water on metal-support catalyst, 192
Samoilov, and free water molecules in solution, 49
Scattering, on catalysts, theory for, 240
Schmelnick and Fiat, and oxygen-isotope measurements, 79
Schulery and Alder, the effects of electrolysis on proton shift, 62
Schuster and Fratiello, 68
Shielding constant, and the Kondo–Yamashito approach, 85
Shift in proton transfer, and interpretation of, 53
Shift, of hydroxyl proton, in mixtures of water, 60
Shifts, due to dilution, 104
  due to hydrogen in solution, 70
  in oxygen, 81
  in solution, 71
Sigmoid shift, and proton-magnetic resonance, 121
Singh and Wood, and proton-transfer theory, 36
Small-angle X-ray scattering, 239
Smith and Kasten, and porosity of carbon blacks, 219
Sodium chemical shifts, in a linear correlation, 101
Sodium shifts, in dimethyl-formamide, 102

Solute, polar organic, 58
Solute shifts, 101
  in solution, 84
Solution, lines broadening of spectra, 109
Solvation, 120
  of ions, 96
  preferential, 117
  in a solution, and entropy, 155
Solvation number, of mechanism ion, 68
  of salts from NMR, 66
Solvent dipoles, libration of, Conway and Gordon, 151
Solvent entropy and solvent hump, 162
Solvent excess entropy, definition, 132
  modelistic approach, 150
Solvent shifts, 95
Solvent–ion overlap, 87
Solvents, aprotic, 98
  other than alcohol, and NMR, 97
Spillover theory, and the Bockris–McHardy work, 193
Spin, and contribution to proton transfer, 47
Spin–lattice effects, 299
Spin–quantum number, in solution, 105
Stearn and Eyring, and absolute reaction rates, 19
Stirling's approximation, 152
Stockton and Martin, and proton shifts with alcohol, 115
Stokes, his radius, hydrodynamic, fraction of, 4
  his transport, 3
Structure, α-helix and hydrogen bond, 2
  of liquids, and hydrogen bonding, 14
  of TaS_2 complex, 288
  of TEA, 284
Structure breaking effects, 24
Structural effects with proton shifts, 116
Sulphate species, and calculation of species in solution, 114
Sulphur-doped, cobaltites, 190
Summary, of electrocatalysts on supports, 257
Supports, other than carbon, 189
Surface area, high, in catalysts, 201
  measurement of catalysts by BET, 217
  of noble blacks, 204
Surface areas, electrochemical, 233
  measurement by electrochemical methods, 233
Surface potential, of water and metal-solution interface, 147
Surface excess entropy, 134
Swift and Sayre, and NMR techniques, 67
Swinehard and Tauber, and early NMR studies in nonaqueous solutions, 95
Symmetry, and proton transfer, 33
Szabo, and transference number of hydrogen, 15

TCNQ salts, 283
Technique, for preparing electrocatalysts, 200
Techniques, electrochemical, for surface-area determination, 227
  for determining surface area of supported catalysts, 224
  for pore-size distribution, 221
Temperature dependence, and entropy of conduction, 285
  of conductivity, 278
  of TTF–TCNQ complex, 287
Temperature-programmed reduction, 207
Templeman and Van Geet, and observations of NMR shifts for sodium solids, 90
Tetrahydrofurane, concentration dependence in solution, 103
Thallium, shifts due to, 106
Theory, thermodynamic, of solvation, 120
Thomas and Marum, and conductance in nonaqueous solutions, 7
Three-state water model, of Bockris and Habib, 140
Tijmstra, and abnormal mobility of pyridinium ion, 3

Titrations, for charge transfer complexes, 290
Tourky and Mikhail, and conductance in nonaqueous solutions, 8
Tracer studies, in solution, 47
Transfer, and equidistances obtained, 35
  of protons in solution, 1
  successive, Goldschmidt and Udby, theory of esterification, 1
Transference number and conductance, 15
Transference numbers, of hydroxol, 13
Trasatti, and absolute calculation of metal solution p.d., 167
  criticized by Frumkin, 171
  and orientation of water, 172
  and the potential zero charge, 171
  and the work function relationship, 170
Tsung and Bevan, and freeze drying, 210
Tungsten carbide, 209
Turkevich and Kim, and preparation of catalytic materials, 202

Underpotential, deposition, 234

Viscosity and conductance, 12
Vishwanathan and Shastri, and pore-size distribution, 220

Walden, and conductance in nonaqueous solutions, 7
  product, in nonaqueous solutions, 11
  product, due to Walden, 9
Wannier, his calculations of proton conductance, 18
Warren, and the geometric means for β, 236
Water, bound, in solution, 71
  free and bound, 79
  structure of, solvent, and solvent mixtures, 48
Water solution, dilute, 57
Water–water repulsion energy, 143
Watts-Tobin and Mott, and model for interfacial water, 137
Weidemann and Zaundel, and quantum mechanical theory, 28
Wicke, proton transfer in solution, 20
Woods, and hydrogen adsorption, 229
Work function, and dipole contributions, 175
Wulff and Hartmann, and water–dioxane mixtures, 7

X-ray analysis, 235
X-ray diffraction, 243
  and catalysts, 207
  on catalysts, 238
  and line broadening, 235
X-ray structure of pyren, diagramed, 274

Yamagati, and shifts in solution, 85
Yeager and Kogawa, and kinetics of oxygen reduction, 188

Zatsepina, and proton-transfer theory, 30
Zeidler, and review of organic ions, 74
Zero point energy, 29
Zhurmann, and change of work function, 175